THE SOCIOLOGY OF FOOD AND AGRICULTURE

As interest has increased in topics such as the globalization of the agrifood system, food security, and food safety, the subjects of food and agriculture are making their way into a growing number of courses in disciplines within the social sciences and the humanities, like sociology and food studies. This book is an introductory textbook aimed at undergraduate students, and is suitable for those with little or no background in sociology.

The author starts by looking at the recent development of agriculture under capitalism and neo-liberal regimes and the transformation of farming from a small-scale, family-run business to a globalized system. The consequent changes in rural employment and role of multinationals in controlling markets are described. Topics such as the global hunger and obesity challenges, GM foods, and international trade and subsidies are assessed as part of the world food economy. The second section of the book focuses on community impacts, food and culture, and diversity. Later chapters examine topics such as food security, alternative and social movements, food sovereignty, local versus global, and fair trade. All chapters include learning objectives and recommendations for further reading to aid student learning.

Michael Carolan is a professor at Colorado State University (USA) and Chair of its Department of Sociology. He is author of *The Real Cost of Cheap Food* (Earthscan, 2011).

"Finally! Amidst the burgeoning literature on agrifood studies a sociological perspective has been sorely missing. In particular, Carolan's book will prove invaluable to those of us who teach in this area and have been looking for a way to weave in sociological theory to examine the wide array of food and agricultural topics." – Carmen Bain, *Assistant Professor in the Department of Sociology, Iowa State University, USA.*

"This highly recommended text is an accessible, lively and up-to-date introduction to the sociological imagination and agrifood studies, filling the gap between the sociologies of food production and food consumption." – Sally Foster, *Senior Lecturer in the Faculty of Health and Social Sciences, Leeds Metropolitan University, UK.*

"Clearly written, but theoretically sophisticated, Carolan uses a wealth of published research to present a more critical 'sociological' analysis of contemporary food system issues that contrasts sharply with the mainstream narratives that dominate in the agricultural and food sciences. The examples are authoritative, fresh, provocative, and speak directly to current social and political debates on farm and food issues." – Douglas Jackson-Smith, *Professor of Sociology, Utah State University, USA.*

"This is critical sociology at its best. In this introductory text, Carolan goes behind the scenes of the global agrifood industry to examine the complex socio-economic and political arrangements that shape food production and consumption. Presenting the latest findings from internationally-based research, the book highlights the structural causes of present-day concerns about hunger, obesity, rural social disadvantage, farmer dispossession, supermarket power and environmental degradation. Oppositional movements challenging the current system of food provision are also discussed in detail.

Carolan is one of the foremost writers in contemporary agrifood studies and he has fashioned a book that provides an up-to-date, informative and highly readable overview of the global agrifood system. The book will have immediate appeal to students, policy-makers and all those concerned about the future of food and farming." – Geoffrey Lawrence, *Professor of Sociology, Head of Sociology and Criminology and Food Security Focal Area Co-Leader at the Global Change Institute, The University of Queensland, Australia.*

THE SOCIOLOGY OF FOOD AND AGRICULTURE

Michael Carolan

LONDON AND NEW YORK

First published 2012
by Routledge
2 Park Square, Milton Park, Abingdon, Oxon OX14 4RN

Simultaneously published in the USA and Canada
by Routledge
711 Third Avenue, New York, NY 10017

Routledge is an imprint of the Taylor & Francis Group, an informa business

British Library Cataloguing in Publication Data
A catalogue record for this book is available from the British Library

Library of Congress Cataloging in Publication Data
Carolan, Michael.
The sociology of food and agriculture / Michael Carolan.
p. cm.
Includes bibliographical references and index.
1. Agriculture–History. 2. Agriculture–Social aspects–History. 3. Food–Social aspects–History.
I. Title.
S419.C366 2012
635–dc23

2011039121

ISBN 978-0-415-69851-1 (hbk)
ISBN 978-0-415-69858-0 (pbk)
ISBN 978-0-203-13679-9 (ebk)

Typeset in Bembo
by Graphicraft Limited, Hong Kong

Printed and bound in Great Britain by
CPI Antony Rowe, Chippenham, Wiltshire

CONTENTS

LIST OF FIGURES, TABLES AND BOXES

Figures

Tables

Boxes

ACKNOWLEDGMENTS

Stone Soup is an old folk story about a hungry stranger who persuades the local people of a town to contribute food to what turns out to be a dish consumed by everyone. Writing this book has reminded me repeatedly of this tale of sharing and cooperation. So many have contributed to its making, though all omissions and mistakes (as any honest chef must admit to) are my responsibility alone. The term "generosity" keeps coming to mind as I search for words to explain the gratitude I feel toward so many: Alison Alkon, Dominique Apollon, Ika Darnhofer, Neal Flora, Jan Flora, Michael Goodman, Julie Guthman, Philip Howard, Yvonne Liu, Linda Lobao, Philip McMichael, Miranda Mirosa, Patrick Mooney, Gyorgy Scrinis, Rachel Slocum, Paul Stock, Curtis Stofferahn, and Julie Zimmerman. From those kind enough to share "in press" works (in a few cases entire book manuscripts) to others who opened up their schedules to read draft sections to still others who shared their tables, figures, and illustrations: thank you. I also own a debt of gratitude to the reviewers. I know there were a lot of you; some whose identity I know because you've since identified yourself to me, others who I will likely never be able to thank personally. For the known and the anonymous reviewers: your efforts were appreciated tremendously – thank you.

Thanks also to the amazing staff at Earthscan. Earthscan's emergence as the premier publisher on subjects related to sustainability, natural resources, and food and agriculture is testament to your hard work. And to Tim Hardwick: an author couldn't ask for a better commissioning editor. Thanks for your prompt responses to my many emails, for giving your time to read draft chapters, and for all the energy you've devoted to this project.

I must also acknowledge Colorado State University and its Department of Sociology for providing a work environment so conducive for writing. Thanks also to my many students who make that environment meaningful and lively.

Lastly: to Nora, Elena and Joey. Your mere presence inspires me. For everything: thank you.

1

INTRODUCTION

KEY TOPICS

- The sociology of food and agriculture has deep roots, from its origins in rural sociology, on the agriculture "side," to its links with cultural studies, on the food "side."
- There are many reasons why food and agriculture ought to be thought about sociologically.
- Questions relating to the book's intended audience are answered and an overview of the rest of the text round out the chapter.

Key words

- Adoption-diffusion
- Agrifood system
- Goldschmidt thesis
- New Rural Sociology
- Organic public sociology
- Present-at-hand
- Readiness-to-hand
- Sociological imagination
- Traditional public sociology

Is anything as simultaneously familiar and distant as food? We are all intimately acquainted with food. The average American spends between 63 and 68 minutes a day eating, and in other countries (France comes immediately to mind) that figure is even higher.[1] Compare that to the amount of time the average American spends talking with their children: 30 minutes.[2] There is little doubt that we all – save for those in abject poverty – have a close relationship with food. And yet, how much do we really know about food? For most consumers in developed economies, rows of crops and annual harvests have been replaced by aisles of shelves and weekly shopping trips. Children as young as five can name all the planets in our solar system (and a few constellations too!), while their older brothers and sisters still don't know that pickles and bacon started their lives as cucumbers and pig, respectively. No; we don't think, *really* think, about the life of food prior to its arrival at our nearest supermarket. This is where sociology comes in, to help us come to grips with this big picture by better understanding the biography of food, from gene to shelf to rubbish bin.

I know from personal experience that among the uninitiated confusion can follow hearing the title "sociology of food and agriculture" for the first time. I can't tell you how many times I've had to field the question, "So what *exactly* does it mean to think about food and agriculture sociologically?" I have no standard boilerplate response. And the two or three sentence answer never seems to do justice to the field. What I tend to do is quickly explain how varied the literature is, in terms of subject matter, theoretical approach, and method. As the reader will quickly learn, one's sociological imagination can truly run wild thinking about food systems. Even a question as seemingly simple as "What is food?" proves to be exceedingly complex once exposed to sociological treatment. As Harris (1986: 13) reminds us, "[w]e can eat and digest everything from rancid mammary gland secretions to fungi to rocks [or cheese, mushrooms, and salt if you prefer euphemisms]." Anyone who has traveled to another country – or spent much time watching, say, the Discovery Channel – knows the label "food" is a terminological box that can be filled with wildly different phenomena depending on culture, circumstances, and time period.

Then there is the entire world "behind" our food. The food *system* is an incredibly (and increasingly) entangled network of people, organizations, states, regulations/laws, ecosystems, and values/beliefs. We've become accustomed to seeing food as "just" food (a state, as later detailed, that is also not without sociological consequence). This is why a sociological treatment of the subject is so revealing. For at its core, sociology is fundamentally about showing us that nothing is *just* as it seems.

A brief historical overview

An explicit sociology of food and agriculture dates back a couple of decades; though its roots extend back a century in the rural sociological literature. Initially inserting its way into the literature as a background variable (subordinate to things like community structure), agriculture's rapidly changing structure during the latter half of twentieth century caused a growing number of rural sociologists to devote more time to its study as a primary (dependent) variable. African-American sociologist W.E.B. DuBois (1898, 1901, 1904), for example, conducted a number of studies at the turn of twentieth century looking at the plight of black rural laborers and farmers in such states as Virginia and Georgia. At around that same time, James

Williams was writing his dissertation at Columbia University. Completed in 1906, "An American Town: A Sociological Study" examines a rural community in New York. The study determined agriculture to be the town's most important economic activity, and as such was viewed by Williams as an essential variable for understanding the community's past, present, and future social trajectories.

C.E. Lively (1928) provides us with one of the first comparative community studies, with agriculture looming large as an important independent variable. Looking at a dairy farming community in Wisconsin and a grain farming community in Minnesota, Lively was able to document that the type of agricultural system can impact community life, as a result of, for example, seasonal fluctuations in farm income, population density, and labor demands. The most famous early comparative study is now known simply as the Goldschmidt study (or Goldschmidt thesis). Walter Goldschmidt's (1978 [1947]) book *As You Sow* (which I discuss in greater detail in Chapter 5) examined two communities in California's upper San Joaquin Valley: one community surrounded by small-scale farms (averaging 23.1 hectares [57 acres]); the other surrounded by large-scale farms (averaging 201 hectares [497 acres]). Goldschmidt then attempted to discern how farm structure shapes community structure. The general thesis to emerge from this research is that large-scale agriculture has, as a whole, a greater number of negative than positive impacts on surrounding rural communities.

The 1950s marked a turning point in rural sociological circles, toward the social psychological. Buttel *et al.* (1990: 44) describe this period as the "social psychological-behaviorist era of rural sociology," which "would remain unchallenged in rural sociological studies of agriculture until the early to mid-1970s" (Buttel *et al.* 1990: 46). This era is defined largely by the level of rural sociological scholarship devoted to the adoption-diffusion research tradition. One of the earliest adoption and diffusion studies came out of Iowa State University in the 1940s. Sociologists Bryce Ryan and Neal C. Gross (1943) examined the adoption of hybrid corn among Iowa farmers. From this they were able to highlight social and psychological variables they believed hastened the adoption of new technologies (which hybrid corn was at the time) throughout a social network.

Because the adoption-diffusion model is primarily concerned with the process by which a technology is communicated through social channels it ignores many other relevant sociological questions, like "Should technology X be adopted in the first place?" and "Who gains (politically, economically, and socially) from adoption and who loses?" The adoption-diffusion model has also been heavily criticized for offering a top-down, expert-driven model of social change (a point I elaborate upon further in Chapter 7). Evidence of its "promotional posture toward technology" (Buttel *et al.* 1990: 46) can be seen in how the model seeks to educate change agents in how to get an innovation adopted. Adoption-diffusion scholars discuss, for example, how important it is "for change agents to work through opinion leaders in a society" and how "a change agent will be more successful if innovations are introduced that match clients' needs" (Rogers *et al.* 1988: 314). But what if the change agents are the ones in need of change? This question began to be asked with greater frequency as the 1960s gave way to the 1970s, particularly as conventional agriculture and its proponent (those aforementioned "change agents") came under increasing scrutiny.

Rural sociology underwent a profound change in the 1970s. This shift, as described by Buttel (2001: 168), who witnessed it first hand, rejected "the 1960s American rural sociology" with "its emphasis on technique, its diffusionism, its lack of attention to rural poverty and deprivation, and its lack of critical imagination with respect to state policymaking and the role of rural sociology in policy." A number of variables lie behind this change. For one, rural sociologists, especially those engaged in international development, were growing increasingly disillusioned with the prevailing paradigms. Some, while conducting research in the developing world, were becoming exposed to theoretical perspectives (often Marxist in orientation) that challenged the then-dominant development orthodoxy (Buttel *et al.* 1990: 75). Related to this was the "discovery" of Marxian thought (and shortly thereafter Durkheim and Weber), which had previously gone largely unrecognized in the sub-discipline. This burst of theoretical activity energized a new generation of rural scholars, causing them to see the world more critically. As Marxism is concerned almost exclusively with issues tied to production, it should not be surprising that New Rural Sociology – the title given to this shift in rural sociology in the 1970s – ushered in an intense interest in agriculture ("farming") and its transformation.

While being pulled theoretically toward agriculture, rural scholars also could not ignore the profound empirical changes that were occurring on the ground. The USA lost farms at a rate of 1.1 percent annually from 1980 to 1995. In Japan, the figure was 2.2 percent, while in the European Union (EU) it was 1.8 percent (Tweeten 2002: 17). This was also the period when the food system – a term used to describe all that lies between seed and grocery shelf – began to take on its characteristic "hourglass" shape, referring to the highly concentrated "middle" that connects farms with consumers (discussed in much greater length in Chapter 3). But even before this, in the early 1970s, a growing chorus of criticism was directed at what was perceived as an unjust, unsustainable system. Critics took aim at things like the green revolution, the environmental impacts of conventional farming, the growing grip the agribusiness complex had over the food system, and the perceived role that land-grant universities had in promoting agricultural technologies to the detriment of the family farm (famously encapsulated in James Hightower's [1973] *Hard Times, Hard Tomatoes*). From this point on, "the sociology of agriculture became an important focus in rural sociologists' research" (Friedland 2002: 353). It is from this point that I pick up discussion in the following chapter.

These changes were of immense interest to social scientists as they were of deep sociological consequence. The shuttering of farms dramatically impacts rural communities, in terms of, among other things, demographics, employment, and community structure. But not just rural communities were (and are) being affected by structural changes to the food system. The globalization of food commodity chains, and I might add feed and fuel chains (recognizing that agriculture also "feeds," among other things, cows, chickens, and cars), is dramatically changing the lives of people around the world. And often these changes are not positive.

Unlike the type of scholarship practiced in, say, the 1980s, the focus today is no longer fixed exclusively on the farm, as scholars increasingly prefer to talk about the agrifood *system* – the term "agrifood" represents a tip of the hat to the fact that this system produces much more than just food. Nevertheless, the sub-discipline's preoccupation with social, economic, and political structures carries on a tradition that dates back to the days of New Rural Sociology. Yet the move away from production, I admit, has been slow. I am reminded of something written by the editors for the book *Hungry for Profit*, a widely read collection of sociological essays critical of corporate agriculture. Regarding farmers' markets as an example of consumer-driven social change they ask "whether this pathway is really a solution to the problems or rather something that will produce only a minor irritant to corporate dominance of the food system" (Magdoff *et al.* 2000: 188). *Real* social change, they countered, involving "a complete transformation of the agriculture and food system, it might be argued, requires a complete transformation of society" (Magdoff *et al.* 2000: 188). Consumption, in other words, just doesn't create sufficient leverage to result in meaningful social change. I investigate further these competing views about whether production or consumption is the most important site for sociologists of food and agriculture to study at various times throughout the book.

I want to be careful here not to overstate this rift between production-oriented and consumption-oriented scholarship. If it exists the rift is diminishing, rapidly. The New Rural Sociology scholarship of the 1980s rarely mentions consumption. Sociologists who studied food consumption, historically speaking, were often by training sociologists of culture. This gave rise to analyses that focused a great detail on the symbolic meanings of food (Beardsworth and Keil 1997). Since the late 1990s a concerted effort has been made to bring these two sociological traditions together and "build a better theoretical bridge between [consumption-oriented] food studies and [production-oriented] agro-food studies" (Goodman and DuPuis 2002: 11).

Why think about food and agriculture sociologically?

One particularly exciting thing about the sociology of food and agriculture is its multidisciplinary makeup, especially in recent decades. C. Wright Mills wrote famously about what he called the sociological imagination. His book, *The Sociological Imagination*, first published in 1959, is arguably still the most widely read book in sociology today. The sociological imagination, to put it simply, involves stepping back and grasping the wildly interconnected character of today's world. Mills contrasts this with what the "mere technician" does, who, while incredibly well trained, possesses an exceeding myopic understanding of the world. According to Mills, you need not be a professional sociologist to have this imagination. A sociological imagination obeys disciplinary boundaries about as well as air obeys geopolitical boundaries. So, what you'll read in this book is agrifood scholarship infused with a sociological imagination. That, to me, is what makes it "sociological." And that, to me, qualifies that scholarship for inclusion in this book.

That being said, *social* variables are given analytic primacy in this text. You will therefore find the following pages populated with ideas and research not only from sociologists (of course), but also from scholars with such professional titles as "anthropologist," "geographer," and "historian". You will not, however, find much attention given to, say, agricultural economics, as much of that literature seeks to explain the world through the lens of economic rationality. Conversely, *political* economy approaches are given substantial consideration, as that framework views markets, consumer "choice," and the like as being shaped markedly by broader sociological variables.

There are few sociological topics hotter at the moment than those dealing with food. Over the last decade we have been fed a steady diet of bestselling books on the subject, such as Eric Schlosser's *Fast Food Nation* (2001) and Michael Pollan's *The Omnivore's Dilemma* (2006). We have also seen in recent years the rise of the Slow Food and Local Food movements. The word "locavore" – which refers to someone who actively seeks out and eats local food – was even declared the word of the year for 2007 by the *Oxford American Dictionary*. While I stand by my earlier statement that we tend not to think – *really* think – about food, I believe an increasing number of people want to.

Philosopher Martin Heidegger makes the distinction between readiness-to-hand and present-at-hand (there is a sociological point to this, I promise). While he uses these terms to talk about how we relate to technology, they can equally be applied to our relationship with food. Heidegger argues that technological artifacts, and I would also argue the systems they are dependent upon, tend only to become visible when they break down. While we may stare intently at a television for hours our attention is usually only drawn to the images and sounds of the movie we are watching. When operating properly, the television, as well as the entire "system" that allows this artifact to work, can be said to have a readiness-to-hand quality as they withdraw from our perceptual field and become transparent. It is only after, say, the television breaks or the power goes out that these previously invisible elements come into focus, at which point they become present-at-hand.

Let's apply this thinking to food. There are many aspects of the food system that are not easily perceived at the point of consumption. This gets back to my earlier point that most don't *really* think about food; though part of the reason for this is because there are many barriers that make it *really* difficult to think systematically about food. Environmental degradation, social injustices, and reductions in biodiversity are just a few of the "absences" that are part of the food system. (And these absences, I might add, make it easier for the system to continue.) Typically, it is only after an aspect of the food system has malfunctioned that we actually actively think about it, like during a massive food recall. Food recalls due to things like mad cow disease or an *E. coli* outbreak invariably get us thinking and talking about the food system. But this state is always temporary. After a while, the system lying "behind" food always manages to recede to the background, becoming once again transparent to the average consumer. Consequently, while we all know that system is there, most of the time it is given little thought.

This is unfortunate, as I firmly believe the food *system* needs to be thought about, discussed, and debated. And I am not saying that the people involved in this massive conversation about the agrifood system need only be professional sociologists. I would prefer it, in fact, if most were not. Better if participants come from all walks of life, so that the collective sociological imagination is informed by perspectives from all strata of society.

Heidegger, I am afraid, is not much help to us beyond this point. While a world-class philosopher, he was not much of a sociologist. I am particularly bothered by his apparent lack of interest in *why* artifacts take on a readiness-to-hand quality. These relational states do not just happen. These presences and absences, rather, are sociological. There are many aspects of the food system that surprise people upon their being made visible. Allow me to offer just a few examples.

- The Food and Agriculture Organization (FAO 2002: 9) of the United Nations (UN) calculates that world agriculture produces enough food to provide everyone in the world with at least 2,720 kilocalories (kcal) per person per day, yet global food insecurity, in absolute numbers, is increasing (Weis 2007: 11).
- The 30 largest countries (economy-wise) spend roughly US$365 billion annually – that's *one billion* dollars a day – on food and agricultural subsidies. Who gets the largest share of those subsidies? Large agribusiness. For instance, in the EU the Dutch firm Campina (a Dutch dairy cooperative that merged with Royal Friesland Foods in 2008) received €1.6 billion in agricultural subsidies between 1997 and 2009. In second place: the Denmark-based firm Arla Foods amba, having cashed checks from taxpayers worth just less than €1 billion between 1999 and 2009 (Carolan 2011: 196).
- At the height of the global food crisis in 2008, millions of Haitians resorted to eating biscuits made out of mud, oil, and sugar (Katz 2008). At the same time, agribusinesses were reporting record profits. For example, during the first quarter of 2008, at the very height of the food crisis, Cargill, ADM (Archer Daniels Midland), and Bunge saw their net earnings rise 86, 55, and 189 percent, respectively (McMichael 2009: 241).
- Approximately 50 percent of all food in the USA is wasted, costing the US economy at least US$100 billion annually (Jones 2005: 2). That "waste" constitutes sufficient nourishment to pull approximately 200 million people out of hunger (Stuart 2009: 82).
- It takes roughly 10 calories of fossil fuel to produce one calorie of food (Frey and Barrett 2007) and 15,500 liters of water to produce 1 kg of intensively reared beef (Hoekstra 2010).

What are the reasons for this state of events, where such inequities and waste are allowed to persist? What are the proposed "solutions" and are they really an improvement over the status quo? Who benefits and who loses when problems like those mentioned above are addressed? Thinking about food and agriculture sociologically forces us to confront questions like these; questions that are admittedly difficult and not easily answered. But if we are serious about wanting to achieve a just and sustainable form of global food security, they are questions that *must* be answered.

In his 2004 Presidential Address to the American Sociological Association, Michael Burawoy (2005) called for a public sociology. This public sociology can take one of two forms. The first he calls a "traditional public sociology." This form of sociology is primarily instrumental, meaning its value lies in advising. The other Burawoy calls "organic public sociology." This is a form of public sociology that is conscience-raising, in that it imparts upon others a sociological consciousness. In actively seeking to make the invisible visible, by getting readers to think imaginatively (sociologically speaking) about food, agriculture, and society, I believe this book can be classified as a work of organic public sociology. But then again, much of the sociology of food and agriculture literature is precisely that: a conscience-raising resource that will cause readers to never see food in the same light again.

Intended audience

While the book's scope makes it suitable for both lower and upper level classes, I try to write in a style that makes the subject matter accessible even to those grappling with these sociological issues for the first time. Questions provided near the end of each chapter are written so they can be answered by either beginner or advanced students. A distinction has been made, however, between lower and upper level students when it comes to suggested reading (which the reader will find immediately following the questions).

The book is also written for an international audience. I have tried to draw upon empirical subject matter from all over the world. The systems that feed us are made up of many millions of faces and places, located in developed and developing countries, residing in cities as well as the countryside. I hope to have done justice to this diversity by not focusing exclusively on one specific country or region.

This is not the same as saying all countries are discussed equally. I will be the first to admit that this book centers heavily on affluent nations (and the USA in particular). But I have my reasons. For one, the vast majority of the sociological agrifood scholarship centers on the USA, UK, (Western) Europe, Australia, and New Zealand. Reviewing this literature, I will be reproducing any regional biases found within it. I would also argue that, from a political economic standpoint, if we really want to understand *why* the food system has taken the shape it has we are going to need to spend some time looking at the policies and practices coming out of countries like the USA and regions such as the EU. This simply cannot be avoided given the pivotal role that these spaces (and actors therein contained) have played in shaping the world food system. But given the importance of food security for the world's growing population, as well as growing demand in China, India, and Brazil, we should as far as possible aim for an international perspective, including the poorest, developing countries.

Brief overview of chapters

The book is broken down into three sections, beginning with Part I, Global Food Economy. In Chapter 2, the subject of the changing structure of agriculture is defined and analyzed, leading to a discussion of such things as agricultural subsidies, the various "treadmills" in agriculture, as well as an overview of some popular agrarian theories from the sociology of agriculture in the 1970s and 1980s. In Chapter 3, discussion expands beyond the farm gate, leading to an overview of the changing structure of the food *system*. In addition to reviewing concentration trends, which has since given the food system its infamous "hourglass" shape, notable frameworks and approaches are reviewed, like the commodity chain analysis and the food regime. In Chapter 4, issues of hunger, malnutrition, and obesity are exposed to a full sociological treatment. By properly contextualizing these issues, some of their real root causes are exposed.

Part II, Community, Culture, and Knowledge, begins with Chapter 5 and an overview of some of the social and public health impacts of the current industrial model of food production, in both affluent and less affluent nations. Chapter 6 elaborates further on the subject of impacts, specifically turning attention to the effect the dominant food system is having on culture. I also take the opportunity in this chapter to discuss the "cultural turn" in the sociology of food and agriculture literature, which includes a "turn" toward issues of gender, race, and ethnicity. In Chapter 7, the link is made between cultural diversity and biodiversity – or what's known as biocultural diversity (in doing this the subject of biopiracy also comes up). I also discuss in this chapter models of knowledge-innovation transfer as they pertain to production agriculture.

Part III, Food Security and the Environment, kicks off, with Chapter 8, with a discussion of food security in light of current demands. These "demands" are examined as are other broader political economic realities that make global food security a problematic goal. In Chapter 9, attention turns to the ecological world; the productive *base*, as it's called, that makes food possible. The subject of agroecology is discussed as is the literature that deals with how "nature" itself is theorized. Frameworks discussed here range from social constructionism to actor network theory. Chapter 10 then looks at the ecological cost of current models of food production, distribution, consumption, and waste.

To conclude, Part IV, Alternatives, takes a critical look at those models, movements, and proposals held up as alternatives to the convention model of food provision. Chapter 11 begins the section by highlighting some of these alternatives, namely organic agriculture, the Slow Food movement, and a peasant movement known as *La Via Campesina*. Chapter 12, through the lens of local food, takes a more critical look at these alternatives. Here the concept of the local trap is discussed, which leads to a discussion of a whole lot of other potential "traps" we would do well to be aware of. Chapter 13, besides concluding the book, takes a look "ahead," in terms of pointing out some gaps in the sociology of food and agriculture literature that need to be filled, and "back," as I attempt to retrospectively map the journey offered in the following pages.

Discussion questions

1 Can you think of examples of how understandings of food in your country have changed over the last century? What might have accounted for this change?
2 Are the terms "rural" and "urban" still relevant? If still relevant, how has "rural" changed over the past century (in terms of things like employment, demographics, community structure, and overall "look")? And how should we think of the terms – as on a continuum, as definitional opposites, or perhaps another way?
3 When sociology first emerged as a discipline in the 1800s the world was predominately agrarian and all its population ate (including sociologists). Why, then, did it take almost another 100 years until its practitioners started studying agriculture and food systems systematically?

Suggested reading: Introductory level

Friedland, William. 1991. Introduction: Shaping the new political economy of agriculture. In *Towards a New Political Economy of Agriculture*, edited by W.H. Friedland *et al.* Boulder, CO: Westview, pp. 1–34.

Suggested reading: Advanced level

Buttel, Fredrick. 2001. Some reflections on late-twentieth century agrarian political economy. *Sociologia Ruralis* 41(2): 165–81.

Friedland, William. 2002. Agriculture and rurality: Beginning the "Final Separation"? *Rural Sociology* 67(3): 350–71.

Lowe, Philip. 2010. Enacting rural sociology: Or what are the creativity claims of the engaged sciences? *Sociologia Ruralis* 50(4): 311–30.

Notes

1 http://www.ers.usda.gov/amberwaves/november05/datafeature/
2 http://www.bls.gov/news.release/atus.t09.htm

References

Beardsworth, Alan and Teresa Keil. 1997. *Sociology on the Menu: An Invitation to the Study of Food and Society*. New York: Routledge.

Burawoy, Michael. 2005. 2004 American Sociological Association Presidential Address: For Public Sociology. *British Journal of Sociology* 56(2): 259–94.

Buttel, Fredrick. 2001. Some reflections on late-twentieth century agrarian political economy. *Sociologia Ruralis* 41(2): 165–81.

Buttel, Fredrick, Olaf Larson, and Gilbert Gillespie. 1990. *The Sociology of Agriculture.* New York: Greenwood Press.

Carolan, Michael. 2011. *The Real Cost of Cheap Food.* London: Earthscan.

DuBois, W.E.B. 1898. The Negroes of Farmville, Virginia: A social study. *Bulletin of the Department of Labor* 3(14): 1–38.

DuBois, W.E.B. 1901. The Negro landholders of Georgia. *Bulletin of the Department of Labor* 6(35): 647–777.

DuBois, W.E.B. 1904. The Negro farmer. In *Negros of the United States.* Washington, DC: US Government Printing Office, pp. 69–98.

FAO (Food and Agriculture Organization). 2002. *Reducing Poverty and Hunger: The Critical Role of Financing for Food, Agriculture, and Rural Development.* Food and Agriculture Organization, International Fund for Agricultural Development, World Food Program, http://www.fao.org/docrep/003/Y6265e/y6265e00.htm, last accessed April, 2010.

Frey, Sibylle and John Barrett, J. 2007. Our health, our environment: The ecological footprint of what we eat. Paper prepared for the International Ecological Footprint Conference, Cardiff, May 8–10. http://www.brass.cf.ac.uk/uploads/Frey_A33.pdf, last accessed September 1, 2010.

Friedland, William. 2002. Agriculture and rurality: Beginning the "final separation"? *Rural Sociology* 67(3): 350–71.

Goldschmidt, Walter. 1978 (1947). *As You Sow.* New York: Harcourt.

Goodman, David and E. Melanie DuPuis. 2002. Knowing food and growing food: Beyond the production-consumption debate in the sociology of agriculture. *Sociologia Ruralis* 42(1): 6–23.

Harris, Marvin. 1986. *Good to eat: Riddles of food and culture.* London: Allen and Unwin.

Hightower, James. 1973. *Hard Times, Hard Tomatoes.* New York: Harper Collins.

Hoekstra, A. 2010. The water footprint of animal products. In *The Meat Crisis: Developing More Sustainable Production and Consumption*, edited by Joyce D'Silva and John Webster. London: Earthscan, pp. 22–33.

Jones, Timothy. 2005. How much goes where? The corner on food loss. *BioCycle* July: 2–3.

Katz, Jonathan. 2008. Poor Haitian resort to eating mud. *National Geographic* January 30. http://news.nationalgeographic.com/news/2008/01/080130-AP-haiti-eatin.html, last accessed February 7, 2011.

Lively, C. 1928. Type of agriculture as a conditioning factor in community organization. *Publications of the American Sociological Society* 23: 35–50.

Magdoff, Fred, John Foster, and Frederick Buttel. 2000. *Hungry for Profit: The Agribusiness Threat to Farmers, Food, and the Environment.* New York: Monthly Review Press.

McMichael, P. 2009. Banking on agriculture: A review of the World Development Report 2008, *Journal of Agrarian Change* 9(2): 235–46.

Mills, C. Wright. 2000 (1959). *The Sociological Imagination.* New York: Oxford University Press.

Pollan, Michael. 2006. *The Omnivore's Dilemma.* New York: Penguin.

Rogers, Everett, Rabel Burdge, Peter Korsching, and Joseph Donnermeyer. 1988. *Social Change in Rural Societies: An Introduction to Rural Sociology*, 3rd edition. Englewoods, CA: Prentice-Hall.

Ryan, Bryce and Neal Gross. 1943. The diffusion of hybrid seed corn in two Iowa communities. *Rural Sociology* 8(1): 15–24.

Schlosser, Eric. 2011. *Fast Food Nation*. New York: Harper.

Stuart, Tristram. 2009. *Waste: Uncovering the Global Food Scandal*. New York: Norton.

Tweeten, Luther. 2002. Farm commodity programs: Essential safety net or corporate welfare. In *Agricultural Policy for the Twenty First Century*, edited by Luther Tweeten and Stanley Thompson. Ames, IA: Iowa State Press, pp. 1–34.

Williams, James. 1906. An American Town: A Sociological Study. PhD dissertation, Columbia University.

Weis, T. 2007. *The Global Food Economy: The Battle for the Future of Farming*. New York, NY: Zed Books.

PART I

Global food economy

2

THE CHANGING STRUCTURE OF AGRICULTURE

Theoretical debates and empirical trends

KEY TOPICS

- There are a number of processes in agriculture that have resulted in the substitution of capital for labor, expansion of farm size, and increased emphasis on a few major crops.
- Agriculture subsidies have had a major role in giving shape to the current structure of agriculture, creating both winners and losers.
- The sociology of agriculture takes a theoretically radical turn in the late 1970s with the application of Marxism to explain, among other things, the family farm's perseverance.
- While originally emerging as an alternative to conventional agriculture, organic agriculture has since taken on a number of its characteristics, giving rise to the conventionalization of organics thesis.

Key words

- Agricultural subsidies
- Agricultural treadmill
- Bifurcation of organics
- Changing structure of agriculture
- Conventionalization of organics
- Decoupled subsidy payments
- Double squeeze in agriculture
- Fertilizer treadmill
- Formal and real subsumption of nature
- Mann–Dickinson thesis
- Marxist political economy of agriculture
- Mechanization revolution in agriculture
- Pesticide treadmill
- Policy path dependency
- Seed treadmill
- World Trade Organization (WTO)
- WTO's subsidy "boxes"

There have been numerous consequential innovations since agriculture's inception: advances in irrigation in the valleys of Mesopotamia and the Nile 3000–5000 Before Present (BP); the development of aquatic rice-growing in the valleys and deltas throughout Asia between 2000 and 3000 BP; and the emergence of cultivation agriculture based on animal-drawn plows, which emerged between the eleventh and thirteenth centuries in Europe. One of the more consequential innovations over the last 200 years was the substitution of capital for labor – what is often referred to as the "mechanical revolution" in agriculture. Examples include: the cotton gin in 1793, which separated the cotton lint from the seed (and initially greatly increased the demand for slave labor in US southern slave states); the threshing machine in 1816, which separated grain from the stalk; the John Deere steel plow in 1837, which turned the soil cleanly thereby minimizing drag that bogged down earlier plows; and Cyrus McCormick's reaper in 1834, which cut grain. Later in the nineteenth century came the tractor, which, unlike the horses it replaced, did not require food, water, or sleep, and allowed the land used previously to raise hay and oats to feed one's horses to be used to produce still more commodities.

All of this labor-saving technology had a clear impact on the structure of agriculture in countries where it was becoming widely adopted. Take, for example, the USA, where the number of human hours required to farm an acre of corn dropped rapidly throughout the first half of the twentieth century: from 38 hours per acre in 1900 to 33 hours in 1920–1924, 28 hours in 1930–1934, and 25 hours in 1940–1944 (and by 1955–1959 the figure had dropped to only 10 hours of labor per acre). A study conducted by Iowa State College (now Iowa State University) in the 1930s concluded that while the average farmer left about 1–2.5 bushels of corn per acre in their field when harvested by hand, a hired husker left 3–5 bushels. By comparison, a mechanical corn picker left 2–3 bushels. In addition, hired huskers charged on average US$2.00 per acre, while custom mechanical pickers cost roughly US$1.25–1.50 per acre. It is not surprising, then, that by 1938 one-third to half of the laborers previously hired out to hand pick corn were replaced by mechanical pickers (Colbert 2000).

Mechanization also encourages specialization and scale increases at the farm level. Though not the only force driving the specialization trends illustrated in Table 2.1, the mechanization of agriculture undoubtedly contributed to the state's shrinking commodity basket. Iowa is not unusual in this regard. Agricultural specialization at the farm level has occurred wherever conventional agriculture is practiced (see also Chapter 4, specifically discussion on the green revolution). To understand one reason for this, take the following example. When a farmer purchases a combine with a corn and soybean head there are certain benefits to expanding the farm's size while reducing the number of commodities produced on it. Investing US$100,000 in such a piece of equipment encourages the farmer to plant corn and soybeans on all (or most of) their land. It also, when possible, encourages our hypothetical producer to acquire more land so still more corn and soybeans can be raised. The rationale for this is simple, as you want to spread the

TABLE 2.1 Number of commodities produced for sale in at least 1 percent of all Iowa farms for various years from 1920 to 2007

1920	*(%)*	*1935*	*(%)*	*1945*	*(%)*	*1954*	*(%)*	*1964*	*(%)*	*1978*	*(%)*	*1987*	*(%)*	*1997*	*(%)*	*2002*	*(%)*	*2007*	*(%)*
Horses	**(95)**	**Cattle**	**(94)**	**Cattle**	**(92)**	**Corn**	**(91)**	**Corn**	**(87)**	**Corn**	**(90)**	**Corn**	**(79)**	**Corn**	**(68)**	**Corn**	**(58)**	**Corn**	**(54)**
Cattle	**(95)**	**Horse**	**(93)**	**Chicken**	**(91)**	**Cattle**	**(89)**	**Cattle**	**(81)**	**Soybeans**	**(68)**	**Soybeans**	**(65)**	**Soybeans**	**(62)**	**Soybeans**	**(54)**	Soybeans	(45)
Chicken	**(95)**	**Chicken**	**(93)**	**Corn**	**(91)**	**Oats**	**(83)**	**Hogs**	**(69)**	**Cattle**	**(60)**	Cattle	(47)	Hay	(42)	Hay	(37)	Hay	(28)
Corn	**(94)**	**Corn**	**(90)**	**Horses**	**(84)**	**Chicken**	**(82)**	**Hay**	**(62)**	**Hay**	**(56)**	Hay	(46)	Cattle	(42)	Cattle	(35)	Cattle	(32)
Hogs	**(89)**	**Hogs**	**(83)**	**Hogs**	**(81)**	**Hogs**	**(79)**	**Soybeans**	**(57)**	**Hogs**	**(50)**	Hogs	(35)	Hogs	(19)	Horses	(13)	Horses	(11)
Apples	**(84)**	**Hay**	**(82)**	**Hay**	**(80)**	**Hay**	**(72)**	**Oats**	**(57)**	Oats	(34)	Oats	(25)	Oats	(12)	Hogs	(11)	Hogs	(09)
Hay	**(82)**	**Potatoes**	**(64)**	**Oats**	**(74)**	Horses	(42)	Chicken	(48)	Horses	(13)	Horses	(10)	Horses	(11)	Oats	(08)	Oats	(03)
Oats	**(81)**	**Apples**	**(56)**	Apples	(41)	Soybeans	(37)	Horses	(26)	Chicken	(09)	Sheep	(08)	Sheep	(04)	Sheep	(04)	Sheep	(04)
Potatoes	**(62)**	**Oats**	**(52)**	Soybeans	(40)	Potatoes	(18)	Sheep	(17)	Sheep	(08)	Chicken	(05)	Chicken	(02)	Chicken	(02)	Goats	(02)
Cherries	**(57)**	Cherries	(34)	Grapes	(23)	Sheep	(16)	Potatoes	(06)	Wheat	(01)	Ducks	(01)	Goats	(01)				
Wheat	(36)	Grapes	(28)	Potatoes	(23)	Ducks	(05)	Wheat	(03)	Goats	(01)	Goats	(01)						
Plums	(29)	Plums	(28)	Cherries	(20)	Apples	(05)	Sorghum	(02)	Ducks	(01)	Wheat	(01)						
Grapes	(28)	Sheep	(21)	Peaches	(16)	Cherries	(04)	Redclover	(02)										
Ducks	(18)	Peaches	(16)	Sheep	(16)	Peaches	(04)	Apples	(02)										
Geese	(18)	Pears	(16)	Plums	(15)	Goats	(04)	Ducks	(02)										
Strawberry	(17)	Mules	(13)	Pears	(13)	Grapes	(03)	Goats	(02)										
Pears	(17)	Ducks	(12)	Redclover	(10)	Pears	(03)	Geese	(01)										
Mules	(14)	Wheat	(12)	Mules	(06)	Plums	(03)												
Sheep	(14)	Geese	(11)	Strawberry	(06)	Wheat	(03)												
Timothy	(10)	Sorghum	(09)	Ducks	(06)	Redclover	(03)												
Peaches	(09)	Barley	(09)	Wheat	(04)	Geese	(03)												
Bees	(09)	Redclover	(09)	Timothy	(04)	Popcorn	(02)												
Barley	(09)	Strawberry	(08)	Geese	(03)	Timothy	(02)												
Raspberry	(07)	Soybeans	(08)	Rye	(02)	Sweet potato	(02)												
Turkeys	(07)	Raspberry	(06)	Popcorn	(02)	Sweetcorn	(01)												
Watermelon	(06)	Bees	(05)	Sweetcorn	(02)	Turkeys	(01)												
Sorghum	(06)	Timothy	(05)	Rasberry	(02)														
Gooseberry	(03)	Turkeys	(04)	Bees	(02)														
Sweetcorn	(02)	Rye	(02)	Sorghum	(01)														
Apricots	(02)	Popcorn	(02)																
Tomatoes	(02)	Sweetcorn	(02)																
Cabbage	(01)	Sweetclover	(01)																
Popcorn	(01)	Goats	(01)																
Currants	(01)																		
n = 34		n = 33		n = 29		n = 26		n = 17		n = 12		n = 12		n = 10		n = 9		n = 9	

Based on data from the US Census of Agriculture, 1920–2007.

cost of that sizable investment over as large an operation as possible (you wouldn't invest in a paint sprayer to paint a single toothpick).

The aforementioned mechanical revolution found particular favor in the rapidly expanding agricultural sector in the USA; especially in the Midwest and Great Plains, where land was plentiful but labor scarce. Pfeffer (1983) argues that the variability of available labor across the USA – specifically looking at family farming in the Great Plains, corporate farming in California, and sharecropping in the South – helps explain the multiple trajectories taken by agriculture in the USA. Throughout much of Europe, conversely, the situation was just the opposite at the turn of twentieth century, where population was increasing but the arable land base remained unchanged, much as it did for centuries.

This chapter focuses primarily on the production "end" of the commodity chain – the farm – in terms of trends and socio-theoretical debates. It also starts the process, which is carried on through several chapters, of looking closer at what is known as the changing structure of agriculture. The term is a catchall of sorts that speaks generally of changes that are occurring throughout the food system, both on and off the farm. Some of these changes include the trend toward fewer and larger farms, the growth in market dominance of food processors and retail chains, and changing food consumption patterns. In this chapter, however, the changes that are examined are those occurring on the farm.

I begin by highlighting some the processes driving these structural changes in agriculture. Yet, with all the changes that have been undertaken on the farm one persistent artifact remains: the family farm. At the chapter's midpoint, I therefore review a flurry of scholarship from the late 1970s and 1980s that attempts to explain the perseverance of the family farm. The chapter concludes by discussing the conventionalization thesis of organic agriculture. The subject of organic agriculture is discussed extensively in Chapter 11. In this chapter, I focus solely on how organic agriculture appears to be reshaping itself in the image of capital and is therefore undergoing many of the same structural changes as conventional agriculture did decades earlier.

Agricultural treadmills

In the 1950s, agricultural economist Willard Cochrane introduced the concept of the "agricultural treadmill." According to Cochrane, farmers are under tremendous economic pressure to adopt new technologies and increase the size of their operations. The logic behind the argument is straightforward. Those first to adopt a technology (especially those technologies that increase yields and/or a farm's efficiency) initially experience windfall profits from increased output – like in the case of hybrid corn, which immediately, upon adoption, doubles a farmer's yield. Soon, other farmers, seeing the relative advantage of the technology and wanting to capture additional profits, choose also to adopt. Yet the cumulative effect of the heightened output eventually exerts downward pressure on prices as the market becomes flooded – lest we forget, new agricultural technologies rarely do anything to directly increase consumer demand. At this point, those yet to adopt the technology now *must* if they wish to remain in the market.

Let's say the price of corn drops 50 percent as a result of overproduction. Those producing 50 percent more corn are now making the *same* as they did prior to adopting the output increasing technology, while those who have yet to adopt are now making 50 percent *less*, even though they are farming exactly the same as they always have. Note the aptness of the "treadmill" metaphor: those who have invested in new technologies are essentially running in place. They might be producing considerably more but because of shrinking profit margins they are likely not profiting wildly from these increases in productivity; they may even be making less than they did prior to the treadmill pickling "speed." At the other end, producers slow to adopt are likely to fall off the treadmill, which, in turn, gives other farmers more land to acquire as they attempt to run still faster – though in reality most farmers are, at best, merely running in place.

For Cochrane, the logic of the treadmill points to inevitable overproduction as supply continually outstrips demand. To bolster his case, Cochrane restates Engel's Law (after the nineteenth century statistician Ernst Engel), noting that whereas demand for food grows mainly with population growth, and is therefore inelastic because the stomach is inelastic, production grows much faster as new technologies increase yields and as additional land is brought into production. One could criticize Cochrane, however, for overstating the fixed nature of food demand, which is understandable given when the treadmill argument was first described over 50 years ago. Since then, the biofuel industry has greatly expanded demand for grains, as has the livestock industry. Finally, the data do not entirely support Cochrane's premise that the human stomach cannot expand. According to the US government funded National Health and Nutrition Examination Survey (NHANES), for example, US men increased their calorie consumption from 1971 to 2000 by an average of 150 calories per day, while women increased their consumption by over 350 calories (see Chapter 4 for additional information on the so-called obesity epidemic) (Taubes 2007: 250).

Global grain prices have not acted quite as predicted by the treadmill argument either. Market prices and profit margins have not been in an unending downward decline since the dawn of twentieth century. It is obvious that, in absolute terms, the price of, say, corn has increased many times over during the twentieth century – in the USA from roughly US$0.30 a bushel in 1931 to around US$6 a bushel in 2008. Yet when indexed (converted to a normalized average that allows for comparisons across time), the price of farm outputs has remained surprisingly stable over the last 35 years.

Conversely, during that same 35-year period, the price of inputs has more than doubled – a phenomenon known as the double squeeze in agriculture (van der Ploeg 2006: 259; Figure 2.1). Unlike most other sectors of the economy, farmers sell their products at wholesale prices but pay retail for their inputs (Magdoff *et al.* 2000: 12). And over the decades the price of those inputs – fuel, seed, pesticides, and fertilizer – have been on the rise. Farmers are therefore "squeezed"

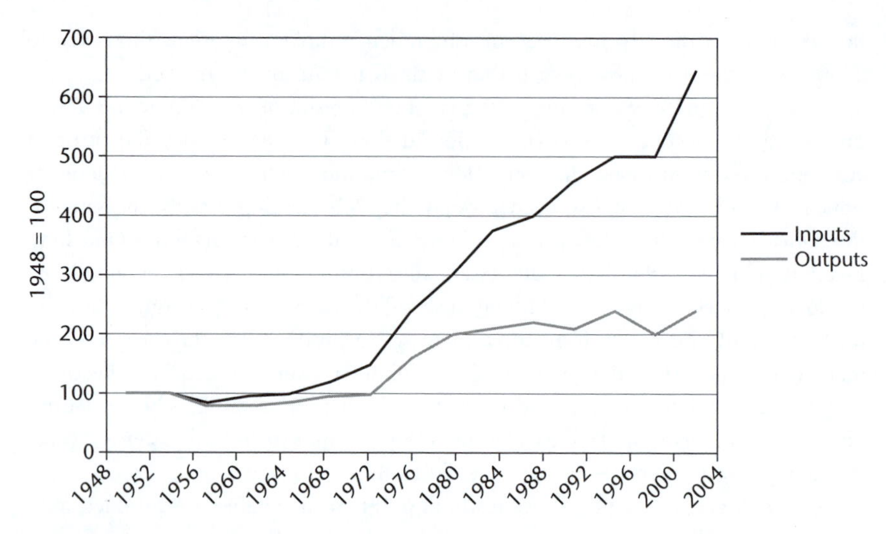

FIGURE 2.1 Price index for agricultural outputs and inputs, 1948–2004 (USA). Data from Fuglie *et al.* (2007: 3)

by both ends, as a result of stable (or declining) farm gate prices and increasing input costs.

The fact that there are far fewer full-time farmers in affluent countries than a couple of generations ago points to the enduring truth of the treadmill logic. Even though the average farm size has increased considerably with agriculture's industrialization, most farmers today can't make enough to survive without supplemental off-farm income. More land; more equipment; more expenses; and more output: all for the same (or less) profit.

Now let's look a little closer at the "squeeze" that farmers experience in terms of their costs. The agricultural treadmill allows us to speak *generally* about these ever-increasing costs. A consequence of being on the agricultural treadmill is the continual pressure to adopt the latest technologies and inputs while at the same time increasing the scale of one's operation. This expands operating costs, requiring farmers to produce more just to break even. Yet there are other, input-specific, treadmills also at work here, which give us a more fine-tuned understanding of why conventional (input-intensive) farming costs as much as it does (Figure 2.2). The "pesticide treadmill" is perhaps the best known of these additional treadmills, having first been discussed back in the 1970s (van den Bosch 1978). This is the phenomenon where insects evolve to become resistant to pesticides, which leads to more applications and/or higher concentrations and/or new chemicals, which leads to still further resistance, and so on. Studies going back to the 1970s note that despite large increases in pesticide use during the 1960s and 1970s, crop losses due to insect pests were actually *increasing* during the same period (Pimentel *et al.* 1978); a trend that continues to this day (Pretty 2004; see also Chapter 9, specifically its discussion on agroecology).

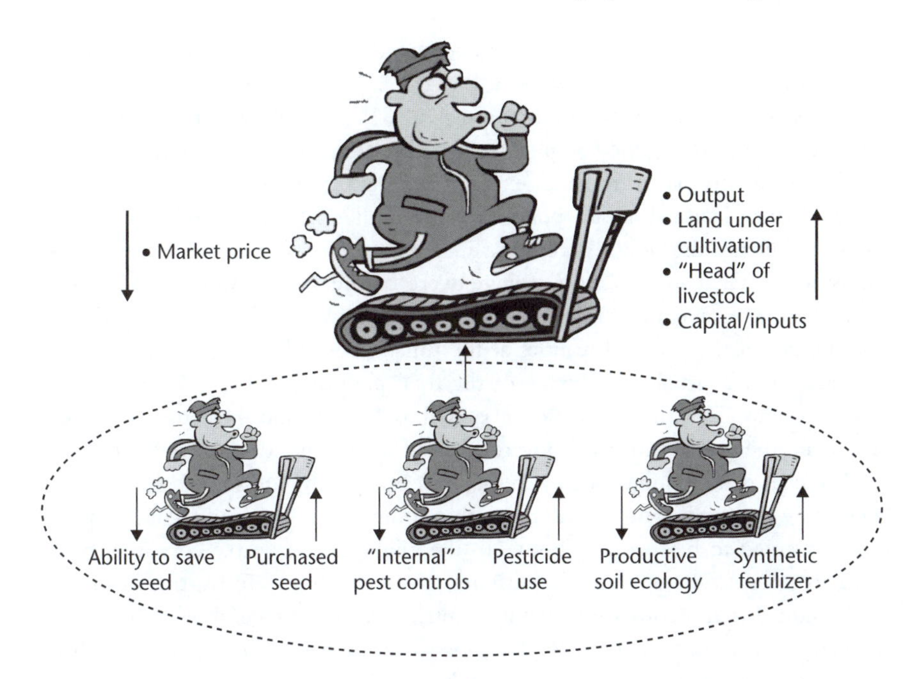

FIGURE 2.2 Agricultural and input treadmills of conventional agriculture

Another input-specific treadmill involves the use of synthetic fertilizers. Synthetic fertilizer use can beget higher application rates of these chemicals. This treadmill is particularly acute when tilth, levels of organic matter, and microorganism activity are diminished due to poor soil management practices and a general "mining" of one's land. Under such a scenario, the only option becomes ever-increasing application levels of synthetic fertilizer in order to maintain one's yield.

Lastly, there is the seed treadmill. A number of protections, as discussed in Chapter 3, are utilized by the seed industry which either discourage or outright prevent farmers from saving their seeds and replanting them the following season. Moreover, as discussed in Chapter 7, the knowledge of how to save and replant seeds is lost when farmers cease practicing seed saving. This creates a dependency upon seed firms. While seed saving continues to be widely practiced in many developing countries, it is coming under increasing threat (Carolan 2010). For example, in the USA the rate of saving corn seed fell from approximately 100 percent at the turn of twentieth century to less than 5 percent by 1960, while rates for soybean saving dropped from 63 percent in 1960 to 10 percent in 2001 (Howard 2009).

Agricultural subsidies

Governments have a long history of supporting agriculture. In 1839, for example, the US Patent Office began importing seeds from around the globe and freely distributing them to US farmers; a practice continued by the US Department of Agriculture (USDA) with its formation in 1862. British Corn Laws are another example. Introduced by the Importation Act 1815 (though later repealed by the Importation Act 1846), these import tariffs were designed to protect domestic corn (and other grain) prices from less expensive foreign imports.

Modern agricultural subsidy policy emerged in the early decades of the twentieth century. Initially, the policies were put into place to act as income redistribution measures, as farm incomes, on average, were considerably lower and less stable than those obtained in other sectors of the economy (Gardner 1987). Let's also be mindful of the political realities at the time. The rural voting electorate in all countries was a sizable percentage of the total population. In the USA in 1930, with 56 percent of the population classified as "urban" and 44 percent classified as "rural," the problem of rural poverty carried significant currency in Washington DC. Similar demographics were also found in Europe, though some countries (e.g. France) were significantly more rural than others (e.g. England). Politicians no doubt wanted to be seen as sensitive to the plight of this sizable constituency.

This justification, at least for subsidies, no longer holds. Today, thanks in part to subsidies, farm incomes in affluent countries are, on average, above those of the average (non-farm) household. But this doesn't mean that farmers are on the whole better off than their non-farming neighbors. What is different today, compared with the 1930s, is the level of income disparity that exists *within* the agricultural sector. And unlike three-quarters of a century ago, when subsidies served more as a safety net to protect those most in need, the overwhelming majority of farm subsidies today are going to the most wealthy.

Between 1995 and 2009, the US government paid out US$211 billion in farm payments (Figure 2.3). Of this, 88 percent (US$186.5 billion) went to 20 percent of all farms, leaving US$24.5 billion to be distributed to the remaining 1,760,000+ farms. And the top recipient? From 1999 to 2009, Riceland Foods, Inc. (Stuttgart, AR) received over US$554 million – that's more than what was paid to all the farmers in Alaska, Connecticut, Hawaii, Maine, Massachusetts, Nevada, New Hampshire, New Jersey, and Rhode Island combined. The top three recipients – Riceland Foods Inc., Producers Rice Mill (Wynne, AR) and Farmers Rice Cooperative (Sacramento, CA) – collectively received during this period over *1 billion* dollars. Perhaps even more stunning is the fact that between 1999 and 2005 US$1.1 billion were paid to US farmers who were no longer living (Peterson 2009: 147).

What about the European Union (EU) and its Common Agricultural Policy? How is their €55 billion annual agricultural subsidy pie distributed? It turns out that the majority goes to large landowners and agribusinesses. The biggest "piece" has been going to the Dutch firm Campina (a Dutch dairy cooperative which merged with Royal Friesland Foods in 2008). Between 1997 and 2009, Campina received €1.6 billion in taxpayers' money. The Denmark-based firm Arla Foods amba (formed by the 2000 merger between Sweden's Arla and Denmark's MD Foods) received the next biggest subsidy. Between 1999 and 2009, their payments totaled €952 million. Rounding out the top three was the London-based company Tate

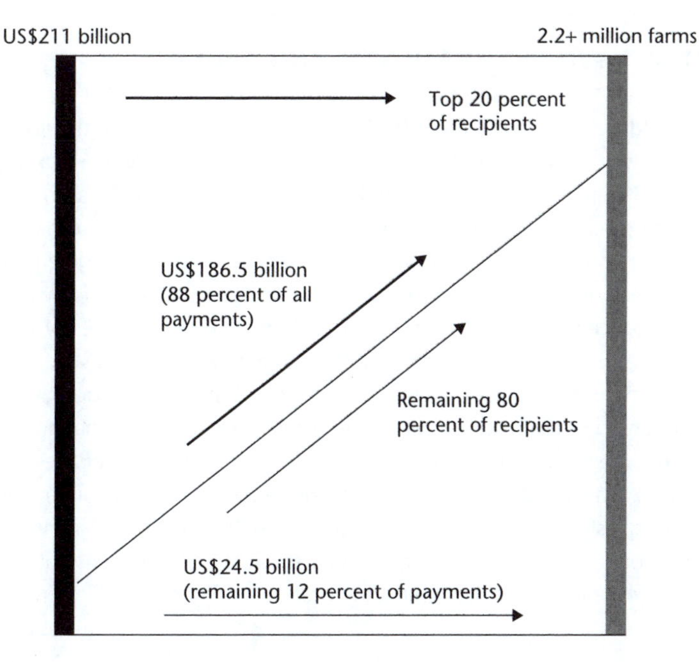

FIGURE 2.3 Asymmetries in US Department of Agriculture farm payments, 1995–2009. Compiled utilizing databank maintained by Environmental Working Group at http://farm.ewg.org/progdetail.php?fips=00000&progcode=total&page=conc

and Lyle Europe, who received €826 million between 1999 and 2009. All totaled: three companies, (roughly) 10 years, and €3.378 billion in agricultural subsidies.

Under the guise of enhancing global commerce, the World Trade Organization (WTO) has entered into the subsidy debate (Box 2.1). One of the goals of the WTO is to minimize policies that are viewed as trade distorting. But isn't that what agricultural subsidies do – send artificial signals to farmers causing them to act in ways that sometimes run counter to those being sent by the market?

To answer this question I'll need to talk a bit about the WTO. Domestic supports are assigned to three "boxes" – amber, blue, and green (Box 2.2). The amber box is the box to be avoided. Policies placed here are deemed trade distorting and must be reduced. Often policies aimed explicitly at increasing production, like input subsidies (e.g. reduced price fertilizer), are targeted for this box. In the blue box are measures designed to limit production, like conservation programs where farmers are paid to pull their land out of production. Blue box measures are not restricted. Finally: the green box. This box is for measures *said* to have no effect on production. These too are not restricted by the WTO. Green measures are the most interesting because in this box you'll find the agricultural subsidies of countries like the USA and regions such as the EU.

The USA, EU, and Japan account for 87.5 percent of the world's total green box expenditures (Maini and Lekhi 2007: 176). Since these policies are based on

BOX 2.1 WORLD TRADE ORGANIZATION

The World Trade Organization (WTO) was organized to oversee the liberalization of international trade. Its official origins date to January 1, 1995, replacing the General Agreement on Tariffs and Trade (GATT) which had been in existence since 1948. In addition to reducing trade barriers the organization provides a framework for negotiating and formalizing trade agreements as well as a dispute resolution process aimed at ensuring that participants adhere to trade agreements and free market principles. The WTO currently has 153 members, which represents roughly 98 percent of the world's population. The organization does not concern itself with developmental or societal "ends." Its primary concern, rather, lies in the establishment of international trade rules that are informed by an unwavering faith in free trade and free market principles. The WTO has perhaps as many detractors as proponents. Many not only questioning the inherent "goodness" of free trade, but are also quick to point out, as in the case of agricultural subsidies, the hypocrisy of affluent nations who fail to practice what they preach.

BOX 2.2 THE WTO'S COLORED BOXES FOR DOMESTIC SUPPORT IN AGRICULTURE

Amber box: Includes all domestic support measures believed to distort production and trade. These measures include policies to inflate commodity prices or subsides that directly encourage production. Member states are required to reduce such support unless current levels fall below certain parameters. Developed countries can keep a total of amber box support that is equivalent to up to 5 percent of total agricultural production, plus an additional 5 percent on a per crop basis. In developing countries their amber box support can go up to an amount that is equivalent to 10 percent of total agricultural production.

Blue box: Measures are directed at limiting production. There are currently no limits on spending on blue box subsidies.

Green box: Subsides that do not distort trade or that at most cause a minimal distortion (like government monies for research and extension). Green box measures can provide support for things like environmentally sound farm management practices, policies defined as being for regional or rural development, pest and crop disease management, infrastructure, food storage, income insurance, and even direct income support for farmers as long as that support is "decoupled" from production. There are no limits on green box subsidies.

Compiled from Rosset (2006: 84–6) and World Bank (2005: 25) "decoupled" payments – payments unrelated to production levels and therefore nontrade distorting – they can exist at unlimited levels. An example of this would be crop insurance, which is paid to farmers regardless of how much they produce (in this case a payment is made due to weather-related crop failure). For further examples of blue and green box subsidies see Box 2.3. Yet doesn't this still shield producers from low prices and therefore distort markets, in addition to clearly

BOX 2.3 EXAMPLES OF BLUE AND GREEN BOX AGRICULTURAL SUBSIDIES

Direct payments: Cash subsidies for producers of crops highlighted for support. Popular commodities include wheat, corn, sorghum, barley, oats, cotton, rice, soybeans, and oilseeds.

Marketing loans: A price-support program that typical guarantees minimum prices for certain crops.

Countercyclical payments: Provides a "safety net" in the event of low crop prices (for certain crops). Payments are issued if the price for a commodity is below the target price for the commodity.

Conservation subsides: Money dispensed to farmers for purposes of "conservation." An example of this would be the Conservation Reserve Program (CRP) in the USA. Created in 1985, the CRP pays farmers to leave idle millions of acres of farmland.

Insurance: "Yield" and "revenue" insurance are available to farmers to manage risk of adverse weather, pests, and low market prices. This subsidized insurance allows farmers to pay roughly one-third the full cost of the policy.

Disaster aid: This is a form of emergency crop relief due to "acts of God" (possibly for crops already covered by subsidized crop insurance).

Export subsides: Government money used to expand a country's agricultural exports.

Agricultural research and statistics: Government money used to support agricultural research, statistical information services, and economic studies. In some instances, this taxpayer-funded research results in patented products that consumers (and farmers) must pay for again in the form of higher commodity (e.g. seed) prices.

Indirect subsides: Subsides that indirectly reduce the cost of production (e.g. subsidized electricity or undervalued water) or increase demand for a commodity (e.g. subsidies for ethanol processing plants or government biofuel mandates).

From the Cairns Group (2010) and Edwards (2009)

advantaging those with crop insurance over those in other countries lacking this safety net?

Why, then, are these policies allowed to continue? And how have we gotten to a place where the 30 most industrialized economies spend roughly a billion dollars a day on food and agricultural subsidies (Peterson 2009)? There are no easy answers to questions like these. One way that's helpful to think about these questions is to note how the benefits of farm subsidies are concentrated (illustrated earlier in Figure 2.3), while their costs are relatively diffuse (spread across all taxpayers). Agricultural producers have more to lose than what any other group would individually gain from the removal of subsidies. Consequently, those who gain from this arrangement will likely spend more to keep the status quo than consumers and taxpayers, for whom subsidies cost them individually very little. And so while agricultural subsidies might change subtly from year to year in any given country the policies themselves will likely not go away anytime soon.

Another useful concept as we think through why subsidies have proven so resilient is policy path dependency: the well-documented phenomena whereby once a policy is enacted it tends to become "locked in." The greater the levels and history of support, the more dependent the farm and farm supply chain becomes on continuing levels of support, which creates tremendous resistance to the removal of these policies. As farm outputs today are utilized by not only traditional food manufactures but also fuel, livestock, and industrial sectors of the economy, there are a lot of parties interested in maintaining the status quo (though as New Zealand has shown, as discussed in Box 2.4, it is possible to roll subsidies back).

Let's investigate further the question, "Who benefits from subsidies?" Starmer and Wise (2007a) look at the links between grain subsidies and the livestock industry in the USA. Between 1997 and 2005, according to their calculations, the livestock sector saved roughly US$3.9 billion annually from being able to purchase their feed (corn and soybeans) at prices below what it cost to raise them. Total savings by industrial hog, broiler, egg, dairy, and cattle operations totaled close to US$35 billion over the 9-year period. Elsewhere, Starmer and Wise (2007b) turn their focus on the swine industry. According to their calculations, grain subsidies gave large-scale hog operations in the USA a 10 percent feed discount before 1996 and a 26 percent discount after this date. For Smithfield Farms, the largest hog-producing company in the USA (with a 31 percent market share), this translates into a collective savings of US$2.54 billion from 1997 to 2005 (or an average savings of US$284 million).

Figure 2.4 illustrates the differences in operating costs between hog-CAFOs (concentrated animal feeding operations) and mid-sized hog farms. Merely adding the costs of full-priced feed negates the price advantage of CAFOs relative to mid-sized producers. As illustrated in the figure, current average operating cost of CAFOs and mid-sized producers is estimated to be US$50.11 and US$45.85 per hundredweight, respectively. If we were to include the additional costs of full-price feed, those figure change to US$45.84 and US$45.85, respectively.

There is a considerable amount of debate around whether to roll back subsidies entirely or to allow them to exist, just in a highly reformed state. At most,

BOX 2.4 NEW ZEALAND'S ROLLBACK OF AGRICULTURAL SUBSIDIES

The New Zealand government's Ministry of Agriculture and Fisheries explains that "no other country in the world has reduced its subsidies for agricultural production to the same extent [. . .] from 34 percent of gross agricultural revenue in 1984 to almost zero in 1995" (as quoted in Le Heron and Roche 1999: 203–4). The level of government protection for agriculture in New Zealand is among the lowest in the developed world. Export subsidies were eliminated and import tariffs were gradually phased out. Fertilizer and other input subsidies were also eliminated, as were free government services for farmers.

Land prices fell immediately on the heels of the reforms, improving the farm industry's international competitiveness; strong support for David Ricardo's almost 200-year-old theory predicting government support for farmers will inevitably drive up the cost of production, thereby further increasing levels of protection, which further drives up the costs of production, and so forth. The elimination of subsidies has led to significant reductions in the use of fertilizers and pesticides as well as in levels of soil erosion, while improving water quality; though the increased use of nitrogen fertilizer, used to feed a growing dairy sector, to some degree negates some of these environmental gains. A steady change of land use has also been recorded since the mid-1980s, from pastoral agriculture (declining from 14.1 million hectares in 1983 to 12.3 million in 2004) to forestry (increasing from 1 million hectares in 1983 to 2.1 million in 2004) (FAO 2006: 232).

Finally, the removal of subsidies is said to be responsible for the further intensification of stocking and crop systems and concentration within certain food sectors, most notably in the dairy industry (Willis 2004).

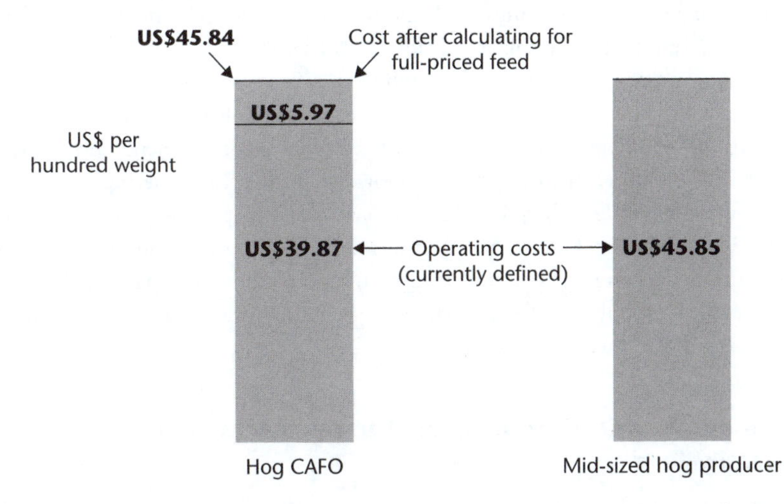

FIGURE 2.4 Operating costs of large-scale and mid-scale hog production (before and after calculating for full-priced feed). Based on data from Starmer and Wise (2007b)

one-tenth of 1 percent of all domestic agriculture subsidies in the USA goes to supporting fruit and vegetable crops (Fields 2004: 822). What's keeping governments from doing more to subsidize healthy foods? As Richard Atkinson, Professor of Medicine and Nutritional Sciences at the University of Wisconsin-Madison, argues: "There are a lot of subsidies for the two things we should be limiting in our diet, which are sugar and fat, and there are not a lot of subsidies for broccoli and Brussels sprouts" (quoted in Fields 2004: 823). Or, at the consumption end, why is additional money not spent on programs that help those of lower socio-economic status buy fresh fruits and vegetables (like food stamps)? Would not such a strategy also help fruit and vegetable growers by further spurring demand for their commodities?

Perhaps it's not a black and white question of whether agricultural subsidies ought to exist or not. Maybe we should be more interested in who are being helped by subsides and the societal ends being served. In suggesting this I am thinking about a recent case involving the African country of Malawi. For years, the country was starving, suffering from persistent famine. In 2005, after a disastrous corn harvest, 5 million of the county's 13 million residents needed emergency food aid. By 2007, the country was turning away food aid. What happen? Two words: fertilizer subsidies. Since the 1980s the World Bank was leaning on Malawi to eliminate its fertilizer subsidies – as input subsides are aimed at increasing production levels they fall into the dreaded amber box. The removal of this subsidy, however, proved disastrous to poorer Malawian farmers. With it, most could no longer afford fertilizer. Yet the gravely depleted soils in the country require inputs, artificial (e.g. chemical fertilizer) and/or otherwise (e.g. manure). Malawi's President, Bingu wa Mutharika, seeing other countries extensively subsidize their own farmers, decided after the 2005 harvest to bring back fertilizer subsidies.

The Malawi case also brings us back to the question, "Who benefits from subsidies?" Even with a fertilizer subsidy, Malawian farmers are at a significant disadvantage in the global marketplace when they have to compete against farms that can receive upwards of hundreds of thousands of dollars annually from their respective governments. Governments of affluent nations spend on average US$6,000–10,000 per farm laborer per year. For comparison, the typical African government spends less than US$10 per farm worker per year on agriculture (Bezemer and Headey 2008: 1350). And as for the link between subsidies and the aforementioned changing structure of agriculture: a study by USDA economists implicates agricultural subsidies for having contributed directly to increases in the average size of US farms (Key and Roberts 2007).

Sociology of agriculture explains the perseverance of the family farm

The previous chapter mentions rural sociology's theoretical "turn" in the 1970s, which gave birth to New Rural Sociology and, in time, to what we know today as sociology of food and agriculture. The earliest contributions associated with this turn took the form of a Marxist political economy of agriculture. Prior to this, Marx was all but ignored in rural sociological circles. There were multiple reasons for this, from the unpopularity of Marxist thought in the traditionally more conservative land-grant system to a general theoretical problem with Marxism that seemed to suggest that this framework did not translate well to agriculture. The latter problem centered on the continued perseverance of the family farm. In most other sectors of the economy the capitalist and labor classes were becoming increasingly distinct. Yet in agriculture the family farm, through the first half of the twentieth century at least, was not only surviving but thriving. The "discovery" of Karl Kautsky's writings were immensely influential upon agrarian-minded sociologists of that era, specifically his book *The Agrarian Question* (published in 1899). The book attempts to not only extend Marx's analysis of capitalism to agriculture but also explains why, as all other sectors of the economy were becoming more capitalistic, the German family farm continued to persist.

The most famous article from that era comes from Susan Mann and James Dickinson. Published in 1978, this path-breaking article, "Obstacles to the development of a capitalist agriculture," rejects the dominant "subjectivist" arguments about why family farms persist in advanced capitalist societies. One popular subjectivist argument at the time was that family farms persist because of their willingness to accept lower levels of wages – self-exploitation – allowing them to overcome the productivity advantages of capitalist farms. Mann and Dickinson develop instead, through Marx, a structural explanation for why capital–labor relations proceed more slowly in agriculture than in other sectors of the economy.

Mann and Dickinson (1978: 467) note that "[c]apitalist development appears to stop, as it were, at the farm gate." That represents, for them, a significant anomaly in relation to Marx's theory of the transitional (which is to say temporary) nature of petty commodity production (the family farm). What then accounts for this anomaly? According to Mann and Dickinson (1978: 466), "the secret of this 'anomaly' lies in the logic and nature of capitalism itself." To put it another way, there's something about nature itself that is holding capitalism at bay when it approaches the farm gate. Drawing inspiration from Marx, Mann and Dickinson identify the following "barriers": the gap between production time and labor time; unproductive production time; and risks and/or invariabilities of production.

The gap between production time and labor time

Marx distinguished between the time it takes to produce a commodity, what he called "production time," and the amount of labor time injected into that production process, what he called "labor time." Often the two are inaccurately conflated. Yet, while "working time is always production time [...] not all time during which capital is engaged in the process of production is necessarily working time" (Marx 1967: 242). In many industries these two times almost imperceptibly overlap, such as in manufacturing. And when there is some divergence between the two, such as in the painting of an automobile, technologies are employed to bring the two times more in line (e.g. utilizing special ovens and/or quick-dry paints). In agriculture, however, production time and labor time are often wildly out of sync.

Factors inherent in the production of living things have historically presented natural barriers that capital has been slow to penetrate. As Mann and Dickinson explain:

> For example, cereal production is characterized not only by a relatively long total production time (as the produce only matures annually), but also by a great difference between production time and labor time; there is a lengthy period when labor time is almost completely suspended as when the seed is maturing in the earth. In this case the reduction of production time is severely restricted by natural factors and thus cannot easily be socially modified or manipulated as occurs in industry proper. Similarly, in stock production, the reproduction of the species is prescribed by definite natural processes. Neither the period of gestation nor the period of growth to economic maturity (i.e. to five-year-old cattle) can be easily shortened.
>
> *Mann and Dickinson (1978: 472)*

And those spheres of agricultural production where labor time and production time diverge the greatest from each other will "prove unattractive to capital on a large scale" (Mann and Dickinson 1978: 473).

In the thirty-plus years since the Mann–Dickinson thesis was first articulated much within agriculture has changed. Most notably, the gap between labor time and production time has been reduced considerably with the help of things such as genetic engineering and hormones. For instance, in the 1930s beef cows did not go to slaughter until they were 4 or 5 years old. In the 1950s, that number was reduced to between 2 and 3 years. Today, the magic number is down to around 14 months. Dairy cows injected with the hormone rBST (recombinant bovine somatotrophin), which helps mobilize body fat to use for energy and diverts feed energy more toward milk production than for tissue synthesis, experience increases in milk production by, on average, 10–15 percent (although increases of 30 percent have been recorded).

Unproductive production time

The divergence between labor time and production time also means there is a lot of unproductive production time. If rent is paid by the farmer that rent covers the entire growing year, regardless of whether they are engaged in value-producing labor. Large-scale, industrial agriculture also involves significant sunken costs, in machinery, buildings, land, and so forth. Many of these items are not only steadily depreciating but can lay idle for many months of the year. A US$100,000 combine, for instance, may be used for only a couple of weeks during the fall harvest. The remainder of the year it lies in the machine shed collecting dust and depreciating. This "[i]dle, constant capital thus appears as a burden to the farmer and something to be avoided by the capitalist" (Mann and Dickinson 1978: 475).

One of goals, then, of any profit maximizing farmer is reducing this unproductive production time. Increasing the efficiency and thus productivity of, for example, a dairy cow increases the number of milkings a dairy farmer performs in a day and therefore reduces the idle time of their milking parlor and milking equipment. Other ways to reduce this unproductive production time is to hire someone and their equipment (known as "custom work") for a specific task to avoid having to purchase for oneself an expensive piece of equipment that is only used a handful of times a year.

Risks and/or invariabilities of production

Food can also be highly perishable, especially compared with typical industrial products, like cars, shoes, and light bulbs. As Marx (1967: 131) wrote, "the more perishable the commodity is and the greater the absolute restriction of its time in circulation as a commodity on account of its physical properties, the less it is suited to be an object of capitalist production." Yet recent changes in the food system have reduced these risks considerably, from improvements in packaging, transportation, and refrigeration to the invention of food additives and the genetic engineering of food to withstand picking and transportation. Uncertainties related to weather have also historically presented considerable risk to farmers, though crop insurance and federal farm subsidies now make the production of certain commodities a far safer bet than was the case a century ago.

When it comes to their natural barriers thesis, Mann and Dickinson are not arguing that capital has not penetrated agriculture. For them, like Marx, any and all barriers are *potentially* penetrable, with sufficient research, capital, and knowledge. The question is more a matter of "when?" than of "why not?" Goodman *et al.* (1987: 20–1) speak succinctly to the point about how capitalism (often quite successfully) has had to treat agriculture differently from most nonfarm industries. In their own words:

> In manufacturing, "nature" is broken down by processing and introducing into the machine as a raw material input, which thus can be adapted to the speed of machine production. By contrast, nature in agricultural production cannot be reduced to an input; indeed it is the "factory" itself. Consequently, instead of restructuring the production process, mechanization effectively represented an implement adapted to the spatial and temporal characteristics of agriculture. Rather than the Copernican revolution of manufacturing whereby nature must circulate around the machine, nature in agriculture maintains its predominance and it is the machine which must circulate.
>
> *Goodman* et al. *(1987: 20–1)*

Harriet Friedmann also contributed to this Marxist renaissance, taking an approach similar to that of Mann and Dickinson. Yet, whereas Mann and Dickinson focus on why capitalists have not been as interested in investing in production agriculture, Friedmann argues that the persistence of family farms rest on their ability to match if not outcompete capitalist firms. Friedmann (1978) documents how, following the global drop in market price for wheat in the late 1800s, family wheat producers in the USA outcompeted capitalist producers from around the world. She attributes their success and persistence to their superior flexibility. This allowed family farms, during this period of depressed market prices, to reduce their level of consumption to a subsistence level and liquidate their assets.

The work of Patrick Mooney from this era is also significant, as not only a notable critic of Mann and Dickenson but also for his attempt to synthesize Marxian and Weberian approaches. Looking at various measures of capitalist penetration in different commodity groups, Mooney (1983) argues that while family farms may not be capitalist in the strictest sense of the word they are still part of a capitalist system and therefore subject to its logics of exploitation. Mooney thus speaks of "detours" to capitalist exploitation rather than immutable "barriers"; though, as Mann and Dickinson (1987) later point out, their "barriers" are not insurmountable.

Among other things, Mooney points out a variety of ways in which capital penetrates agriculture. As Mooney highlights, the family farm–capitalist farm dichotomy is a tremendously oversimplified lens through which to understand exploitation in agriculture. More problematic still, as illustrated in Figure 2.5, Mooney shows just how difficult it is to place those involved in agriculture in the pure class locations of "proletariat," "capitalist," and "petty bourgeoisie" (the latter category being where Marxist thought typically locates the family farm). Mooney explains some of these contradictory class relations as follows:

> In the usual case, we will find that the tenant, debtors, or contract producers who hire labor also engage in manual production tasks alongside their worker. Like capitalist farmers, but unlike simply commodity producers, they redistribute surplus value to the creditor. Like proletarians, but unlike capitalist, they are active in the production of alienated surplus value. [. . .] With the capitalists, they control the labor power of others. With the proletariat, they control over production processes, perhaps even the workers they hire, is immediately circumscribed by capital in the terms of credit, lease, or contract. [. . .] Domination of the market by monopoly capital eliminates an essential characteristic of the petty bourgeois mode of production, free exchange on an open market.
>
> *Mooney (1986: 232)*

Mooney argues that many of these indeterminate class locations, or "contradictory classes" (from Wright 1978), reflect a "new petty bourgeoisie."

Yet I would argue that Mooney's most valuable contribution is his attempt to open a Marxist political economy of agriculture up to account for nonlinearity – to be,

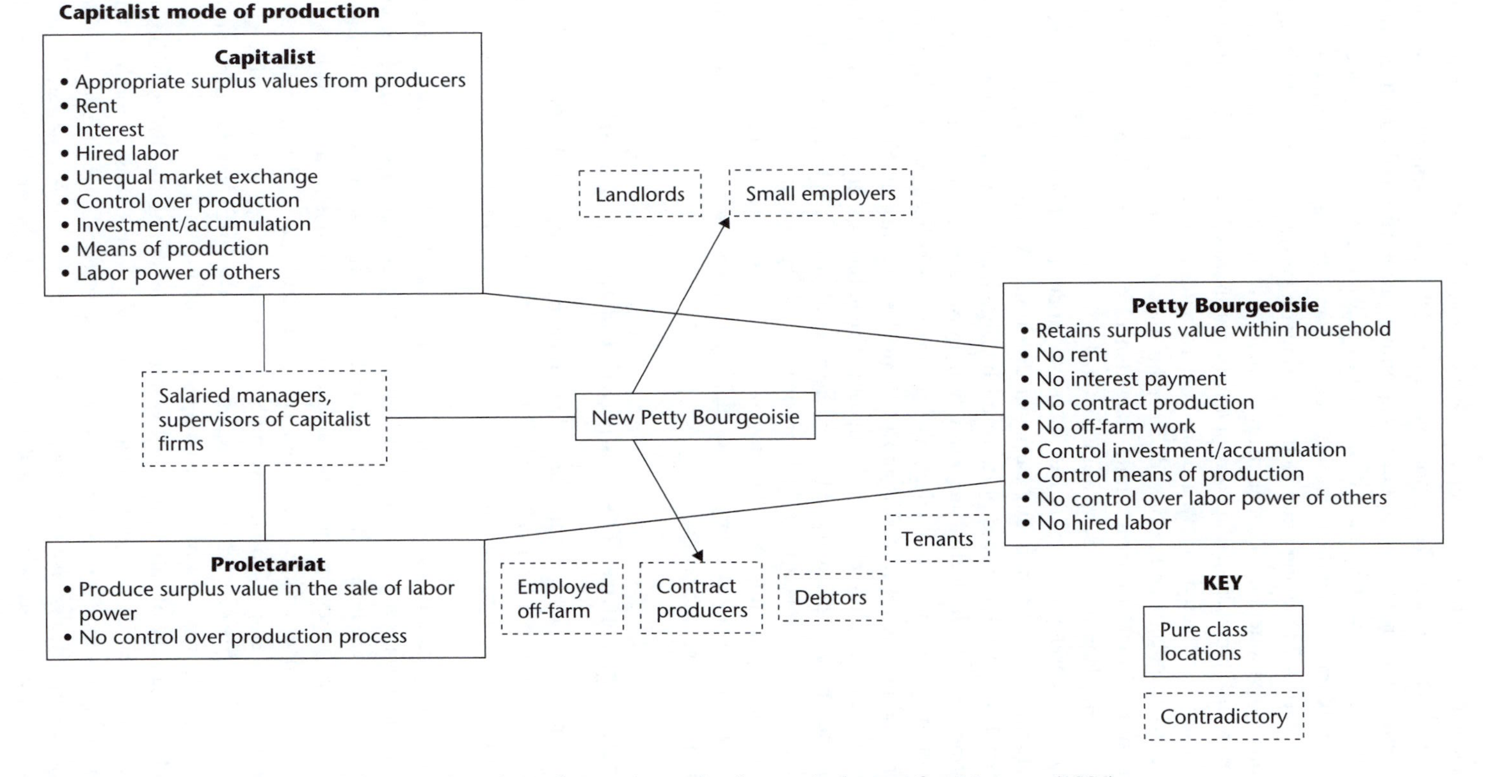

FIGURE 2.5 Mooney's "map" of the complex class relations in production agriculture. After Mooney (1986)

in other words, explicitly nondeterministic. Devoid of any understanding of subjective experience and human action, Mann and Dickinson paint themselves into a theoretical corner when it comes to explaining historical variability. If the omnipresent logics of capitalism are all we have to explain agriculture, why does it look so very different around the world, even in spaces that share the same so-called natural barriers? Drawing upon Weber's distinction between formal and substantive rationality, Mooney argues that many farmers are motivated by more than just profit (formal rationality). Many are equally (if not more so) motivated by substantive rationality, as reflected in, say, a desire to have autonomy over their work, to be stewards of the land, and to provide a certain quality of life (that can't be obtained with money alone) for their family. As evidence of his rejection of linear theorizing, Mooney later moved away from his initial objective of theorizing the "detours" to capitalist development to his later desire to "undermine the notion that capitalist development need flow in any particular direction" (Mooney 1987: 293).

Boyd *et al.* (2001) extend portions of this argument when they propose a distinction between the formal and real subsumption of nature. Doing this highlights the distinct ways that biological systems are industrialized and often made to operate as productive forces in and of themselves. Formal and real subsumption allow for analytic distinctions to be made between biologically based (e.g. cultivation) and nonbiologically based (e.g. extractive) industries. Under real subsumption, capital circulates *through* nature (albeit unevenly) as opposed to *around* it, as in the case of formal subsumption. This move allows for a more forceful argument to be made about the social agency of capital itself, as having the power to literally transform nature into its own image.

Organic agriculture: The "conventionalization debate"

The concept of "conventionalization" to reference trends in the organic sector was first introduced by Buck *et al.* (1997) to describe the infiltration of capital into the organic vegetable industry in Northern California. As agribusiness capital became increasingly successful in "finding ways to re-shape the particularities of organic agriculture to its own advantage" (Buck *et al.* 1997: 17) the original spirit of the organic movement became marginalized as ecological or social concerns went by the wayside. Another key term in this literature is "bifurcation," which is said to be the result of conventionalization. As agribusiness enters organics, the organic system essentially splits into two: one made up of larger conventional operations that employ input substitution strategies directed at monoculture production of high value crops for distant markets; the other consisting of smaller farms growing a variety of products for local markets.

The originality of this argument lies not in questioning the ability of organic agriculture to transform conventional agriculture. As the previous section highlights, scholars by this time were very careful not to interpret change – even seemingly transformative change like the growth of the organics sector – as necessarily being antithetical to the interests of capital. Buck *et al.* were, however, the first to research systematically into the structural trends taking place in organics; a research project that has since been thoroughly extended by others.

The conventionalization of organics was helped along considerably by the formation of a uniform organic standard for the entire USA. Prior to this, the regulatory environment was a patchwork, with considerable variability across the country. A single organic standard benefited agribusinesses. It meant they only needed to invest according to that single standard, after which they would then have access to high price premiums across the entire country. The standards eventually adopted in the USA – as throughout much of the world – allow for input substitution. That means, just like in conventional agriculture, "internal" (agroecological) controls can be substituted for external inputs. The resulting agriculture thus differs from the conventional systems only in the *type* of inputs used. There is therefore nothing about the "organic" label that says it cannot come from a system that is energy-intensive, based on large-scale monocultures, and inattentive to social justice and animal rights issues. There are unquestionably still smaller organic operations out there, just as there are organic farmers who treat their workers and animals with great care. Yet they are frequently relegated to those local markets that do not interest large agribusinesses, like farmers markets, roadside stands, and subscription farming (such as community supported agriculture). While these local arrangements might seem to contradict the conventionalization thesis, according to proponents of the thesis, they are nothing more than default choices for growers with few alternatives.

The conventionalization thesis is not without its critics. Some have questioned the apparent implied inevitability of conventionalization. Research from New Zealand has led some to conclude that small-scale organic growers can in fact thrive under the current arrangement, even with the obvious infiltration of capital into organics (see, e.g., Coombes and Campbell 1998; Campbell and Coombes 1999). While these studies acknowledge that organics is undergoing a level of conventionalization and bifurcation, they disagree that the trends are inevitable or inherently negative. Take, for example, the government programs in Europe that explicitly support small-scale, artisan, and (often) organic agriculture. And national certification standards: do they not benefit *all* organic producers – large and small – by increasing the legitimacy of the entire organic sector?

In response to critics, and specifically the argument that small-scale organic growers need not worry about becoming marginalized by larger producers, Guthman (2004a) points to the regulatory success had by agribusiness in the formation of federal organic standards in the USA. According to Guthman, it is not just that corporate actors have been able to extract significant profits from organics. More importantly, their very participation in organics "alters the conditions under which all organic growers participate in the sector by unleashing the logic of intensification" (Guthman 2004b: 304). In other words, all organic growers, small and large alike, are essentially now participating in a game whose rules were written by (and thus favor) agribusiness.

The last decade has not done much to settle the debate, as evidence for both (and sometimes neither) "sides" is accumulating. For example, a study of organic farmers in Ontario, Canada, had mixed results (Hall and Mogyorody 2001). While finding signs of conventionalization and bifurcation among crop farmers, organic fruit and vegetable operations showed signs of neither, tending overall to consist of small, diverse farms that market directly and locally to consumers. Research from Australia and New Zealand, pointing to the expansion of organic markets as an example of "opportunistic corporate greening" (Lyons 1999: 262; see also, e.g., Lockie *et al.* 2000), lends support for the conventionalization thesis. Another study, in an attempt to better understand the local and/or regional context of the conventionalization of organics in Denmark and Belgium, finds various institutional factors that shape the structure of organic farming systems (Lynggard 2001). A more recent study focusing on organic dairy producers in upstate New York also has trouble coming down firmly on either "side" of the debate (Guptill 2009). On the one hand, its findings lend some support to the conventionalization hypothesis, noting that organic milk has long been treated as a commodity and the federal regulations governing organic dairies have facilitated making the industry into a fully fledged commodity-based system. Yet it is also found that some organic dairies are responding to these pressures to conventionalize by becoming, in certain respects, "more" organic, "by going deeper into the alternative organic model, forging more direct and local relationships along the value chain and embracing principles of the organic movement" (Guptill 2009: 29).

Transition . . .

The focus of this chapter has remained (mostly) confined to the farm gate. This limited scope in part mirrors the scope of sociology of agriculture during its early days, from its inceptions in the 1970 up through the 1980s – though, as reflected by the conventionalization thesis debate, the farm remains a popular site of study for scholars up to this day. The 1990s, however, mark a noticeable change in the literature, as scholarship expands beyond (while still interested in what goes on within) the farm gate. And with that, the sociology of not just agriculture but also of food (systems) begins to take shape.

Discussion questions

1 Besides mechanization, what else might explain the specialization trend depicted in Table 2.1?
2 What role (if any) should government subsidies have in our food system?
3 Applying the Mann–Dickinson thesis to agriculture today, what "natural barriers" do agricultural scientists seem most interested in overcoming? Are any such barriers insurmountable in your estimation?
4 Retail giants like Walmart have recently started selling local foods. Does this mean local food is becoming "conventionalized?"
5 How might Willard Cochrane's agricultural treadmill argument inform our understanding of the conventionalization thesis?

Suggested reading: Introductory level

Guthman, Julie. 2003. Fast food/organic food: Reflexive tastes and the making of yuppie chow. *Social and Cultural Geography* 4(1): 45–58.

Strange, Marti. 2008 (1988). *Family Farming: A New Economic Vision*, second edition. Lincoln, NE: University of Nebraska Press.

Suggested reading: Advanced level

Allen, Patricia and Martin Kovach. 2000. The capitalist composition of organic. *Agriculture and Human Values* 17(3): 221–32.

Boyd, W., W. Prudham, and R. Schurman. 2001. Industrial dynamics and the problem of nature. *Society and Natural Resources* 14(7): 555–70.

Guptill, Amy. 2009. Exploring the conventionalization of organic dairy: Trends and counter-trends in upstate New York. *Agriculture and Human Values* 26(1–2): 29–42.

Mann, Susan and James Dickinson. 1978. Obstacles to the development of capitalist agriculture. *Journal of Peasant Studies* 5(4): 466–81.

Mooney, Patrick. 1983. Toward a class analysis of Midwestern agriculture. *Rural Sociology* 48(4): 563–84.

References

Bezemer, D. and D. Headey. 2008. Agriculture, development and urban bias. *World Development* 36(8): 1342–64.

Boyd, W., W. Prudham, and R. Schurman. 2001. Industrial dynamics and the problem of nature. *Society and Natural Resources* 14(7): 555–70.

Buck, Daniel, Christina Getz, and Juile Guthman. 1997. From farm to table: The organic vegetable commodity chain of northern California. *Sociologia Ruralis* 37: 3–20.

Cairns Group. 2010. *Export Subsidies, Fact Sheet.* Cairns Group, Cairns, Australia, www.cairnsgroup.org/factsheets/export_subsidies.pdf, last accessed September 22, 2010.

Campbell, Hugh and Brad Coombes. 1999. Green protectionism and organic food exporting from New Zealand: Crisis experiments in the breakdown of fordist trade and agricultural policies. *Rural Sociology* 64(2): 302–19.

Carolan, Michael. 2010. *Decentering Biotechnology: Assemblages Built and Assemblages Masked.* Burlington, VT: Ashgate.

Colbert, Thomas. 2000. Iowa farmers and mechanical corn pickers, 1900–1952. *Agricultural History* 74(2): 530–44.

Coombes, Brad and Hugh Campbell. 1998. Dependent reproduction of alternative modes of agriculture: Organic farming in New Zealand. *Sociolgia Ruralis* 38(2): 127–45.

Edwards, C. 2009. *Agricultural Subsidies*. Washington, DC: CATO Institute, www.downsizinggovernment.org/agriculture/subsidies, last accessed September 22, 2010.

FAO. 2006. Livestock's Long Shadow: Environmental Issues and Options, Food and Agriculture Organization of the United Nations, Rome, Italy, ftp://ftp.fao.org/docrep/fao/010/a0701e/a0701e.pdf, last accessed March 2, 2010.

Fields, S. 2004. The fat of the land: Do agricultural subsidies foster poor health? *Environmental Health Perspectives* 112(14): 820–3.

Friedmann, Harriet. 1978. World market, state, and family farm: Social bases of household production in an era of wage labor. *Comparative Studies in Society and History* 20: 545–86.

Fuglie, Keith, James McDonald, and Eldon Ball. 2007. Productivity Growth in US Agriculture, Economic Brief Number 9, United States Department of Agriculture, Economic Research Service, Washington, DC, http://www.ers.usda.gov/publications/EB9/eb9.pdf, last accessed September 21, 2010.

Gardner, Bruce. 1987. Causes of US farm commodity programs. *Journal of Political Economy* 95(2): 290–310.

Goodman, David, Bernardo Sorj, and John Wilkinson. 1987. *From Farming to Biotechnology: The Industrialization of Agriculture*. New York: Blackwell.

Guptill, Amy. 2009. Exploring the conventionalization of organic dairy: Trends and countertrends in upstate New York. *Agriculture and Human Values* 26(1–2): 29–42.

Guthman, Julie. 2004a. *Agrarian Dreams: The Paradox of Organic Farming in California*. Berkeley, CA: University of California Press.

Guthman, Julie. 2004b. The trouble with 'organic lite': A rejoinder to the 'conventionalization' debate. *Sociologia Ruralis* 44(3): 301–16.

Hall, Alan and Veronika Mogyorody. 2001. Organic farms in Ontario: An examination of the conventionalization argument. *Sociologia Ruralis* 41(4): 399–422.

Howard, Philip. 2009. Visualizing consolidation in the global seed industry: 1996–2008. *Sustainability* 1: 1266–87.

Key, N. and M. Roberts. 2007. Commodity Payments, Farm Business Survival and Farm Size Growth, Economic Research Report, No ERR-51, Economic Research Service, USDA, Washington, DC, www.ers.usda.gov/Publications/ERR51/ERR51ref.pdf, last accessed August 11, 2011.

Le Heron, Richard and Michael Roche. 1999. Rapid reregulation, agricultural restructuring, and the reimaging of agriculture in New Zealand. *Rural Sociology* 62(2): 203–18.

Lockie, Stewart, Kristen Lyons, and Geoffrey Lawrence. 2000. Constructing 'green' foods: corporate capital, risk and organic farming in Australia and New Zealand. *Agriculture and Human Values* 17: 315–22.

Lynggard, Kennet. 2001. The farmer within an institutional environment: Comparing Danish and Belgian organic farming. *Sociologia Ruralis* 41(1): 411–27.

Lyons, Kristen. 1999. Corporate environmentalism and the development of Australian organic agriculture: The case of Uncle Tobys. *Rural Sociology* 64(2): 251–65.

Magdoff, Fred, John Foster, and Fred Buttel. 2000. An overview. In *Hungry for Profit: The Agribusiness Threat to Farmers, Food, and the Environment*, edited by Fred Magdoff, John Foster, and Fred Buttel. New York: Monthly Review Press, pp. 7–21.

Maini, Kusum and R.K. Lekhi. 2007. Implications of World Trade Organization on dairy sector of India. In *Emerging Dimensions of Global Trade*, edited by R.S. Jalal and Nandan Sighn Bisht. New Delhi: Sarup and Sons, pp. 174–80.

Mann, Susan and James Dickinson. 1978. Obstacles to the development of capitalist agriculture. *Journal of Peasant Studies* 5: 466–81.

Mann, Susan and James Dickinson. 1987. One furrow forward, two furrows back: A Marx–Weber synthesis for rural sociology. *Rural Sociology* 52: 264–85.

Marx, Karl. 1967. *Capital*, Vol. 2. Moscow, Russia: Progress Publishers.

Mooney, Patrick. 1983. Toward a class analysis of Midwestern agriculture. *Rural Sociology* 48(4): 563–84.

Mooney, Patrick. 1986. Class relations and class structure in the Midwest. In *Studies in the Transformation of US Agriculture*, edited by A. Eugene Havens, with Gregory Hooks, Patrick Mooney, and Max Pfeffer. Boulder, CO: Westview Press, pp. 206–51.

Mooney, Patrick. 1987. Desperately seeking. One-dimensional Mann and Dickinson. *Rural Sociology* 52(2): 286–95.

Peterson, E. Wesley. 2009. *A Billion Dollars a Day: The Economics and Politics of Agricultural Subsides*. Malden, MA: Wiley-Blackwell.

Pfeffer, Max. 1983. Social origins of three systems of farm production in the United States. *Rural Sociology* 48(4): 540–62.

Pimentel, D., J. Krummel, D. Gallahan, J. Hough, A. Merrill, I. Schreiner, P. Vittum, F. Koziol, E. Back, D. Yen and S. Fiance. 1978. Benefits and costs of pesticide use in United States food production. *BioScience* 28(772): 778–84.

Pretty, Jules (ed). 2004. *Pesticide Detox: Towards a More Sustainable Agriculture*. London: Earthscan.

Rosset, Peter. 2006. *Food is Different: Why We Must Get the WTO Out of Agriculture*. New York: Zed Books.

Starmer, Elanor and Timothy Wise. 2007a. Feeding at the Trough: Industrial Livestock Firms Saved $35 Billion from Low Feed Prices, Policy Brief No. 07-03. Global Development and Environment Institute, Tufts University, Medford, MA.

Starmer, Elanor and Timothy Wise. 2007b. Living High on the Hog: Factory Farms, Federal Policy, and the Structural Transformation of Swine Production, Working Paper No. 07-04, Global Development and Environment Institute, Tufts University, Medford, MA.

Taubes, Gary. 2007. *Good Calories, Bad Calories: Challenging the Conventional Wisdom on Diet, Weight Control, and Disease*. New York: Knopf.

van den Bosch, Robert. 1978. *The Pesticide Conspiracy*. Garden City, NY: Doubleday.

van der Ploeg, Jan. 2006. Agricultural production in crisis. In *Handbook in Rural Studies*, edited by Paul Cloke, Terry Marsden, and Patrick Mooney. Thousand Oaks, CA: Sage, pp. 258–77.

Willis, Richard. 2004. Enlargement, Concentration and Centralisation in New Zealand Dairy Industry, *Geography* 89(1): 83–88.

World Bank. 2005. *Agriculture Investment Sourcebook*. Washington, DC: World Bank.

Wright, Erik. 1978. *Class, Crisis, and the State*. London: New Left Books.

3

UNDERSTANDING THE FOOD SYSTEM

Past, present, and future

KEY TOPICS

- The food commodity chain has undergone tremendous change in recent decades as certain sectors have become highly concentrated, giving rise to what has come to be known as the "hourglass" shape of the food system.
- Nowhere is this market concentration higher than in the seed industry.
- Changes to the food system have caused sociologists to look beyond the farm gate, as evidenced by two frameworks (still popular to this day) which emerged in the 1980s and early 1990s: the commodity systems and food regime approach.

Key words

- Appropriationism
- Commodity chain analysis
- Commodity network analysis
- Commodity studies
- Contract farming
- *Diamond* v. *Chakrabarty*
- Exit power
- Filière approach
- Firm concentration ratios
- Food regime
- Food system "hourglass"
- Genetic use restriction technology
- Global commodity chains
- Horizontal–vertical concentration
- Monopoly
- Monopsony
- Product of nature doctrine
- Regulation theory
- Substantial equivalence principle
- Substitutionism
- Terminator technology

A discernible shift occurs in the sociological literature in the late 1980s, after which the title sociology of *agriculture* no longer adequately describes the empirical and theoretical focus of scholarship of this subdiscipline. Increasingly, terms like "food" and "food system" began to be inserted in the title to reflect practitioners' and scholars' budding interest in issues beyond the farm gate. In this chapter, I review some of the frameworks – most notably the food regime approach and commodity systems analyses – responsible for redirecting sociologists away from an almost exclusively productivist orientation to one more open to the entire food chain. Before doing this, however, I think it would be useful to acquaint the reader with some of the issues that were (and still are) causing sociologists to look "beyond" (and in some cases "prior to," as in the case of the seed industry) the farm. One of the more interesting developments to occur in the food system has to deal with its concentration in recent decades, giving it the shape of an "hourglass." After making some general points about concentration in the food system, attention turns to the seed industry in particular to explore some of the reasons why, at least within this industry, those trends are occurring.

The food system hourglass: Market concentration

A popular measure of market concentration is the four firm concentration ratio – or simply CR4. The CR4 is defined as the sum of market shares of the top four firms for a given industry. A standard rule of thumb is that when the CR4 reaches 20 percent a market is considered concentrated, 40 percent is highly concentrated, and anything past 60 percent indicates a significantly distorted market. With this rule of thumb in mind, note in Table 3.1 the CR4 (unless otherwise noted) for certain agrifood markets in the USA.

TABLE 3.1 Concentration of US agricultural markets

Sector	*CR4* (%)*
Beef packers	83.5
Steer and heifer slaughter	79.0
Pork packers	67.0
Pork producers	37.3
Broilers	58.5
Turkeys	55.0
Soybean crushing	80.0
Corn seed	CR1 80.0
Soybean seed	CR1 93.0
GE cotton seed	CR1 96.0

*Unless otherwise stated.

Source: Developed from Domina and Taylor (2010), Hendrickson and Heffernan (2007), Paarleberg (2010: 130), and Wise (2010).

The CR4 statistic is a measure of horizontal concentration; concentration at one "link" in the food commodity chain. Horizontal integration occurs when firms in the same industry and at the same stage of production merge and monopolize a market. Yet the food system has undergone tremendous vertical concentration too – a phenomenon describing when companies are united throughout the supply chain by a single owner. Smithfield Foods, for example, has captured (CR1) a 31 percent market share of the US pork packing industry. Additionally, they raise 19.7 percent of all hogs in the USA. This duel concentration – horizontal and vertical – gives firms unique advantages that cannot be had in more open markets. For example, as millions around the world in 2007 and 2008 were rioting in the streets to protest rising food prices, agribusinesses were riding a wave of near-record profits. In 2007, Cargill's, ADM's (Archer Daniels Midland), and Bunge's – who collectively control 90 percent of the world's grain – profits rose 36 percent, 67 percent, and 49 percent, respectively (Box 3.1). Their net earnings rose still further during the first quarter of 2008: 86 percent, 55 percent, and 189 percent, respectively (Carolan 2011).

BOX 3.1 APPROPRIATIONISM AND SUBSTITUTIONISM

Two processes said to underpin agribusiness expansion are "appropriationism" and "substitutionism" (Goodman *et al.* 1987; Goodman and Redclift 1991). Appropriationism refers to the replacement of elements of the production process with industrial (or manufactured) artifacts. Off-farm industries appropriating on-farm processes and practices allow the former to further extract profits out of the latter. Put simply (and to harken back to the Mann–Dickinson thesis in Chapter 2), capital appropriates what it can when it can. Examples of appropriationism include: bringing the machine (e.g. combine) to nature (e.g. field); speeding up elements of production time with technology (such as feeding growth hormones to livestock); and making farmers purchase their seed every spring rather than allowing them to use what they saved.

Substitutionism speaks to the replacement (or substitution) of costly agricultural end-products with seemingly cheaper industrial ones. This occurred with cane sugar, which in recent decades has been widely replaced by high fructose corn syrup. As Friedmann (1991: 74) once pointed out, what food processing firms want "is not sugar, but sweeteners; not flower or cornstarch, but thickeners; not palm oil or butter, but fats; not beef or cod, but proteins. Interchangeable inputs, natural or chemically synthesized, augment control and reduce cost better than older mercantile strategies for diversifying sources of supply." The ability to substitute one sweetener (or thickener, fat, protein, etc.) for another gives firms tremendous power, as it allows them to bypass particular farmers, commodities, and entire regions when sourcing ingredients (Lawrence and Grice 2004).

Most are familiar with the term "monopoly." It refers to a level of seller dominance at which point they can effectively choose the price at which to sell something. To put it simply, monopoly is a state of seller power. But that's not what we have in the food system, at least not among food processors and manufactures. What they possess (the big ones at least) is buyer power – or what is also known as "monopsony." Under monopsony conditions, market concentration has reduced the number of potential buyers to the point that the seller has few options other than accept the price dictated by the buyer.

Farmers are particularly susceptible to buyer power given the perishable nature of most of what they produce. Meat and dairy producers are particularly vulnerable to buyer power. Livestock producers rely on selling their animals at optimum weight, which, in the case of cattle, is a window of just a couple weeks. In the case of hogs, a timely sale is also essential in order to clear room for the next litter. For dairy producers, the bulk tank must be emptied daily (sometimes multiple times a day), making it impossible for the farmer to hold out for a better price. Livestock producers also have an incentive to avoid, when at all possible, distant markets, which makes seeking out competitive bids from far away unlikely. Shipping live animals long distances can be prohibitively expensive, increase animal mortality and carcass shrinkage, and result in a drop in the quality of their meat (which will negatively affect the price they receive).

Another important component to all of this is the rise of contract farming. In livestock production, where contracts are most widely utilized, the contractor (the processing firm) owns the animals. Producers, conversely, are responsible for building and managing the facilities – usually to the contractors' specifications – who in turn receive all inputs from their buyer: animals, feed, veterinary services, and transportation when it is time for the animals to be slaughtered. Yet there are a number of problems with this relationship for those at the producer end.

First, farmers have little negotiating power. The aforementioned narrowing of the commodity chain means that while processors have their pick of producers, farmers may have only one or two processors from whom to obtain a contract. This means producers lack "exit power" – the power to walk away from the negotiating table knowing that others are out there to purchase the product. Contracts also place considerable – many say "too much" – risk on the shoulders of the producer. Chicken producers can invest as much as US$500,000–1,000,000 in facilities that have a 20–30 year economic life with no practical alternative use. So once such a facility is built the producer is under tremendous pressure to obtain and maintain their contract with processors, no matter how unfairly it might be structured. And then there are the required upgrades, which can cost a poultry grower an additional US$50,000–100,000. A medium-sized finishing hog operation with six 1,100 head hog houses costs on average US$600,000–900,000 to build. Between 1995 and 2009, poultry growers in the state of Alabama, USA, witnessed a negative net return in 10 of those 15 years (Taylor and Domina 2010: 9). This translates to average cumulative losses for each grower in excess of US$180,000.

Large processors can also control market access through the setting of standards. For instance, together Nestlé and Parmalat forced more than 50,000 Brazilian dairy farmers out of business by buying out milk cooperatives in the 1990s and changing the standards for handling and storing milk prior to purchase (Birn *et al.* 2009). With no other buyers (including cooperatives) left, farmers had little choice but to accept these new demands. Parmalat and Nestlé, after establishing their market dominance, placed the burden of meeting these standards, which involved, for instance, purchasing and installing expensive refrigeration units, squarely on the shoulders of the dairy farmers. The expense was simply too great for many of the small-scale producers, who had minimal capital reserves and little access to credit. Consequently, many were forced out of business.

Rising concentration in the retail sector has also shaped the structure of the food system. Large retail firms like Walmart and Kroger are dealing increasingly with just a handful of very large packers, bypassing the wholesale sector entirely. Retail firms do this in part to exploit the buyer power held by large processing firms. Large retailers, in pursuit of the best price possible, are usually able to get a better deal from large firms like Smithfield Foods than from smaller processors who do not wield the same influence over producers. The processors then pass the tighter margins on to producers, rather than absorbing them, which might explain the growing gap between what producers are paid and retail prices for those products. A study from 2004, for example, found that the difference between the price paid farmers and the prices paid by consumers *increased* by 149 percent between 1970 and 1998 (Marsh and Brester 2004). Buyer power for retailers, it turns out, does not necessarily translate into lower prices for consumers.

Retail sector concentration varies considerably around the world. Independent grocers, for example, still account for 85 percent of all retail sales in Vietnam and for 77 percent in India (von Braun 2007). At the other extreme is Australia, whose retail sector has a CR2 statistic of 80 percent (Wardle and Baranovic 2009). CR3 ratios for the retail sectors in Sweden, the Netherlands, France, Spain, Greece, and Italy are 95 percent, 83 percent, 64 percent, 44 percent, 32 percent, and 32 percent, respectively (Lang *et al.* 2009).

Large retail firms wield tremendous buyer power thanks in part to the sheer volume they deal in. Walmart's global sales for 2010 topped, for the third year in a row, US$400 billion (up from approximately US$250 billion in 2003). If Walmart was a country, its revenues would push Saudi Arabia out of the top 25 for world's largest economies. The company topped the Fortune 500 list of all the world's companies with the highest gross revenue in 2010 (they were followed by three oil companies). Large retailers also prefer to deal directly with manufacturers and processors, cutting out traditional wholesalers entirely. General Mills and Kraft Foods, for instance, generate about 19 percent and 15 percent, respectively, of their revenue through Walmart retail sales (Bloomberg 2009). When so much revenue is dependent upon one contract this gives the buyer tremendous negotiating power. Walmart possesses such buyer power that they can often simply tell manufactures and wholesalers what they are willing to pay and the sellers must find a way to make that price work (see, e.g., Frontline Transcript 2004). This has forced sellers looking to retain their Walmart contract to either go overseas in search of cheaper labor, in the case of manufactured goods, or squeeze farmers even further, in the case of food.

When looked at from farm-to-fork, the demographics of the food system take on an hourglass figure. The system is relatively wide at both "ends," representing farmers and consumers, with a severely truncated "middle," where the food processors and manufacturers and food and beverage retailers reside. Adding the even less competitive seed industry (as detailed earlier in Table 3.1), a better metaphor might be an hourglass hanging by a thread. Figures 3.1 and 3.2 illustrate this market concentration for both the USA and New Zealand, respectively.

Consolidation in the global seed industry

The seed industry, with its remarkable level of concentration, deserves a closer look. The top four seed firms control 56 percent of the global brand-name seed market. When it comes to genetically engineered seed, the level of concentration is even greater. In the USA, for example, the largest three seed firms control 85 percent of transgenic corn patents and 70 percent of noncorn transgenic plant patents. Of all the land cultivated in the USA in genetically engineered seed, 85 percent of the corn and 92 percent of soybean acreage holds seeds that are the proprietary technology of Monsanto, far and away the world's most profitable seed company (Figure 3.3).

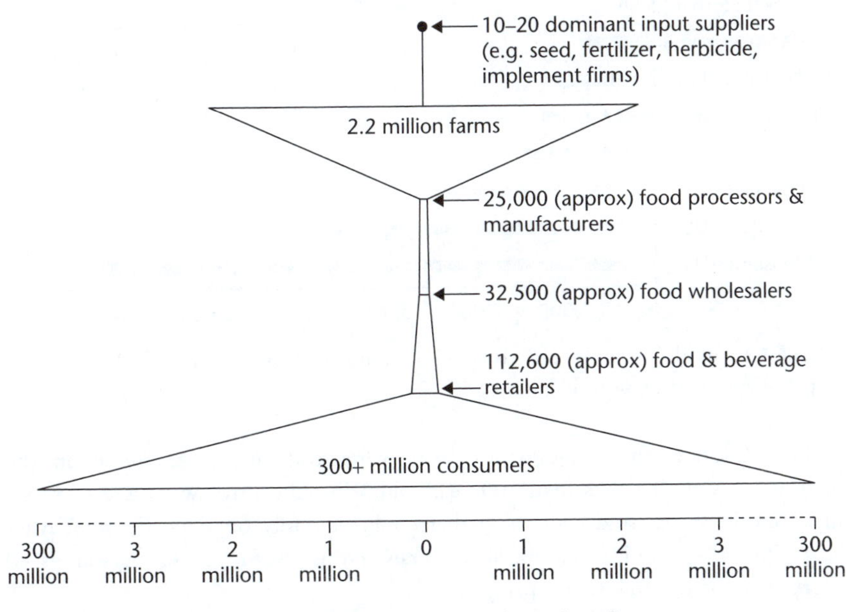

FIGURE 3.1 US food system "hourglass" (hanging by a thread)

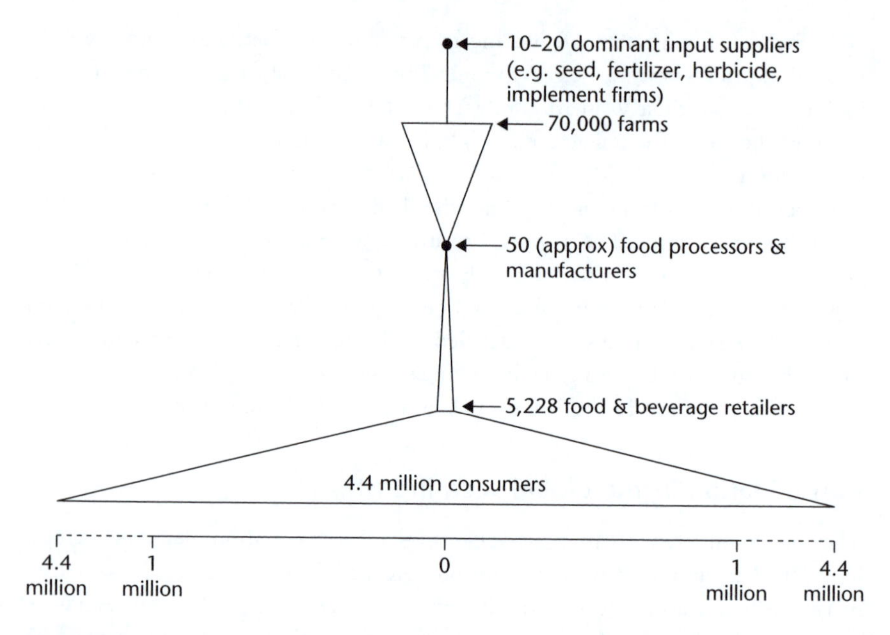

FIGURE 3.2 New Zealand food system "hourglass" (hanging by a thread). Compiled by Michael Carolan, with assistance from Paul Stock and Miranda Mirosa

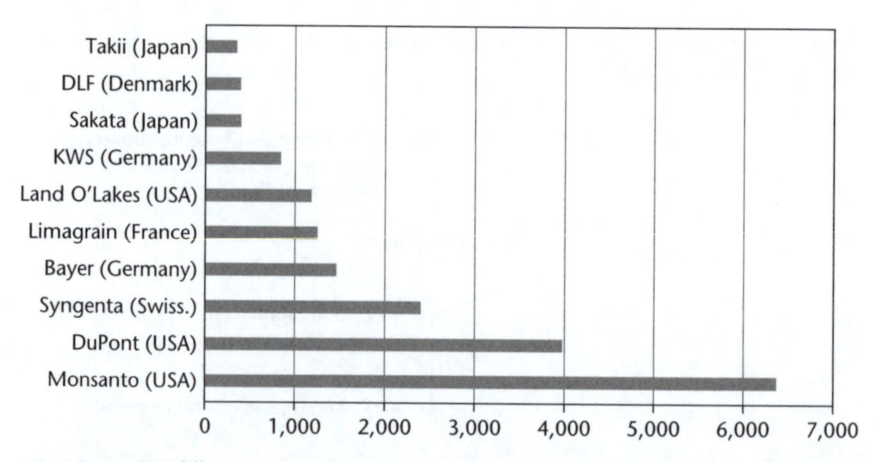

FIGURE 3.3 World's top 10 seed companies (annual sales in US$ millions). Data from http://www.etcgroup.org/en

In addition to this consolidation being a relatively recent phenomenon, the seed industry itself is less than 100 years old. What incentive was there to invest in an industry that sold something that could be easily obtained for free? After all, people had for millennia managed to raise crops just fine by saving some seed every year for the following spring.

Hybrid corn changed things considerably. By producing progeny that does not breed "true" (which is to say they do not yield at the same level as their parents), hybrid corn essentially forced farmers back to their local seed dealer every spring. And the reason why farmers were willing to suddenly pay for something that had previously been free: higher yields. As Kloppenburg explains:

> Despite a reduction of 30 million acres on which grain corn was harvested between 1930 and 1965, the volume of production increased by over 2.3 billion bushels [. . .]. And hybrid corn's 700 percent annual return on investment remains the much cited and archetypical examples of the substantial returns society enjoys from agricultural research.
>
> *Kloppenburg (1988: 92)*

Yet this praise for hybrid corn must be tempered. It has been argued that had an equal investment been made utilizing more conventional methods breeders could have created high-yielding varieties of corn that still bred true, in which case farmers would have been able to save and replant harvested seeds (Dutfield 2008). The adoption of hybrid corn was also helped along with mechanization and increases in farm size that accompanied this movement away from animal and human labor. With mechanization farmers suddenly *needed* crops that matured at an even rate allowing stalks to be harvested at the same time. Also required were plants vigorous enough to remain standing by harvest so the ears were accessible to the mechanical corn picker. Stalks lying on the ground were missed by the picker's head.

As a de facto patent, hybrid vigor forced farmers to come back year after year to purchase seed. Once adopted, farmers needed to purchase this previously free input. Yet hybridization does not work for crops like wheat. And with no law keeping farmers from replanting their seeds, breeders and seed companies continued to seek legal (e.g. patents) and later biological (e.g. so-called terminator technology) protections for their labors, even after the discovery of hybridization.

Not surprisingly, the hybrid seed corn industry was first to consolidate, as they had the earliest viable market. Seed industry consolidation has been rapid since the 1970s, helped along considerably by plant and biotech patents. Philip Howard developed the illustration depicted in Figure 3.4. The figure shows the global seed industry structure (in 2008), noting also that this sector is dominated by more than just seed firms ("other" companies tend to be biotechnology firms). Full ownership is represented with a solid line; partial ownership is represented with a dashed line. The figure also shows connections between key firms through joint ventures.

Let's look more closely at Monsanto's place at the top of the seed industry; a cherished position as it represents the first link in the food chain. In the 1990s alone, Monsanto spent close to US$9 billion accumulating new biotech and seed companies. The company projects that seeds and licensed traits will provide roughly 85 percent of gross profits by 2012 (Howard 2009). The protections that patents provide allow firms like Monsanto tremendous flexibility when setting the price of their product. There is no free market when it comes to patented seeds and thus no chance for what economists call "price discovery" – lest we forget, patents are veiled state-sanctioned monopolies. Take the added cost that the Roundup Ready trait in

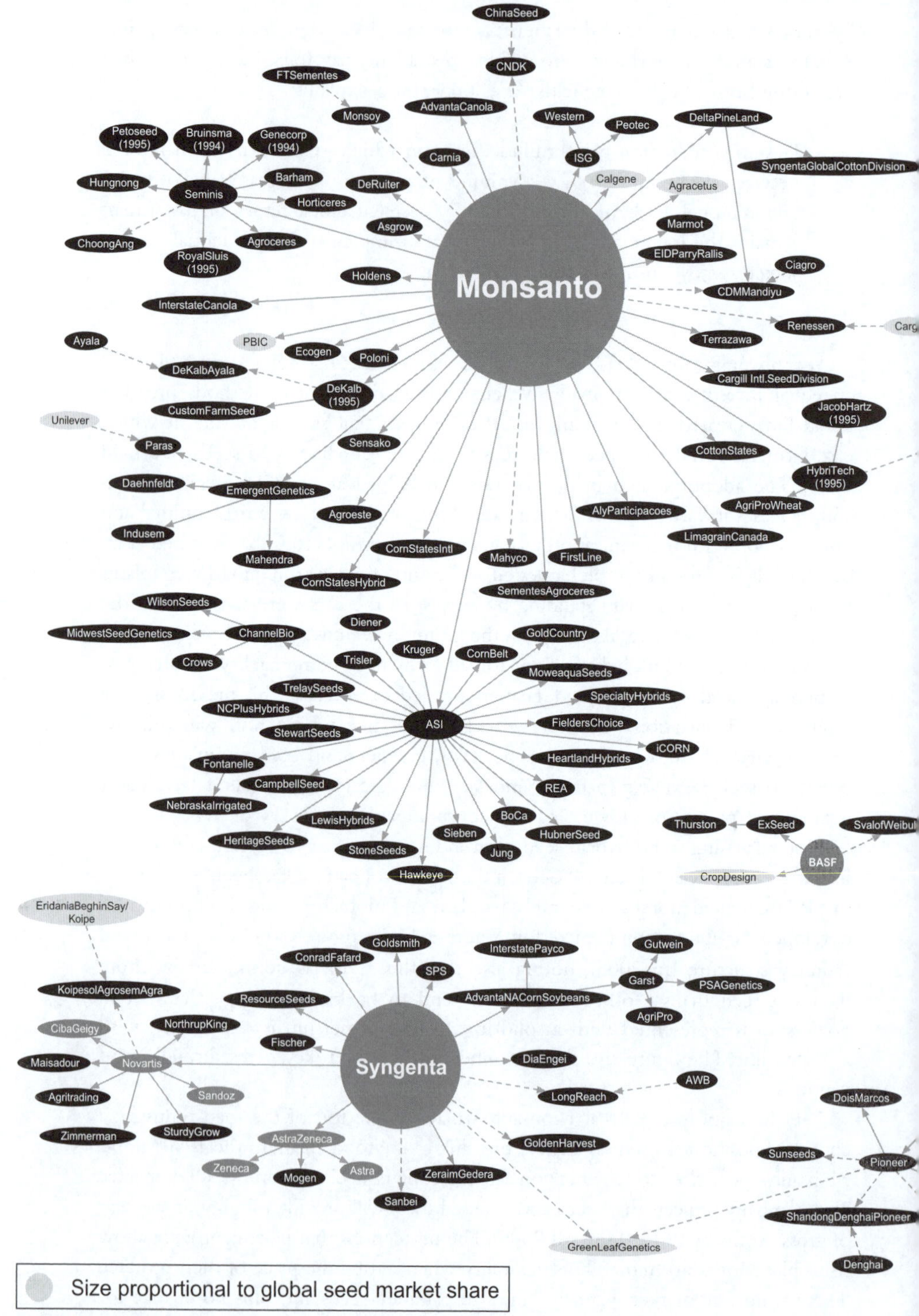

FIGURE 3.4 The global seed industry structure (in 2008)

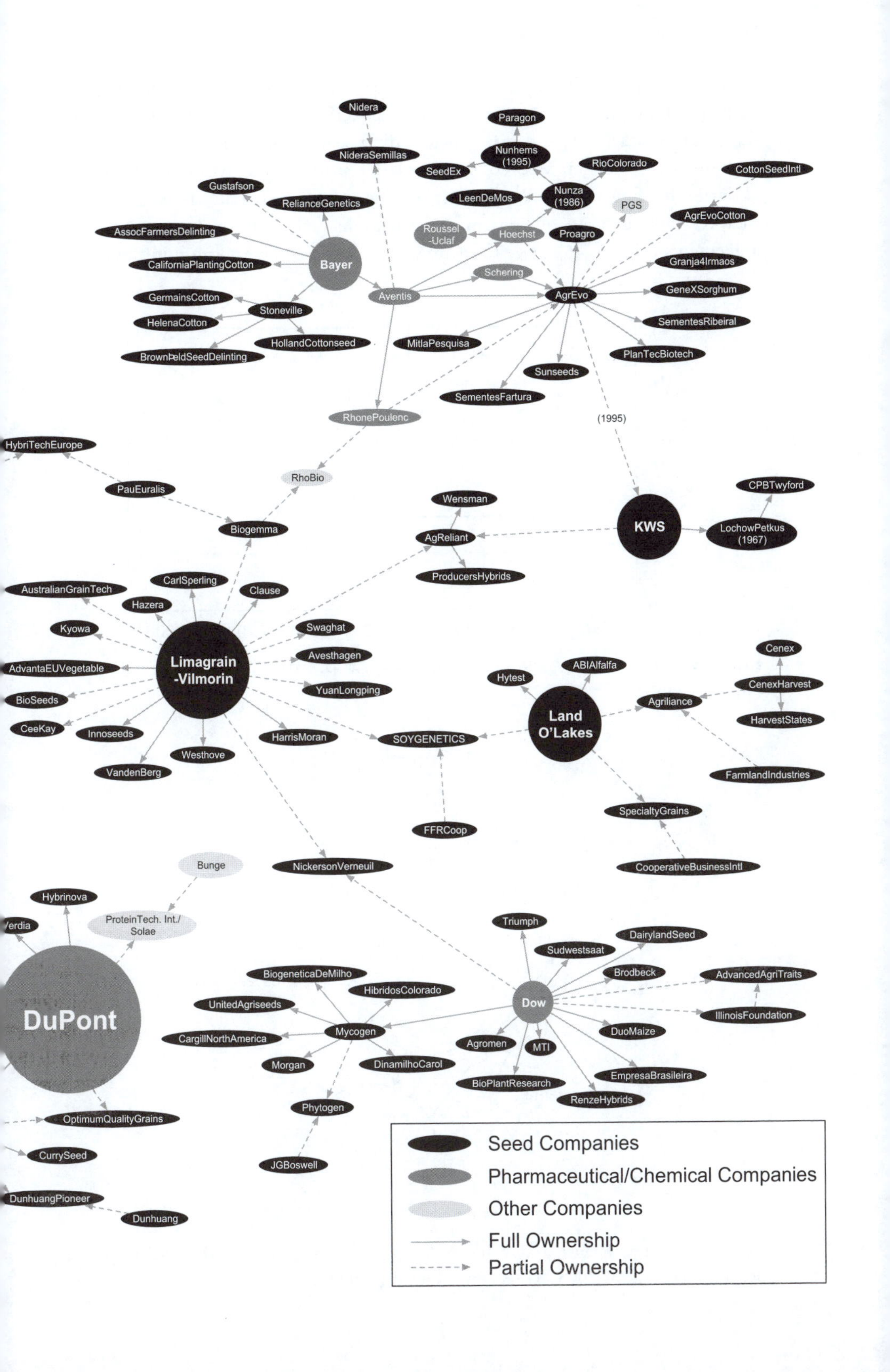

Nidera
NideraSemillas
Paragon
Nunhems (1995)
SeedEx
RioColorado
CottonSeedIntl
Nunza (1986)
LeenDeMos
PGS
Gustafson
RelianceGenetics
AssocFarmersDelinting
Roussel -Uclaf
Hoechst
Proagro
AgrEvoCotton
Bayer
CaliforniaPlantingCotton
Schering
Granja4Irmaos
GermainsCotton
Aventis
AgrEvo
GeneXSorghum
Stoneville
HelenaCotton
SementesRibeiral
HollandCottonseed
MitlaPesquisa
PlanTecBiotech
BrownfieldSeedDelinting
Sunseeds
SementesFartura
RhonePoulenc
(1995)
HybriTechEurope
RhoBio
PauEuralis
CPBTwyford
Wensman
Biogemma
KWS
AgReliant
LochowPetkus (1967)
ProducersHybrids
AustralianGrainTech
CarlSperling
Clause
Hazera
Swaghat
Kyowa
Cenex
Limagrain -Vilmorin
Avesthagen
ABIAlfalfa
AdvantaEUVegetable
Hytest
CenexHarvest
YuanLongping
Agriliance
BioSeeds
Land O'Lakes
HarvestStates
CeeKay
Innoseeds
HarrisMoran
SOYGENETICS
Westhove
VandenBerg
FarmlandIndustries
SpecialtyGrains
FFRCoop
NickersonVerneuil
CooperativeBusinessIntl
Bunge
Hybrinova
ProteinTech. Int./ Solae
Triumph
DairylandSeed
Sudwestsaat
Brodbeck
BiogeneticaDeMilho
AdvancedAgriTraits
HibridosColorado
UnitedAgriseeds
Dow
IllinoisFoundation
DuPont
CargillNorthAmerica
Mycogen
DuoMaize
Agromen
MTI
Morgan
DinamilhoCarol
EmpresaBrasileira
BioPlantResearch
RenzeHybrids
Phytogen
OptimumQualityGrains
CurrySeed
JGBoswell
DunhuangPioneer
Dunhuang
Seed Companies
Pharmaceutical/Chemical Companies
Other Companies
Full Ownership
Partial Ownership

soybeans brought to farmers (a trait that makes these plants resistant to the popular herbicide Roundup). In 2000, the trait added US$6.50 to each bag. The fee for this trait is now $17.50 per bag. In other words, a farmer who plants Roundup Ready soybeans has seen their seed costs nearly triple in the last 10 years (Figure 3.5).

Firms holding patents on genetically engineered seed traits are also forming industry links in ways other than consolidation, such as through cross-licensing agreements (Figure 3.6). Here too Monsanto holds a central position in the global

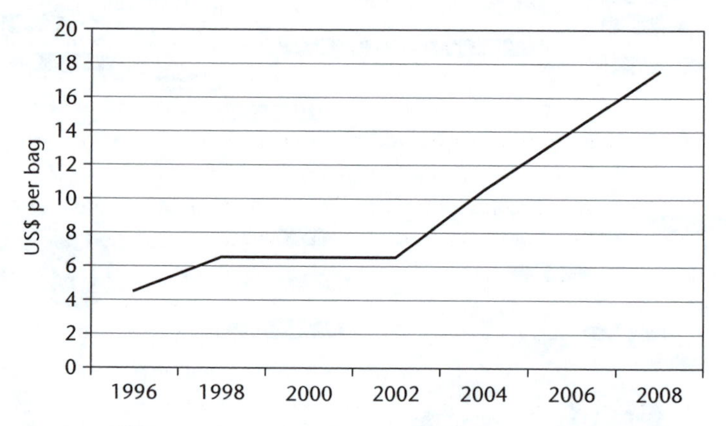

FIGURE 3.5 Roundup Ready soybean technology fee, 1996–2008 (US$). Based on Hubbard (2009)

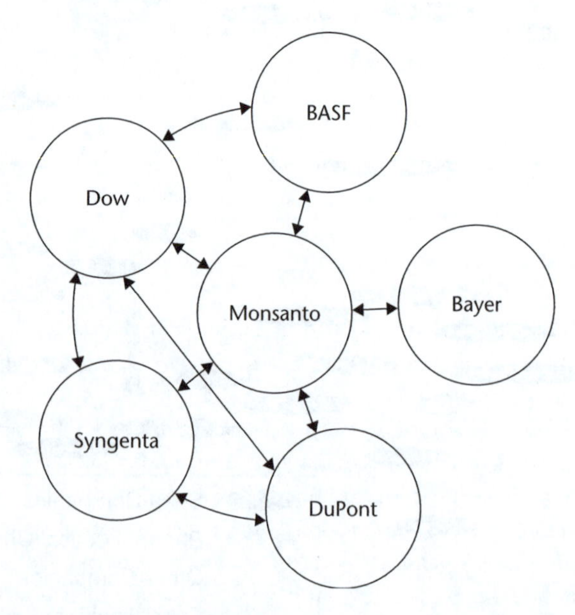

FIGURE 3.6 Cross-licensing agreements for patented traits among the top six patent holding firms (as of 2008). Based on Howard (2009)

seed industry structure, as it is the only firm to have agreements with each of the other major patent holding firms. Cross-licensing agreements have been growing in popularity as of late as firms are beginning to stack multiple patented traits within a single seed. For example, in 2009 Monsanto and Dow announced that they had received approval to introduce their new eight-trait genetically engineered corn "SmartStax" into Canada and the USA; though apparently more is not necessarily better, as early harvest data compiled by Monsanto showed some of its SmartStax corn are yielding less than its older technology triple-stack corn seeds (Kaskey 2010).

Tensions: Legal protections and regulations

Patents give seed firms a 20-year monopoly over their products. Patents have proven a key component to helping seed firms like Monsanto shore up their market dominance. They also increase firms' seller (monopoly) power over farmers as patents give patent holders the power to set the price of their products. But the right to patent seeds has not always existed. And in obtaining that right tensions have emerged between patent laws and food safety regulations, with the former arguing the seed is unique and the latter arguing it is not.

In *Ex parte Latimer*, the US Commissioner of Patents rejected an application on a fiber made out of Southern Pine. This fiber had been "eliminated in full lengths from the silicious, resinous, and pulpy part of the pine needles [of Southern Pine] and subdivided into long, pliant filaments adapted to be spun and woven" (1889 Comm'r Dec.: 123–4). In rejecting the patent, the Commissioner emphasized the naturalness of the fiber. Specifically, he likened Latimer's invention to "wheat which has been cut by a reaper or by some new method of reaping" as well as to a middle ear bone that had been removed "in its natural condition" (1889 Comm'r Dec.: 126–7). The Commissioner argued that Latimer's discovery of this "natural product" did not entitle him to patent protections "any more than to find a new gem or jewel in the earth would entitle the discoverer to patent all gems which should be subsequently found" (1889 Comm'r Dec.: 127). The Commissioner also raised concerns over what might happen if the patent were granted, noting that other "patents might be obtained upon the trees of the forest and the plants of the earth" (1889 Comm'r Dec.: 125–6). Thus, while "the alleged invention is unquestionably very valuable" and of "immense value to the people of the country [...] the invention resides, I am compelled to say, exclusively in the process and not at all in the product" (1889 Comm'r Dec.: 127). And so was born the "product of nature doctrine"; a legal barrier that kept patents from being issued to any and all organisms – including seeds – throughout much of the twentieth century.

A property right is only valuable if it can be enforced. The first step towards enforcement is the ability to specify the identity of the property. While relatively easy to accomplish with a tract of land, specifying the identity of a living organism was long considered problematic given the lack of knowledge about biological identity (e.g. blood types, DNA). The identity-specifying tools born out of the genetic revolution were not yet available to nineteenth and early twentieth century scientists, inventors, and judges. This left patent law ill-equipped to see independent "objects" when it looked at biological artifacts, even those altered by human hands. This is why, for example, the production of nature doctrine disqualified all but asexually reproducing plants from gaining patent protection under the 1930 Plant Patent Act; because only those plants looked sufficiently "fix" (unchanging). An asexually reproduced plant – like a rose propagated through a "cutting" – is genetically identical to the plant it was derived from.

The legal landscape changed entirely in 1980 with the US Supreme Court ruling of *Diamond* v. *Chakrabarty*. In 1972, Ananda Chakrabarty, a biochemist for General Electric Company, filed a patent for a bacterium that had been genetically altered to consume oil slicks. Initially, Chakrabarty's patent application was denied for reasons relating to the product of nature doctrine. In the end, however, the US Supreme Court granted Chakrabarty his patent claims in a 5 to 4 vote. The Court ruled that the bacterium was human-made, that it was a new composition of matter, and, finally, that whether an invention is alive is not a legitimate legal question. In supporting their ruling, the Court wrote the now famous passage that explains the virtually limitless subject matter potentially eligible for patent protection:

> The Patent Act of 1793, originally authored by Thomas Jefferson, defined statutory subject matter as "any new and useful art, machine, manufacture, or composition of matter, or any new or useful improvement." [. . .] In 1952, when the patent laws were recodified, Congress replaced the word "art" with "process," but otherwise left Jefferson's language intact. The Committee Reports accompanying the 1952 Act inform us that Congress intended statutory subject to *include anything under the sun that is made by man.*
>
> Diamond *v.* Chakrabarty *(447 U.S. 303 [1980]: 308–9) [my emphasis]*

What the Court essentially argued was that Chakrabarty's involvement in making this bacterium disqualified it from being labeled "natural"; a fact that in turn exempted it from the product of nature doctrine. In the Court's own words: "His discovery is not nature's handiwork, but his own; accordingly it is patentable subject matter under § 101" (*Diamond* v. *Chakrabarty*, 447 U.S. 303, 206 USPQ 193 [nhj1980]). Yes; the patented bacteria were still living organisms. But what the Court chose to focus on instead was *Chakrabarty's involvement* in making them. This involvement, essentially, was sufficient to give them a non-natural identity, therefore qualifying this invention for patent protections. And so was born bio-*technology*.

Yet this identity is not fixed, as reflected in the tensions that have arisen between patent laws and food safety regulations. Proponents of biotechnology, when discussing issues of regulation, routinely make statements such as "The genes introduced through GM [genetic modification] are not qualitatively different from those genes introduced by conventional breeding from exotic sources or from novel genes produced through mutation" (Trewavas and Leaver 2001: 456). How can this be? Do not genetic patents rest upon the very assumption that biotechnology is qualitatively *different* from their non-GM counterparts? In short: how can they be both natural and unnatural?

When the goal is minimizing government regulation and winning over public opinion, biotechnology is given a different identity. The goal among proponents now becomes to make these objects appear "natural" by directing attention away the social networks, people, capital, and technology that brought them into existence. Instead, connections to the biophysical realm, which were earlier ignored when the argument was about patents, are now elevated. *Bacillus thuringiensis* (*Bt*) corn provides a good example of this. This corn has been genetically modified by transfer of a gene from the soil bacterium *Bacillus thuringiensis* that has insecticidal properties. To minimize government regulation and win over public sentiment, proponents of *Bt* corn like to point out that *Bt* corn is natural given that this bacterium occurs naturally in soils (see, e.g., DeGregori 2002). Another "naturalizing" technique is what is known as the substantial equivalence principle.

The principle of substantial equivalence is widely used by national and international agencies, such as the Canadian Food Inspection Agency, Japan's Ministry of Health and Welfare, the US Food and Drug Administration, the United Nation's Food and Agriculture Organization, the World Health Organization, and the Organization for Economic Cooperation and Development (OECD). The concept was first articulated by the OECD, in the seminal publication *Safety Evaluation of Foods Derived by Modern Biotechnology: Concepts and Principles* (OECD 1993). The principle requires regulators to compare so-called genetically engineered foods with their nearest known ("natural") counterpart. If the engineered food is found to be substantially equivalent to its nonengineered counterpart, through such techniques as metabolic profiling and protein profiling, it is said to be at least as safe as that counterpart. Evaluating the substantial equivalence of a new food involves measuring the bioavailability and concentration of important nutrients in the food, such as proteins, carbohydrates, vitamins, minerals, and fats. Such biochemical profiling is conducted to ensure that engineered foods fall within the normal range of variability as found in their nonengineered counterparts. If no discernible differences are noted, then the food is said to be of an "equivalent composition" (Novak and Haslberger 2000). At this point, these objects have had their identity reversed back to *bio*-technology (Carolan 2008).

Biological protections: Terminator technology

In addition to legal protections, like patents, seed firms are also actively looking to use technological protections to secure their market dominance. When seed firms talk about technological protections they are talking about what is commonly known as terminator technology – or the less pejorative term genetic use restriction technology (GURT). GURT refers to genetically engineered seeds that germinate into plants that produce sterile seed. These come in two forms: Trait GURT (T-GURT) and Varietal GURT (V-GURT). T-GURT produces viable seed but restricts gene expression of a certain trait which can be switched on with the application on a specific input. V-GURT, which is particularly controversial, restricts the use of the entire plant because they render seeds sterile. Since V-GURT have seen the greatest commercial application, in part because they are viewed as offering a more complete biological protection system than T-GURT, I will focus my discussion on this form.

V-GURT work by producing a toxin protein during seed germination that yields plants or seeds that can be harvested, eaten, and used as feed but which are themselves sterile. When first publicized in 1998 there was enormous public outrage over the technology, from farmers' rights groups, to environmentalists, religious organizations, and seed breeders. Monsanto, having received the brunt of this criticism due to its position as a leader in agro-biotechnology, responded by publically announcing that it would not commercialize this technology until its impacts had been thoroughly studied and understood. Yet by 2003, in a paper co-authored by Monsanto scientists and released by the International Seed Federation, the benefits of GURT were being lauded for their ability to protect the proprietary interests of seed firms. Terminator technology is still being pursued today; though, understandably, without much fanfare so as to minimize attention to these activities. Delta and Pine Land Company won new GURT patents in both Europe and Canada in October 2005. On June 1, 2007 the US Justice Department approved Monsanto's US$1.5 billion takeover of Delta and Pine Land Company, thus making Monsanto a leader in GURT.

One popular criticism is how GURT will "raise costs and lock farmers into tightly controlled marketing and licensing agreements" (Christian Aid 1999: 1). This is particularly worrisome for farmers in developing countries who live in economic systems vulnerable to disruptions, which can occur due to war, civil disorder, and natural and technological disasters (Van Dooren 2007: 82). Such dependency, it is feared, will force farmers to "return to the seed corporations every year and will make extinct the 12,000 year tradition of farmers saving, adapting, and exchanging seed in order to advance biodiversity and increase food security" (ETC 2002: 2). Consequently, "GURT varieties may displace locally adapted genetic material rather than integrate with it, and this could affect the resilience and long term productivity of the low input farm systems" (Van Wijk 2004: 133–4). Moreover, while this technology produces sterile seed, it can still generate fertile pollen with the ability to carry the terminator trait to sexually compatible plants and render them sterile – a particularly troublesome outcome if that now-sterile seed was saved for the following year's planting (Pendleton 2004: 16).

While often compared to hybridized corn, which has been around for generations and also (as mentioned earlier) forces farmers back to their seed dealer year after year, the reasoning that claims terminator seed its equivalent is problematic. For example, with hybridized seed, breeders can still retrieve genetic material from second generation plants for purposes of research. While hybrid varieties do not breed "true" they still produce fertile progeny. Furthermore, hybridization was developed to *increase* yields and plant vigor. The nonviability of second generation seed from hybrid crops is ultimately an ancillary effect; though, admittedly, had it not been present hybridization would likely not have been pursued with such vigor. GURT, conversely, is designed solely for facilitating monopoly control over the seed market.

Food regime

By the late 1980s, rural studies was said to be suffering from a "hangover" (Marsden 1989: 313) thanks to a decade of literature giving analytic superiority to the point of production – in other words, to agriculture *itself*. Perhaps the term "hangover" overstates things a bit. At the very least, we can say sociologists of agriculture were drinking from more than just the Mann–Dickinson thesis bottle, which helps explain why the debate began to grow old by 1989 – the year Friedmann and McMichael's seminal *Sociologia Ruralis* paper on food regimes was published. Almost a decade earlier we can already find studies that look beyond the point of production, which recognize agriculture as but one point in a complex network (see, e.g., Friedland *et al.* 1981). As the 1980s gave way to the 1990s, agrifood scholars were calling for approaches that left analytic room for dynamics, processes, and practices beyond the farm gate, like those associated with supply chains and at the point of consumption. There was also concern that the theoretical arguments that dominated much of sociology of agriculture scholarship in the 1980s were – or at least could be interpreted as – suggesting *linear* trajectories of agrarian change. Much of the "agrarian question" literature from the 1980s ignored issues of historical, spatial, and social contingency, leaving us with a single-arrow understanding of social change.

Yet the arrows of history point in all directions when it comes to agrifood systems. For example, commercial farms in Japan average 1.9 hectares, while in the USA the average farm (both commercial and noncommercial) is 182 hectares. I've already written earlier in this chapter about the tremendous variability in market concentration across national retail sectors. Or take global tomato processing: the USA and EU account for 42 percent and 34 percent of world production, respectively. While Australia, Canada, Chile, China, Latin America (minus Chile), the Middle East and Africa, and Turkey make up 1 percent, 1 percent, 3 percent, 4 percent, 5 percent, 4 percent, and 4 percent, respectively, of the world's share of tomato processing (Pritchard and Burch 2003: 253). The agrarian theories of the 1980s are not terribly helpful for understanding this *global systemic* variability.

And so the field made a dramatic turn in the late 1980s. Fred Buttel, famous for his encyclopedic knowledge of the literature (among other things), summed up this "turn" as follows:

> But only five years after the seminal piece – Friedmann and McMichael's 1989 *Sociologia Ruralis* paper on food regimes – was published, the sociology of agriculture had undergone a dramatic transformation. [. . .] [T]he Friedmann and McMichael *Sociologia Ruralis* article on food regimes was arguably the seminal piece of scholarship in the abrupt shift away from the new rural sociology, and "regime-type" work has proven to be one of the most durable perspective in agrarian studies since the late 1980s, in large part because it is synthetic and nuanced.
>
> *Buttel (2001: 171, 173)*

"Regime-type" research seems to be as popular today as in 2001. I concede that as this research tradition is now over 20 years old I cannot possibly do it justice through brief review. There are some characteristics shared by all food regime studies, which I will attempt to highlight. Just to be clear, regime-type theorizing is not a rejection of agrarian political economy but an extension of it (Borras *et al.* 2010: 575). It is also debatable whether we should think of food regime as a theory *per se*. Some (like McMichael) think it is best understood as an approach to the contingent history of capitalist relations; a way, in other words, of ordering how we think about the structuring of the world food order.[1]

The aforementioned 1989 Friedmann and McMichael piece views the global food system as driven as much by historically specific *political* variables as economic logic. Regime-type research is careful not to divorce agriculture and the food system more generally from state agricultural policies and the capitalist world-economy. That is because, ultimately, "the food regime concept is a key to unlock not only structured moments and transitions in the history of capitalist food relations, but also the history of capitalism itself" (McMichael 2009b: 281). The term *food* regime is therefore a bit of a misnomer, as it "is not about food *per se*, but about the relations within which food is produced, and through which capitalism is produced and reproduced" (McMichael 2009b: 281).

A key component of the food regime is the "rule-governed structure of production and consumption of food on a world scale" (Friedmann 1993: 30–1). Food regimes are defined by periods of relatively stable sets of rule-governed relationships, separated by unstable periods shaped by political contests over a new way forward (Friedmann 2005: 228). This focus on rules reflects a long-standing influence by what is known as "regulation theory." A significant focus of the regulationist literature is on detailing the modes of control (based on rules, culture, and political arrangements) that lie behind periods of capitalist growth, which, in regulation theory, have historically been separated into two distinct epochs: mid-20th century Fordism (e.g. mass production, undifferentiated mass market) followed by post-Fordism (e.g. flexible production, diversification of commercialized cultures and lifestyles). An influential critique by Goodman and Watts (1994) caused scholars to rethink Fordist-type concepts to describe epochs in the political economy of agriculture; though regulation theory lives on through, for example, revised concepts like "global post Fordism" (e.g. Bonanno and Constance 1996). Among other things,

Goodman and Watts argue convincingly that agriculture is sufficiently different from industry, which precludes the two sectors from being conflated into one. The critique did not, however, stop scholars from seeing "significant world-historical periodisation anchored in the political history of capital" (McMichael 2009a: 144). However, the Goodman and Watts piece appears to have pushed regime scholars to make their frameworks more open to historical, spatial, and social contingency. Evidence of this can be seen in McMichael's (2009a) and Friedmann's (2005) recent iterations on the subject, where they include social movements in their theorizing.

Food regime scholars are in agreement about what are known as the first and second food regimes. The first food regime (approximately 1870–1930s) was rooted in British imperialism. Its networks consisted of colonial tropical imports to Europe and basic grain and livestock imports from settler colonies, all to help fuel Britain's image as the "workshop of the world." Under this regime, "development" was understood as an articulation between *both* agricultural and industrial sectors; the policies and actions located in the former spurred growth in the latter sector (but only in settler states, which explains why the USA became the "development model" in the twentieth century). This occurred, for example, by way of states' and firms' ability to reduce the cost of labor by the mass production of food staples like grain, meat, sugar, and coffee. Harriet Friedmann's (1978) study of the late-nineteenth century North American family farmer details how the growth of a world wheat market reduced grain prices and thus the cost of labor.

The second food regime (approximately 1950–1970s) was organized around the disposal of agricultural surplus from the key settler state (namely, the USA). Within this food regime, excess food was utilized as a strategic resource in an attempt to win the Cold War. Food aid helped secure loyalty against communism while aiding in opening up markets to US products. The second food regime was thus organized around the reproduction of US hegemony. All of the country's highly subsidized cheap food (food surplus) following World War II helped feed cheap labor in less affluent nations. Cheap food lowered labor costs in economically and politically strategic nations, while helping to pacify urban labor forces in the developed North (e.g. if food is cheap people won't demand to be paid so much) (McMichael 2009b: 285).

While there is remarkable consensus among food regime scholars, there are also points of debate. For example: where are we today – the second food regime, a third food regime, or perhaps somewhere in between? There seems to be an emerging consensus that a third food regime is either emerging or well underway. Friedmann (2005: 229), who can be placed in the former camp, claims to have identified certain contradictions that point to an emerging "green capitalism" regime: "a new round of accumulation [. . .] based on selective appropriation of demands by environmental movement, and including issues pressed by fair trade, consumer health, and animal welfare activists." Campbell (2009: 309) has recently echoed this argument in his description of an emerging "food from somewhere" regime, which he characterizes as "having denser ecological feedbacks and a more complex information flow in comparison to the invisibility and distanciation characterizing earlier regimes as well as contemporary "'Food from Nowhere'." Campbell notes that, while this regime does offer an opportunity for changing in important ways the character of certain key food relations and their impacts on the environment, its legitimacy, somewhat paradoxically, relies on the continuing existence of the global-scale, unsustainable regime which it is viewed as an alternative to. "The key question," according to Campbell, is "whether new food relations can open up spaces for future, more ecologically connected, global-scale food relations" (Campbell 2009: 309).

McMichael is less tentative in his assertions about where we are at, arguing that we have entered a third food regime – what he calls the "corporate food regime." Under this regime, the goal is no longer about providing labor with cheap food but "rendering [...] the peasantry increasingly unviable" with the goal of eliminating all remaining barriers – to land, nature, labor, capital, etc. – to agribusiness (McMichael 2009b: 285). David Burch and Geoffrey Lawrence (2009), to give a third example, write about a financialized third food regime. This describes the increasing trend of "finance institutions becoming increasingly involved in the agri-food system while agri-food companies come increasingly to behave like financial institutions" (Burch and Lawrence 2009: 277).

There are two ways to view this ambiguity over where, regime-wise, we are currently at. On the one hand, critics could use this to question the explanatory value of regime theory. Conversely, and the position that I prefer, is to see this debate among food regime theorists as evidence that the framework is not linear and that it is sufficiently flexible to see, simultaneously, multiple future trajectories.

Food regimes theory, along with the commodity systems approach (discussed in the following section), have been labeled agrifood theory (Buttel 1996; McMichael 1994). The research agenda driving these beyond-the-farm-gate frameworks reflects a not-so-subtle departure from the more farm-oriented agrarian political economy frameworks that commanded such attention in the 1980s. The impact they have had on the subdiscipline is unmistakable. They helped illuminate, unlike, say, the Mann–Dickinson thesis (Chapter 2), why the food *system* is sociologically significant.

Commodity systems approach

The commodity systems approach emerged in the 1980s as part of the "new" sociology of agriculture. Over a decade later, it had grown into one of the "major foci of theory and research" for agrifood scholars (Buttel 2001: 171). Commodity systems analyses include but cannot be reduced to what are called "commodity studies." For example, de Janvry *et al.* (1980) provide an early example of a tomato commodity study. They focused on one issue as it related to one segment of the commodity chain: mechanization at the site of production and its social consequences. Such limited breadth justifies locating this study in the commodity *studies* camp. Compare this to an early lettuce study (Friedland *et al.* 1981), which more fully embodies the spirit of a commodity *systems* approach. In order to fully understand the social consequences that came with mechanical harvesting the authors looked beyond the site of production and included in their analysis lettuce processors as well as nonagricultural aspects of life, such as community well-being.

The difference between commodity studies and commodity systems studies (and other related approaches; Box 3.2) are often matters of degree rather than of kind.

BOX 3.2 COMMODITY CHAINS AND NETWORKS, FILIÈRES, AND GLOBAL COMMODITY CHAINS

Even among the initiated (I speak from personal experience), the commodity-type literature can be head-scratchingly complex. You have, for example, commodity chain analyses, commodity network analyses, the filière approach, and the global commodity chain literature. Not helping matters, while some view these approaches as each offering something unique, others prefer to "use 'commodity systems,' 'commodity chains,' and 'filières' interchangeably [. . .] [as] all have a similar meaning and refer to the methodology for studying a specific commodity from its origins in production to consumption" (Friedland 2004: 5). With this in mind, I will do my best to distinguish these approaches in a few sentences.

- *Commodity chain*: illuminates the connections between consumers, producers, and workers, giving particular attention to the unequal distribution of power between actors and the social relationships they possess to capital.
- *Filière approach*: the filière approach can follow any number of schools of thought or research traditions (each with its own theoretical underpinnings), including, for example, systems analysis, institutional economics, management science, and Marxist economics. Most filière studies can be distinguished from commodity chain analyses by their attention to those public institutions charged with creating a smooth flow of agricultural commodities (Raikes *et al.* 2000).
- *Global commodity chain*: increasingly, commodity chains are extending around the globe. Researchers interested in global commodity chains (GCC) place particular importance on issues of governance. This concept highlights that GCC rarely operate in either a free market or through a series of arm's-length market-based transactions. In GCC, rules, laws, and organizations (e.g. WTO) strongly influence actors along the entire commodity chain.
- *Commodity network*: "The commodity network approach is more inclusive and wider-reaching than the GCC analysis on which it is based," by also looking at the values actors draw on during market transactions (Klooster 2006: 545).

For example, while Wells' (1996) *Strawberry Fields* spends a great deal of time examining labor relations space is also given to issues like ethnicity, worker–grower relations, and corporate power. Conspicuously absent from her analysis, however, are issues related to strawberry distribution, processing, and marketing. Morgan's (1980) *Merchants of Grain* omits just the opposite, looking almost entirely at the agribusinesses that dominate the global trade of grain while ignoring grain farmers. While both studies look at a number of variables, the general consensus seems to be that neither looks at enough to justify the label commodity *systems* study (see Friedland 2001: 83).

William Friedland (1984) published a seminal paper detailing five basic foci for commodity systems research: production practices or labor process; how growers organize themselves with respect to other actors; the character of the labor market and supply, and the ways workers organize themselves with respect to production; how scientists are mobilized and conduct their research with respect to food production; and the marketing and distribution of commodities beyond the farm gate. Since then, others have added to Friedland's original list. Dixon (2002), for example, argues that attention must also be paid to product design and regulatory politics. Dixon has also argued that Friedland's original methodology unwisely omits the site of consumption, noting the importance of such variables as food access, the eating environment, and the eating experience. Scholars calling for the inclusion of consumption in commodity system analyses, however, still reject the belief of consumer sovereignty. By nature of its focus on the *system*, this approach can easily show how little control exists at either "end" (farm or fork) of the commodity chain. Power, rather, lies in the middle of the hourglass, with giant agribusinesses (like Cargill and ADM), supermarket retailers, and fast food producers. Making room for consumption in commodity system analyses, however, does allow researchers to examine the influence of such important actors as nutritionists, market researchers, and specialists, who engage in symbolic manipulation and thus give shape to consumption patterns.

The commodity systems literature, on the whole, tends to be heavily empirical (Buttel 2001: 174). While not atheoretical, commodity systems scholars tend to let empirical concerns – for instance, "How should I go about studying strawberries?" – guide their research. Lacking a shared theoretical grounding allows different "flavors" of commodity systems scholarship to emerge, based upon the theoretical inclinations of the analyst.

One such "flavor" comes from commodity system scholars guided by actor network theory, also known as ANT (discussed further in Chapter 9). To put it simply, ANT privileges nothing (or everything). What matters, according to ANT, are patterned networks consisting of human, nonhuman, living, and inanimate artifacts. Who (or what) make up those networks are given equal explanatory weight. This separates ANT from most social theories, where humans (and human action) are given analytic superiority. Lawrence Busch and Arunas Juska were among the first to offer a sustained ANT-inspired commodity systems approach (see, e.g., Busch and Juska 1997; Juska and Busch 1994). They examined rapeseed.

This seed had limited market and agricultural value until a scientific network was mobilized – including mobilization of rapeseed *itself* – and certain negative qualities associated with the seed removed. The success of this network is evidenced by rapeseed's transformation from a marginal global commodity to something traded to the tune of US$40 billion a year.

Other commodity system "flavors" include those interested in describing the multiscalar, and often "lumpy," dynamics of globalization. I would place the book *Agri-food Globalization in Perspective* (Pritchard and Burch 2003) in this camp; a book that has been said to offer "probably the most comprehensive of commodity systems analyses" (Friedland 2004: 8). Analyzing growing, processing, and marketing networks from around the world, stretching from the USA to Canada, Australia, Europe, Thailand, and China, the book finds that what "passes for 'the global food system' consists of a set of heterogeneous and fragmented processes, bounded in multiple ways by the separations of geography, culture, capital and knowledge" (Pritchard and Burch 2003: xi). When viewed through this lens, the globalization of food is "understood as an intricate set of processes operating at many scales, and on many levels, rather than a unilateral shift toward a single global marketplace" (Pritchard and Burch 2003: xi).

It is difficult to adequately define the boundaries of the commodity systems literature; a point that speaks to the impact the approach has on the subdiscipline as a whole. One would be hard-pressed today to find an analysis that does not, when examining a commodity or a category of commodities, follow the artifact as it moves through space and time within an ever-changing food system. In many cases, studies might read like a commodity systems analysis without formally taking up such an identity (see, e.g., de la Pena 2010; Freidberg 2009). This is perhaps the greatest testament of all to this approach: that it is now commonsense when studying food to never just look at food *itself* but also at the actors, institutions, and rules and regulations that connect farm-gate and kitchen table.

Transition. . .

Chapters 2 and 3 spent a significant portion of time discussing changes to the structure of agriculture and the food system more generally. Building on this discussion, Chapter 4 speaks of additional reasons for these changes, particularly at the international level. It questions specifically certain conventionally held beliefs about food security and investigates whether changes to global agriculture directed by these beliefs produced the ends promised.

Discussion questions

1 What are some potential benefits associated with the earlier mentioned market concentration in the food commodity chain? What are some of the costs?
2 What is the financial incentive for seed companies like Monsanto, who by definition (as for-profit entities) seek profits, to engage in breeding programs

designed to produce seed for the world's poor (an endeavor that would likely *lose* the company money)?

3 How do the food regime and commodity system approaches differ? What characteristics do they share?

4 In what ways are the food regime and commodity systems approaches an improvement over a Mann–Dickinson-type analysis (as discussed previously in Chapter 2)? Does a Mann–Dickinson-type analysis (classic agrarian theory) still have value?

Suggested readings: Introductory level

Goodman, David and Michael Redclift. 1991. The origins of the modern agri-food system. In *Refashioning Nature: Food, Ecology and Culture*, edited by David Goodman and Michael Redclift. London and New York: Routledge, pp. 87–132.

Heffernan, William D. 2000. Concentration of ownership and control in agriculture. In *Hungry for Profit*, edited by Fred Magdoff, John Foster, and Fredrick Buttel. New York: Monthly Review Press, pp. 61–75.

Suggested readings: Advanced level

Bonanno, Alessandro and Douglas Constance. 2001. Globalization, Fordism, and Post-Fordism in agriculture and food: A critical review of the literature. *Culture and Agriculture* 23(2): 1–18.

Friedland, William. 2001. Reprise on commodity systems methodology. *International Journal of Sociology of Agriculture and Food* 9(1): 82–103.

Glenna, L. and D. Cahoy. 2009. Agribusiness concentration, intellectual property, and the prospects for rural economic benefits from the emerging biofuel economy. *Southern Rural Sociology* 24(2): 111–29.

McMichael, Philip. 2009. A food regime genealogy. *Journal of Peasant Studies* 36(1): 139–69.

Pechlaner, Gabriela and Gerardo Otero. 2010. The neoliberal food regime: Neoregulation and the new division of labor in North America. *Rural Sociology* 75: 179–208.

Note

1 Taken from personal communication with Phil McMichael, June 18, 2011.

References

Birn, Anne-Emanuelle, Yogan Pillay, Timothy Holtz, and Paul Basch. 2009. *Textbook of International Health: Global Health in a Dynamic World*. New York: Oxford University Press.

Bloomberg 2009. Wal-Mart's store-brand groceries to get new emphasis. *Bloomberg* February 19, http://www.bloomberg.com/apps/news?sid=afVJJxZ4oCtY&pid=newsarchive, last accessed September 26, 2010.

Bonanno, A. and D. Constance 1996. *Caught in the Net: The Global Tuna Industry, Environmentalism and the State*. Lawrence, KS: University Press of Kansas.

Borras, S., P. McMichael, and I. Scoones. 2010. The politics of biofuels, land and agrarian change: Editor's introduction. *Journal of Peasant Studies* 37(4): 575–92.

Burch, David and Geoffrey Lawrence. 2009. Towards a third food regime: behind the transformation. *Agriculture and Human Values* 26(4): 267–79.

Busch, Lawrence and Arunas Juska. 1997. Beyond political economy: Actor networks and the globalization of agriculture. *Review of International Political Economy* 4(4): 688–708.

Buttel, F. 1996. Theoretical issues in global agri-food restructuring. In *Globalisation and Agri-food Restructuring: Perspectives from the Australasia Region*, edited by D. Burch, R. Rickson, and G. Lawrence. Avebury: Aldershot, pp. 17–44.

Buttel, Fredrick. 2001. Some reflections on late-twentieth century agrarian political economy. *Sociologia Ruralis* 41(2): 165–81.

Campbell, H. 2009. Breaking new ground in food regime theory: Corporate environmentalism, ecological feedbacks and the "food from somewhere" regime? *Agriculture and Human Values* 26: 309–19.

Carolan, Michael, S. 2008. From Patent Law to Regulation: The Ontological Gerrymandering of Biotechnology, *Environmental Politics* 17(5): 749–65.

Carolan, Michael. 2011. *The Real Cost of Cheap Food*. London: Earthscan.

Christian Aid. 1999. Selling suicide, May: 1–2, http://christian-aid.org.uk/in-depth/9905suic/suicide2.html, last accessed February 12, 2011.

DeGregori, Thomas. 2002. *Bountiful Harvest*. Washington, DC: The Cato Institute.

de Janvry, Alain, Philip Leveen, and David Runsten. 1980. *Mechanization in California Agriculture: The Case of Canning Tomatoes*. Berkeley, CA: University of California.

de la Pena, Carolyn. 2010. *Empty Pleasures: The Story of Artificial Sweeteners from Saccharin to Splenda*. Chapel Hill, NC: University of North Carolina Press.

Dixon, Jane. 2002. *The Changing Chicken: Chooks, Cooks, and Culinary Culture*. Sidney, Australia: University of South Wales Press.

Domina, David and C. Robert Taylor. 2010. The debilitating effects of concentration markets affecting agriculture. *Drake Journal of Agricultural Law* 15(1): 61–108.

Dutfield, Graham. 2008. Turning plant varieties into intellectual property: The UPOV Convention. In *The Future Control of Food*, edited by Geoff Tansey and Tasmin Rajotta. London: Earthscan, pp. 27–47.

ETC (Erosion, Technology and Concentration Group). 2002. Terminate terminator in 2002, February 19: 1–3, http://www.etcgroup.org/documents/terminatorbrochure02.pdf, last accessed March 2, 2011.

Freidberg, S. 2009. *Fresh: A Perishable History*. Cambridge, MA: Harvard University Press.

Friedland, William. 1984. Commodity systems analysis: An approach to the sociology of agriculture. In *Research in Rural Sociology and Development: A Research Annual*, edited by Harry Schwarzweller. Greenwich, CT: JAI Press, pp. 221–35.

Friedland, William. 2001. Reprise on commodity systems methodology. *International Journal of Sociology of Agriculture and Food* 9(1): 82–103.

Friedland, William. 2004. Agrifood globalization and commodity systems. *International Journal of Sociology of Agriculture and Food* 12: 5–16.

Friedland, William, Amy Barton, and Robert Thomas. 1981. *Manufacturing Green Gold: Capital, Labor, and Technology in the Lettuce Industry*. New York: Cambridge University Press.

Friedmann, Harriet. 1978. World market, state, and family farm: Social bases of household production in an era of wage labor. *Comparative Studies in Society and History* 20(4): 545–86.

Friedmann, Harriet. 1991. *Changes in the International Division of Labor: Agrifood Complexes and Export Agriculture*. Edited by W. Friedland, L. Busch, F. Buttel, and A. Rudy. Boulder, CO: Westview Press, pp. 65–93.

Friedmann, Harriet. 1993. The political economy of food: A global crisis. *New Left Review* 197: 29–57.

Friedmann, Harriet. 2005. From colonialism to green capitalism: Social movements and the emergence of food regimes. In *New Diections in the Sociology of Global Development*, edited by Fredrick Buttel and Philip McMichael. Oxford: Elsevier, pp. 229–67.

Friedmann, Harriet and Philip McMichael. 1989. Agriculture and the state system: The rise and fall of national agricultures, 1870 to the present. *Sociologia Ruralis* 29: 93–117.

Frontline Transcript. 2004. Is Walmart good for America? Public Broadcasting Service, http://www.pbs.org/wgbh/pages/frontline/shows/walmart/etc/script.html, last accessed September 26, 2010.

Goodman, David, Bernardo Sorj, and John Wilkinson. 1987. *From Farming to Biotechnology: The Industrialization of Agriculture.* New York: Blackwell.

Goodman, David and Michael Redclift. 1991. *Refashion Nature: Food, Ecology and Culture.* London: Routledge.

Goodman, David and Michael Watts. 1994. Reconfiguring the rural or fording the divide? Capitalist restructuring and the global agro-food system. *Journal of Peasant Studies* 22(1): 1–49.

Hendrickson, Mary and William Heffernan. 2007. Concentration of agricultural markets, 2007, Department of Rural Sociology, University of Missouri, Columbia, MO, http://www.foodcircles.missouri.edu/07contable.pdf, last accessed September 22, 2010.

Howard, Philip. 2009. Visualizing consolidation in the global seed industry: 1996–2008. *Sustainability* 1: 1266–87.

Hubbard, Kristina. 2009. Out of hand: Farmers face the consequences of the seed industry. Farmer to Farmer Campaign on Genetic Engineering, December, National Family Farm Coalition, http://farmertofarmercampaign.com/Out%20of%20Hand.FullReport.pdf, last accessed March 13, 2011.

Juska, Arunas and Lawrence Busch. 1994. The production of knowledge and the production of commodities: The case of rapeseed technoscience. *Rural Sociology* 59(4): 581–97.

Kaskey, Jack. 2010. Monsanto says triple-stack corn beating some SmartStax. *Bloomberg Business Week*, September, 20, http://www.businessweek.com/news/ 2010-09-20/monsanto-says-triple-stack-corn-beating-some-smartstax.html, last accessed March 14, 2011.

Kloppenburg, Jack. 1988. *First the Seed.* New York, NY: Cambridge University Press.

Klooster, Dan. 2006. Environmental certification of forests in Mexico: The political ecology of a nongovernmental market intervention. *Annals of the Association of American Geographers* 96(3): 541–65.

Lang, Tim, David Barling, and Martin Caraher. 2009. *Food Policy: Integrating Health, Environment, and Society.* New York: Oxford University Press.

Lawrence, Geoffrey and Janet Grice. 2004. Agribusiness, biotechnology and food. In *A Sociology of Food and Nutrition*, edited by G. Germov and L. Williams. New York: Oxford University Press, pp. 77–95.

Marsden, Terry. 1989. Restructuring rurality: From order to disorder in agrarian political economy. *Sociologia Ruralis* 29(3/4): 312–17.

Marsh, John and Gary Brester. 2004. Wholesale-retail marketing margins behavior in the beef and pork industries. *Journal of Agricultural and Resource Economics* 29(1): 45–64.

McMichael, P. 1994. Introduction: Agro-food restructuring: unity in diversity. In *The Global Restructuring of Agro-food Systems*, ed. P. McMichael. Ithaca, NY: Cornell University Press, pp. 1–18.

McMichael, Philip. 2009a. A food regime genealogy. *Journal of Peasant Studies* 36(1): 139–69.

McMichael, Philip. 2009b. A food regime analysis of the "world food crisis." *Agriculture and Human Values* 26: 281–95.

Morgan, Dan. 1980. *Merchants of Grain*. New York: Penguin.

Novak, W. and A. Haslberger. 2000. Substantial equivalence of antinutrients and inherent plant toxins in genetically modified novel foods. *Food and Chemical Toxicology* 38(6): 473–83.

OECD. 1993. *Safety Evaluation of Foods Derived by Modern Biotechnology*. Paris: Organization for Economic Cooperation and Development.

Paarlberg, Robert. 2010. *Food Politics: What Everyone Needs to Know*. New York: Oxford University Press.

Pendleton, Cullen. 2004. The peculiar case of "terminator" technology: Agricultural biotechnology and intellectual property protection at the crossroads of the third green revolution. *Biotechnology Law Review* 23: 1–29.

Pritchard, Bill, and David Burch. 2003. *Agri-food Globalization in Perspective: International Restructuring in the Processing Tomato Industry*. Aldershot, UK; Burlington, VT: Ashgate.

Raikes, Philip, Michael Friis Jensen, and Stefano Ponte. 2000. Global commodity chain analysis and the French filière approach: Comparison and critique. *Economy and Society* 29(3): 390–417.

Taylor, Robert and David Domina. 2010. Restoring economic health to contract poultry production. Report prepared for the Joint US Department of Justice and US Department of Agriculture/GIPSA Public Workshop on Competition Issues in the Poultry Industry, May 21, 2010, Normal, AL, http://www.dominalaw.com/ew_library_file/Restoring%20Economic%20Health%20to%20Contract%20Poultry%20Production.pdf, last accessed September 24, 2010.

Trewavas, A. and C. Leaver. 2001. Is opposition to GM crops science or politics? *EMBO Reports* 2(6): 455–9.

Van Dooren, Thom. 2007. Terminated seed: Death, proprietary kinship and the production of (bio) wealth. *Science as Culture* 16(1): 71–93.

Van Wijk, Jeroen. 2004. Terminating piracy or legitimate seed saving. *Technology Analysis and Strategic Management* 16(1): 121–41.

von Braun, Joachim. 2007. The world food situation: New driving forces and required actions. Food Policy Report, December, International Food Policy Research Institute, Washington, DC, http://www.ifad.org/events/lectures/ifpri/pr18.pdf, last accessed September 26, 2010.

Wardle, Jon and Michael Baranovic. 2009. Is lacking of retail competition in the grocery sector a public health issue? *Australian and New Zealand Journal of Public Health* 33(5): 477–81.

Wells, Miriam. 1996. *Strawberry Fields: Politics, Class, and Work in California Agriculture*. Ithaca: Cornell University Press.

Wise, Timothy. 2010. Monopolies are killing our farms, Editorial. *East Texas Review* April 13, http://www.ase.tufts.edu/gdae/Pubs/rp/WiseMonopoliesAndFarms13Apr10.pdf, last accessed September 22, 2010.

4

MALNUTRITION

Hidden and visible

KEY TOPICS

- Rather than the result of there being insufficient food in the world, famines reflect a distribution problem.
- As fundamentally a product of bad policy, global food insecurity cannot be eradicated simply with more technology, higher yielding varieties, and greater use of inputs.
- The so-called global obesity epidemic is the product of an array of social, environmental, economic, and political variables.

Key words

- Apolitical ecologies
- Body mass index
- Famine
- Food aid
- Food crisis, 2007–2008
- Food dependence
- Food desert
- Food riots
- Global obesity epidemic
- Governance
- Green revolution
- International Monetary Fund
- Infant formula controversy
- Marketization of food security
- McLibel
- Micronutrient malnutrition
- Nutritionism
- Obesity–hunger paradox
- Structural adjustment policies
- Trade liberalization
- World Bank

Malnutrition is the result of having an insufficient (calorie-wise) and/or deficient (micronutrient-wise) diet. Since the mid-twentieth century, it has been common to frame hunger as a technological problem; one that can only be resolved with higher yields, more inputs, and all-around better technology. When looked at sociologically, the real complexity of malnutrition, in all of its forms – hunger as well as obesity – is revealed in this chapter. Long-term solutions to food security are located in neither a can (e.g. more inputs) nor patent application (e.g. biotechnology) but in novel social, political, and economic arrangements. And it's an issue that, as illustrated in Figure 4.1, we simply cannot afford to ignore.

It might surprise the reader to find a chapter on such subjects as famine, hunger, and obesity located in this Part and not, say, in Part III, Food Security and the Environment. I've placed the chapter here to emphasize that these problems are rooted in global social, economic, and political structures. We must keep these roots in mind when attempting to draw up solutions to them.

Amartya Sen on famine

According to Amartya Sen (1981), modern food crises are less related to the absence of food as to the inability to buy it. Examining the 1943 Bengal famine, Sen found there to be plenty of food around. The problem, which a *Life* magazine

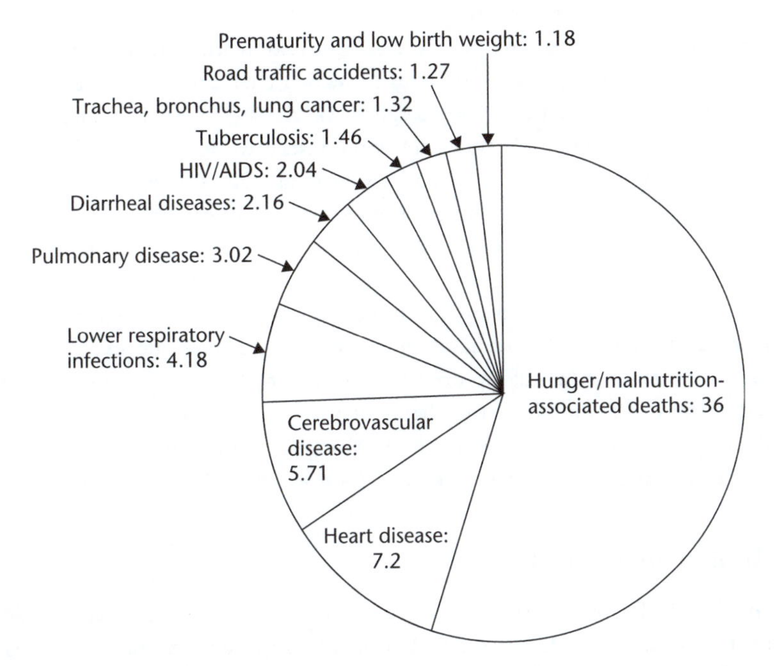

FIGURE 4.1 Top 11 leading causes of death worldwide (in millions). Based on data from the World Health Organization

reporter at the time estimated was taking the lives of more than 50,000 Bengalese a week (Fisher 1943: 16), was not a lack of food but a lack of *available* food. Those who had food were likely to hoard it, knowing that this action would only increase its market price and make them more money in the not-too-distant future. The Bengal famine revealed the limits of the market for delivering food during times of crisis. The British (the famine occurred during the British Raj) bet, incorrectly, that the market would justly and efficiently distribute food to the poor.

Sen notes that similar variables can explain the 1974 Bangladesh famine. Floods put farm laborers out of work and therefore drastically reduced their incomes. The floods created expectations of impending rice shortages, which led to hoarding among those who could afford to do so. This panic buying pushed the price of rice beyond the reach of the poor.

Sen sees democracy – not biotechnology, advantages in plant breeding, or more fertilizer – as offering the surest path to a world free of famine. As Sen (1981: 16) famously wrote: "no famine has ever taken place in the history of the world in a functioning democracy." Famines tend to only affect the poorest segment of a country's population. In nondemocratic states, then, those most likely to experience famine are likewise the least likely to have political access. Their famine-related suffering is therefore allowable to the elites, as it does little to threaten their grip on power. Among other things, democracy gives those living on the margins a voice, so even a country's political elite can feel the pain of famine, even though they and their families remain well fed.

For Sen, it is not enough to have a regulated market. China experienced the Great Leap famine (1958–1961), during which over 15 million lives were lost (Xizhe 1987: 639), with heavily regulated markets. Conversely, India managed to avoid famine in 1965 and 1966 despite two consecutive years of failed crops due to massive monsoon rains. In a democracy, elected leaders have a vested interest in responding promptly during times of food scarcity. In the case of India in 1965 and 1966, officials quickly looked to outside sources for food aid (receiving 10 million tons of grain from the USA), while simultaneously investing in the public food distribution system to ensure that those who needed food received it. In sum: famines are, at their root, a *political* problem, which can be cost effectively alleviated. Famines only stricken a small proportion of the population; a figure that rarely exceeds 5 percent. And since the shares of income and food of those stricken usually do not exceed 3 percent of the total for the nation, it is not difficult (or expensive) to fill in whatever income and food has been lost (Sen 2004).

Political economy of hunger

In April 2008, Haitian protestors, objecting to rising food prices, attempted to storm the presidential palace gates at Port-au-Prince. They were protesting against the rising price of basic foods like rice, beans, and condensed milk, which had increased by more than 50 percent in less than 12 months. A 50 percent price increase would have been hard to swallow in an affluent nation but in a country where 80 percent survive on less than US$1 a day, its effects were devastating. A mother working in Port-au-Prince told a reporter that before the food crisis US$1.25 could buy a small amount of vegetables, some rice, 10 cents of charcoal, and a little cooking oil. At the height of the crisis, that US$1.25 made it impossible to "even make a plate of rice for one child" (as quoted in Quigley 2008).

After the dictator Jean-Claude "Baby Doc" Duvalier was overthrown in 1986, the International Momentary Fund (IMF) provided Haiti with a US$24.6 million loan (Box 4.1). Since Duvalier had pillaged the treasury before fleeing, the country desperately needed cash. Government officials thus accepted the loan. But it came with strings attached – or, in international developmental parlance, it required the country to undergo certain "structural adjustments."

BOX 4.1 IMF AND THE WORLD BANK

Known collectively as the Bretton Woods Institutions after the small community in rural New Hampshire (USA), where they were founded in July 1944, the World Bank and the International Monetary Fund (IMF) are charged, in different ways, with directing and supporting the structure of the world's economic and financial order. The simplest way to separate the two institutions is to say that the World Bank is a bank (primarily a lending institution), whereas the IMF is not.

The World Bank's first loans were granted following World War II to help finance the reconstruction of Western Europe. As these nations recovered, the Bank turned increasingly to the world's poorer countries, in part to help them "develop" by folding them into the world economy and thus lure them away from Communism (as this was at the height of the Cold War). The World Bank's statement mission is to promote economic and social progress (though the Bank believes the latter only comes *after* the former) in developing countries.

The IMF's role is essentially to keep the international monetary system running smoothly by making sure the unpredictable variations in the exchange values of national currencies that helped to trigger (and unduly extended) the Great Depression in the 1930s do not happen again. Toward this end, the IMF does provide loans; though, as mentioned, it is more than just a lending institution.

Both institutions wield considerable power over international affairs, as their loans come with substantial strings attached. These "strings" are said to benefit less affluent nations by enriching the global community as a whole. Yet, to date, this wealth has overwhelming enriched only a handful of nations. For this reason, both institutions have come under intense criticism in recent decades for making matters worse for the world's poorest economies rather than better.

These "adjustments" are attached to all loans provided to developing nations by agencies like the IMF. In Haiti's case, acceptance of the loan required the country to reduce tariff protections for their rice and other agricultural products and to open up their markets to imports. Coincidently, as Haiti was forced to reduce its government support for agriculture the USA was doing just the opposite (direct US taxpayer subsidies to the rice sector have averaged $1 billion a year since 1998).

Prior to 1986, Haiti imported very little rice. The agricultural sector was heavily protected through tariffs and import prohibitions. Trade liberalization drastically reduced these protections, leading, not surprisingly, to a flood of cheap rice into the country, most coming from the USA. In 1985, Haiti imported only 7,000 metric tons of rice. By 2004, the country was importing 225,000 metric tons. All those cheap imports did make, at least in the short term, food more affordable for the urban poor. Estimates by the World Bank (2002: 43) indicate that the real price of rice to consumers was reduced by about 50 percent after trade liberalization in Haiti. But this cheap food came at a steep price.

Trade liberalization policies have had a largely negative effect on the macroeconomic stability of the country. In the early 1960s, the agricultural sector employed 80 percent of the labor force and made up 90 percent of all exports. When trade liberalized those export revenue streams dried up. Moreover, income tax receipts dropped as the country's farmers fell further into poverty. And so began a vicious circle. As foreign investments dropped, the country's infrastructure fell into disrepair. This, in turn, further undermined trade capacity as the country became even less attractive to foreign investors, which further reduced income tax receipts and investment in infrastructure. And so the downward cycle continues to this day.

As noted by a prominent globalization scholar (as well as Representative to the 14th Congress of Republic of the Philippines), when it came to curbing hunger the one-time Philippines dictator Ferdinand Marcos had a better track record than either the World Bank or the IMF:

> To head off peasant discontent, the regime provided farmers with subsidized fertilizer and seeds, launched credit schemes, and built rural infrastructure. During the 14 years of the dictatorship, it was only during one year, 1973, that rice had to be imported owing to widespread damage wrought by typhoons. When Marcos fled the country in 1986, there were reported to be 900,000 metric tons of rice in government warehouses. Paradoxically, the next few years under the new democratic dispensation saw the gutting of government investment capacity. As in Mexico, the World Bank and IMF, working on behalf of international creditors, pressured the Corazon Aquino administration to make repayment of the $26 billion foreign debt a priority.
>
> *Bello (2008: 451)*

Like other countries, the Philippine government was instructed by the IMF to abandon the practice of surplus storage. The elimination of surplus grain inventories occurred because the Bank and IMF (who control the purse strings of many of these nations) believed countries suffering crop failures could always import the food they need. A problem with this reasoning, however, is that these countries were also instructed (by these same institutions) to reduce expenditures so more money would be freed up to service their debt. Yet this gutted the very domestic policies that previously helped secure food security for these nations, such as subsidies to small farms for things like fertilizer and seed. Rolling back the presence of the state also eviscerated the infrastructure of these countries and with that their organizational capacity to effectively handle and distribute the massive levels of grain coming in as food aid during times of crisis. The global trade liberalization policies of the last half a century have been instrumental in transferring responsibility for food security from the state to the market – a process that has been called the "marketization of food security" (Zerbe 2009: 172).

The net result of these policies has been *increased* food dependence, which translates into hunger and malnutrition, particularly during times of heightened global grain prices (such as during the recent food crisis of 2007–2008). Between 1950 and 1970, less affluent nations went from importing no grain to accounting for almost half of all world imports (Friedmann 1990: 20). Before World War II, no nations in Africa, Latin America, or South Asia imported wheat. As recently as the 1960s, countries like Nigeria were almost entirely food *independent*. Yet, by 1983, one-quarter of the country's *total earnings* went to importing wheat (Friedmann 1992). So when prices, say, double, as they did for certain commodities during the food crisis of 2007–2008, food imports suck an enormous amount from the public coffers of countries like Nigeria.

This dependency started with cheap imports from the USA. After World War II, the USA had large stocks of surplus wheat thanks to New Deal price support programs and the government was looking to rid itself of this surplus without harming market prices. It was decided this could be accomplished through "concessional sales" to developing countries. These sales were highly subsidized under Public Law 480 ("food aid"). During the 1950s, the US share of wheat exports grew from one-third to over half of the global market, with the majority destined for the developing world. Between 1950 and 1976, per capita consumption of wheat increased in the developing world over 60 percent, while per capita consumption of cereals (minus wheat) increased 20 percent and per capita consumption of root crops (a tradition staple in many developing nations) decreased by 20 percent.

Directing monies away from sectors whose products can be readily substituted by cheap imports makes sense, in the short term. Unfortunately, weather, population growth, biofuels, a growing livestock industry, and poor food access mean that the market can never be entirely trusted to deliver low cost agricultural commodities forever – a point made painfully clear in 2007 and 2008. Prior to the food riots of 2008, approximately half of Haiti's hard currency was used to purchase food. So when prices of basic foods rose 50 percent during the food crisis of 2007–2008, it is obvious why mud biscuits became for all intents and purposes the national dish.

Food aid

The USA is, and has long been, the largest food aid donating country. In 1954, the US government created Public Law 480 with the passing of the Agricultural Trade Development and Assistance Act. The Marshall Plan (a large-scale economic recovery program offered by the USA to aid Europe's recovery following World War II) showed how foreign aid could be used to increase trade dependence despite a country's lack of money. There was also a belief that food could be used strategically to promote US interests abroad. To quote Senator Hubert Humphrey, in remarks to the Agriculture and Forest Committee of the US Senate in 1957: "[I]f you are looking for a way to get people to lean on you and to be dependent on you, in terms of their cooperation with you, it seems to me that food dependence would be terrific" (as quoted in Weis 2007: 66). We must also remember that food aid arose during the height of the Cold War; a war that was not just about winning over hearts and minds but also stomachs.

Some believed that the ideology most likely to prevail – democratic capitalism *v.* centralized communism – would be the one that could offer a country's citizens the most food. General Lucius Clay, while military governor of West Germany immediately after World War II, said in 1946 that food shortages meant "no choice between coming a Communist on 1,500 calories and a believer in democracy on 1,000 calories" (quoted in Major 1997: 236). Anxiety that West Germans would side with the Soviets simply to obtain more food precipitated the Berlin airlifts, where food and other goods were flown into Berlin from June 1948 to May 1949. Soon after Public Law 480 passed and was used to similar effect as the Berlin airlifts, to stop Communism from spreading in poor countries on account of calories. It also helped bolster the Eisenhower administration's claim that "free men eat better," later echoed by the Kennedy administration by their claim that "wherever Communism goes, hunger follows" (quoted in Cullather 2007: 363).

There is considerable debate, however, about how much food aid actually helps recipient nations. Part of the issue lies in how it is distributed. When an emergency arises, surplus food from affluent nations is donated to relief organizations working closely with the United Nations World Food Program. Another option is to donate food directly to the governments in need. The receipt governments then sell the food at below market prices to raise money for general development programs. In either case, farmers residing in countries receiving food aid typically get less for their products as the food aid inevitably drives market prices downward (Peterson 2009: 115). Consequently, most donating countries, except for the USA, give cash to the World Food Program rather than surplus grain. From 1995 to 2005, for example, the USA accounted for approximately 60 percent of all food aid. The remainder came from the EU (25 percent), Japan (6 percent), Canada (5 percent), Australia (3 percent), Norway (less than 1 percent), and Switzerland (less than 1 percent) (Hanrahan and Canda 2005: 5).

Giving grain to later be sold to generate cash – instead of just giving cash to begin with – is inefficient on a number of levels. For one thing, when the food aid is sold it typically sells well below the market price – 30–50 percent below according to some estimates (IATP 2005). Why sell something for a fraction of what it is worth, which has been paid for at full price with taxpayer dollars, to generate aid revenue when the money originally used to buy the grain could be given directly to those in need? Furthermore, a sizable amount of money spent on food aid is used to transport that heavy grain over oceans and deep into continents. The US Agency for International Development (USAID) food aid budget in 2005 was US$1.6 billion. Of that, only 40 percent (US$654 million) went to paying for food. The rest was spent on overland transportation (US$141 million), ocean shipping (US$341 million), transportation and storage in destination countries (US$410 million), and administrative costs (US$81 million) (Dugger 2005). The US government further requires that 75 percent of all food aid be transported by US flag carriers, regardless of cost. The rationale is national defense. Having a significant presence of US flag carriers out on the open sea is said to enhance US military readiness; though in actuality many of the US flag carriers are owned by foreign companies with little actual military value (Martin 2010). The US government also requires that 25 percent of all food aid must pass through Great Lakes ports, regardless of where it originates from. These requirements cost US taxpayers an additional $70 per ton of food aid shipped in 2007 (Martin 2010). They also slow delivery times substantially. Aid can take as long as 6 months to get from a US grain storage facility to a foreign village (Martin 2010).

The green revolution

The green revolution was a series of research and technology transfer initiatives that occurred immediately after World War II and lasted until the late 1970s. The revolution centered primarily on the development of high-yield varieties of a handful of grains coupled with the expansion of the necessary irrigation infrastructure and input supply chains (fertilizer, pesticides, seeds, etc.). It started in the late 1940s, as international agricultural research centers, funded by large US private foundations like Ford and Rockefeller, were developing high-yield varieties of rice, wheat, corn, and soybeans.

The benefits of the green revolution have been far from evenly distributed. Those with access to fertile ground, adequate water, and credit (to buy the necessary costly inputs and technology) gained the most. Other parts of world, like much of Africa, lacking water, capital, and arable land, missed out almost entirely from the movement (Table 4.1). For example, most of the high-yielding crops developed require copious amounts of water. The green revolution had a measureable positive impact (at least in terms of yields) in Asia, in part because the region receives considerable rainfall and has an extensive river system to support extensive irrigation systems. Contrast this to dry Sub-Saharan Africa. Eighty-five percent of Kenya's arable land, for example, cannot be irrigated (too far from any consequential water source) and must rely on rainfall that may, in

TABLE 4.1 Population and land use by region (2006)

Region	*Population*	*Ag. land (ha)*	*Arable (%)*	*Permanent pasture (%)*
Sub-Saharan Africa	750,500,000	947,000,000	17.0	80.8
China and India	2,480,300,000	736,000,000	41.2	55.9
Asia (other)	1,503,000,000	635,000,000	24.8	68.7
Australia/New Zealand	24,700,000	457,000,000	10.8	88.7
Europe	588,100,000	267,000,000	60.4	34.0
Russia	143,200,000	216,000,000	56.7	42.5
Latin America/Caribbean	565,000,000	726,000,000	19.7	77.6
North Africa	408,800,000	458,000,000	19.1	78.4
North America	335,500,000	477,000,000	45.9	52.3
World	6,592,800,000	4,973,000,000	28.2	69.0

Source: Data from Peterson (2009: 54).

any growing season, never come. A widely cited review from 1995 of over 300 peer-reviewed studies published between 1970 and 1989 evaluates the green revolution's success. Eighty percent of the articles conclude that the green revolution *increased* inequality at both farm and regional levels (Freebairn 1995).

The green revolution sought to export conventional agricultural methods to other parts of the world at the same time as conventional agriculture was itself becoming increasingly energy intensive. A team of researchers concluded in a 1973 *Science* article that agriculture was using an equivalent of 80 gallons of gasoline to produce an acre of corn (Pimentel *et al.* 1973). Furthermore, they note that while the production of corn per acre increased 2.4 times from 1945 to 1970, the input of fuel rose 3.1 times. In other words, yields in corn energy relative to fuel input *declined* 26 percent during this period. Beyond obvious environmental implications, this move toward an energy-intensive agriculture caused a number of green revolution critics to openly wonder if the movement wasn't more interested in building markets for expensive agribusiness inputs (Kloppenburg 1988).

The green revolution, as a solution to hunger and food security, is terribly narrow in its scope. Focusing so extensively on production, for example, does nothing to resolve the problems of food access mentioned earlier. Nor does it alter the tightly concentrated distribution of market power (the "hourglass") discussed in the previous chapter. And it has done little if anything to put money in the pockets of poor farmers; farmers who, with high yielding seeds in hand, now need to purchase expensive petrochemical inputs and install irrigation systems (if farming in area where rains are infrequent and/or insufficient).

Micronutrient malnutrition: Hidden hunger

The Food and Agriculture Organization (FAO) of the United Nation calculates that world agriculture produces enough food to provide everyone in the world with at least 2,720 kilocalories per person per day (FAO 2002). World agriculture is to be commended for producing such an abundance of calories. But a nutritional diet is not built on calories alone.

Even with all these available calories, micronutrient malnutrition plagues billions around the world. As explained in a report prepared by the United Nations Children's Fund (UNICEF), World Health Organization (WHO), and United Nations World Food Program (WFP):

> Deficiencies of micronutrients are a major global health problem. More than 2 billion people in the world today are estimated to be deficient in key vitamins and minerals, particularly vitamin A, iodine, iron and zinc. Most of these people live in low income countries and are typically deficient in more than one micronutrient.
>
> *UNICEF, WTO, and WFP (2007: 1)*

Children are a particularly sensitive population to micronutrient malnutrition. Approximately two-thirds of all deaths of children globally are attributable to nutritional deficiencies (Caballero 2002: 3–4). Growth stunted by malnutrition during early years of childhood is difficult to make up. Children who were between 12 and 24 months of age at the peak of the drought in rural Zimbabwe in 1994–1995 had average heights well below that of comparable children not impacted by this event when measured at ages 60–72 months (Hoddinott and Kinsey 2001).

There are many factors to blame for micronutrient malnutrition, like poor food access and extreme poverty. Another factor is the green revolution. There is no disputing that the green revolution has negatively impacted dietary diversity. Crops bred for traits associated with improved yields or to withstand mechanization have displaced traditional crops that are higher in iron and other micronutrients. In South Asia, for example, cereal production has increased more than fourfold since 1970, while production of pulses (a high protein legume) has dropped 20 percent (Gupta and Seth 2007: 436; Welsh and Graham 1999: 2). There are also a number of studies that link green revolution cropping systems with a decline in the density of iron in the diets of people in South Asia (Seshadri 2001; Welsh and Graham 2002). In addition, the cereals of the green revolution are processed and consumed in such a way that further negatively impacts the dietary health of millions. Rice particularly (and to a lesser degree cereals) is consumed after milling – a process that removes precious micronutrients. Pulses, conversely, are traditionally consumed whole after cooking; a process that leaves their micronutrient profile largely intact (Box 4.2). Moreover, whole cereal grains contain relatively high levels of antinutrients, which are known to lower the absorption of micronutrients, and lower levels of substances that promote the bioavailability of micronutrients (Cordain 1999; Welsh and Graham 1999).

The USAID estimates that iron deficiency costs India and Bangladesh about 5 and 11 percent, respectively, of their annual GDP (Sanghvi 1996). It has been

BOX 4.2 BOLIVIAN QUINOA AND THE UNINTENDED CONSEQUENCES OF EXPORT

Offering an exceptionally balanced level of animal acids, fiber, and an array of vitamins and minerals, quinoa arguably has no equal in the plant and animal world for the level of nutrients it contains. The National Aeronautic and Space Association (NASA) discovered this decades ago, declaring quinoa the perfect food to bring along during long-term human space missions. A centuries-old food staple in Bolivia, only recently has quinoa been perceived as little more than a curiosity for people living beyond the Andes. In the last decades, this "crop of the Incas" has been exported to affluent foreign markets in increasing quantities, raising the incomes of Bolivian quinoa growers substantially. But there has been a trade-off to these rising quinoa prices. Though prices have tripled for quinoa farmers in the last 5 years they have likewise tripled for everyone who eats this cultural staple. Consequently, consumption of quinoa in Bolivia has fallen 34 percent over the same period. And what are they eating in its place? Consumption of less expensive processed foods are on the rise, causing nutritionists to worry about rising rates of malnutrition, obesity, and other diet-related health risks (Romero and Shahriari 2011).

estimated that for India, 4 million healthy life years are lost annually due to iron deficiency, an additional 2.8 million due to zinc deficiency, and still 2.3 million more due to vitamin A deficiency. When collectively assigned a monetary value, these deficiencies have been estimated to reduce the country's gross national income by 0.8–2.5 percent (Stein 2006).

Fortunately, micronutrient malnutrition has been receiving more attention over the last 20 years. As evidence of this, note the rise of what is known as biofortification – the breeding (and genetic engineering) of plants with the aim of higher micronutrient content. Golden Rice – rice genetically engineered to contain high levels of vitamin A – is a well-known example of biofortification. Yet as a solution to a diet deficient in beta carotene, Golden Rice seems to sidestep larger sociological questions. Vitamin A is fat-soluble, which means its uptake within the body is dependent upon a level of fat in the diet. In less developed parts of the world, however, levels of dietary fat are often insufficient. And simply increasing a population's daily intake of rice, by itself, provides none of this dietary fat. Of greater concern, however, is the fact that Golden Rice glosses over entirely the deeper question: why is vitamin A deficiency increasing throughout the developing world thus creating the need for such rice in the first place? I've already given the answer: the green revolution has helped change dietary patterns, which, for some populations, has led to diets dangerously low in vitamin A.

Such irony: the green revolution, through biofortification, is cast as the solution to . . . itself! Golden Rice is directed at symptoms, not the root cause of vitamin A deficiency. And we can't expect a technological fix to solve a problem that is fundamentally social, political, and economic in nature.

Biofortification also reflects a broader trend taking place throughout the food system, where "nutrition" is being reduced to the absolute or relative quantities of certain nutrients in food. Scrinis (2008: 44) refers to this as the ideology of nutritionism: a "quantitative logic" that "obscures the broader cultural, geographical, and ecological contexts in which foods, diets, and bodily health are situated." Scrinis claims all the attention being paid to nutrients in isolation – rather than understanding nutritional health as a complex socio-ecological process – deliberately distracts consumers from asking "deeper" questions about their food. Focusing on the fact that a food has, say, 100 percent of the recommended daily amount of vitamin C or folic acid draws attention away from, among other things, the overall nutrient profile of a food. Let us not forget, a nutritionally fortified Twinkie is still a Twinkie. Similarly, reducing food to fundamental chemical components makes it hard for us to talk about food *systemically* and therefore blinds us to, say, a food's ecological footprint or to questions related to whether it was raised and processed in a socially just way.

Political economy of obesity

Whether starving or not, how and what we eat is always conditioned by sociological (which includes political and economic) structures. Let's start by discussing how the conditions of trade do not treat all food equally. Since 1974, trade in processed products for developed countries as a whole has outpaced trade in primary agricultural products. A variety of reasons lie behind this trend.

One reason relates to foreign direct investment (FDI). One condition of trade liberalization is to open one's economic boarders up to FDI. But what sorts of investments are large multinational food processors most interested in? Increased FDI in the food sector of developing economies by multinational food firms has stimulated growth, not surprisingly, in the production and distribution of processed foods. US food processing firms increased their investment in foreign food processing firms (which often meant buying them outright) from US$9 billion in 1980 to US$36 billion in 2000. This sudden influx of revenue into the food processing sector of developing countries helps explain why processed food sales rose almost 30 percent during that time period, compared with a much less dynamic 7 percent growth recorded in affluent nations. In Thailand, for example, thanks in part to FDI from such US firms as PepsiCo (which also owns Frito-Lay), annual retail sales of so called "junk" food (highly processed sugary, salty, and fatty food) increased almost 70 percent between 1999 and 2004 (Friel and Lichacz 2010: 119–20).

These trends are not confined to developing nations. A US study from 2004 examined how many calories US$1 could buy. Not surprisingly, one could maximize calorie purchasing power in the middle of the store, among the walls of soda, chips, candy, and other nonexpirable goods (Drewnowski and Specter 2004). Similar results were obtained in a French study. Controlling for total caloric intake the researchers found that a 100-gram increase in fats and sweets are associated with a 6–40 cent decrease in daily diet costs, whereas a 100-gram increase in fruits and vegetables are associated with a 22–35 cent increase (Maillot *et al.* 2007).

Researchers examining foods at major retail supermarket chains in Seattle, Washington, found nutrient density to be negatively associated with energy density and positively associated with cost (Monsivais *et al.* 2010). Between 2004 and 2008, the timeframe of the study, the price disparity between healthy and unhealthy foods *increased* in favor of the latter, which is to say junk food became less expensive relative to more nutrient-dense foods. This follows a general trend for the country as a whole, where fats and sweeteners have been getting progressively cheaper while fruits and vegetables are becoming more expensive (Figure 4.2).

This gives some perspective to what is known as the "obesity–hunger paradox"; a term used recently in a *New York Times* article (Dolnick 2010) to describe hunger-related problems in South Bronx, where "the hungriest people in America today, statistically speaking, may well be not sickly skinny, but excessively fat." The obesity–hunger paradox reminds us that access to cheap, processed food and food security are not the same things. A diet rich in calorically dense, yet otherwise nutritionally shallow, foods does not promote health.

The concept "food desert" is helpful here. A food desert refers to any area "with limited access to affordable and nutritious food, particularly such an area

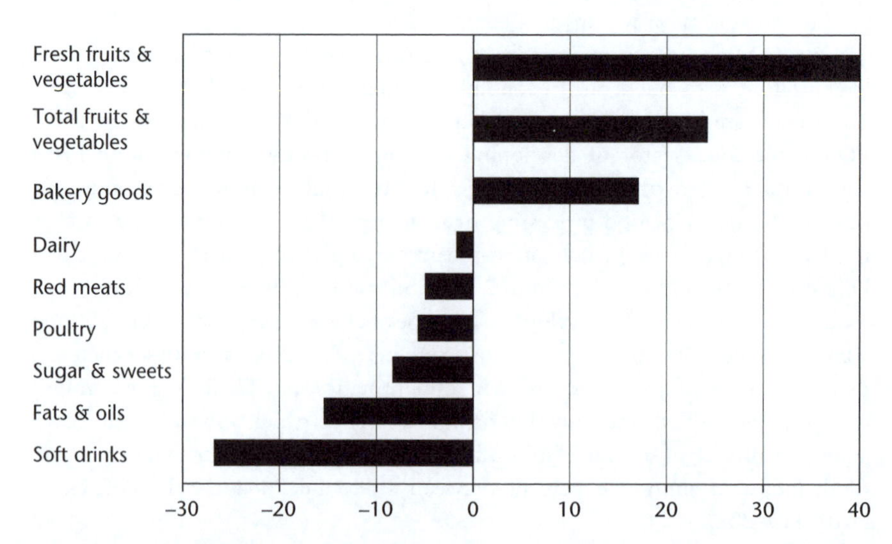

FIGURE 4.2 Change in US food prices, 1985–2000 (converted to real 2002 US$). Based on data from Putman *et al.* (2002)

composed of predominately lower income neighborhoods and communities" (2008 US Farm Bill, Title VI, Sec. 7527). Fewer supermarkets tend to be located in low income neighborhoods than more affluent neighborhoods. Consequently, as a result of this diminished competition, the price of goods in poorer neighborhoods tends to be higher (Morland *et al.* 2002). The presence of supermarkets in a neighborhood has also been shown to be inversely correlated with levels of obesity, while the availability of convenience stores is associated with a higher prevalence of obesity among residents of a community (Morland *et al.* 2006). Research has also found a link between proximity to a supermarket and diet. The closer residents in low-income neighborhoods live to a supermarket the higher their fruit and vegetable consumption (Rose and Richards 2004). A similar relationship has been reported among pregnant women (Laraia *et al.* 2004).

Food deserts also tend to disproportionately impact people of color (Morland *et al.* 2002; Zenk *et al.* 2005). Grocery stores in African-American neighborhoods, for example, have been reported to stock fewer fresh fruits and vegetables and healthier versions of standard foods, even though residents of the community report wanting greater access to these goods (Sloane *et al.* 2003). In another study, fruit and vegetable consumption increased 32 percent for each additional supermarket located in predominately African-American neighborhoods (Morland *et al.* 2002).

And what are we to make of the so-called global obesity epidemic? It is clear, from an empirical standpoint, that people, on the whole, are getting larger in developed (and increasingly in developing) countries (Carolan 2011). Yet does this translate into an "obesity epidemic?" While often used by individuals with good intensions, the term itself should not be immune to sociological scrutiny. That is to say, it needs to be looked at with just as much thoughtful skepticism as we would bring to the study of any food-related phenomenon.

Let's look first at body mass index (BMI). As a convenient, internationally accepted, cost-effective measure to access adiposity (fatness), BMI has become a widely popular tool among epidemiologists and health professionals. BMI is calculated as weight (kg)/height $(m)^2$. Among adults, there are four BMI categories: underweight (less than 18.5); normal weight (18.5–24.9); overweight (25–29.9); and obese (greater than 30). Talk of a global obesity epidemic reflects a steady shift in populations from "lower" to "higher" BMI categories. Estimates classify 23 percent of the world's population as overweight and an additional 10 percent as obese. It has further been calculated that by 2030, 58 percent of the world's population will be obese (Kelly *et al.* 2008). Yet it is worth highlighting that this growth in BMI reflects a small movement to the right of the population distribution of BMI (Campos *et al.* 2006). In others words, what occurred in many countries was a large population of people resting on the heavy side of lower BMI categories gaining a little weight in recent years and pushing their country's average into a new, higher BMI category.

Critics of obesity epidemiology highlight also that there are many metrics and determinants of health and that the relationship between BMI, nutrition, physical activity, and long-term health is complex and still not perfectly understood. Yet, in fairness, this uncertainty cuts both ways, leading just as easily to an *underestimation* of the risks associated with adiposity (Lockie and Williams 2010). Nevertheless, arguments critical of obesity epidemiology cannot be dismissed out of hand. The cultural link, for example, between "fatness" and individual fault – as a cultural statement about, say, one's laziness or lack of control – corresponds with a larger societal shift toward neoliberal forms of what French philosopher Michel Foucault called "governance" (Carolan 2005; Guthman and DuPuis 2006). Foucaultian governance, to greatly simplify a concept that is quite nuanced, refers to the reduction of social problems to an aggregate of individual problems. Doing this skillfully transfers responsibility for developing solutions from society as a whole onto individuals; a move that, in turn, supports the very system responsible for their ill health in the first place. As good consumer-citizens we are therefore taught to shop our way to safety, happiness, and sustainability by buying, in the case of environmental problems, eco-friendly products, and, in the case of the obesity epidemic, weight loss pills and diet plans.

Guthman (2011) develops this point extensively, noting how, in the interest of economic growth, capitalism has helped to create obesity as a material phenomenon (adiposity) and then made it into a moral problem (e.g. obesity = laziness) that can only be resolved through capitalism. She likens current treatments of obesity to prescriptions to combat climate change. In both cases, the treatment involves, in her estimation, a disproportionate focus on individual consumption choices rather than on creating policies that would enforce corporate accountability or to mitigate their effects on those most harmed. Obesity and climate change are thus seen as "apolitical ecologies," which is to say that our explanations for each do not fully account for the asymmetries in power that first create them materially and then later define them as "problems" (and in fact these *asymmetries* are the problem). Instead, both are viewed as the product of bad decision making at the individual level, which can be remedied with such things as education. This move effectively absolves the system of any blame.

Another compelling sociological look at the current obesity epistemic comes from Dixon and Broom (2007). They reject as too simplistic the popular eat less and/or exercise more recommendation coming from many health care professions. The want to know, for example, what it means to eat "right," why we are not eating more nutrient-dense foods, and what barriers might be keeping people from becoming more physically fit? These questions point to a host of often-ignored variables that need to be better understood and included in discussions about weight gain and health, like leisure time (or lack thereof), the commodification of food preparation, the changing structure of our food system, food policy, car dependence, over-aggressive marketing of nutritiously shallow food products, changing parental pressures, and barriers to self-powered forms of personal transportation. Doing this shows weigh gain to be a sociological problem rather than a product of individual weakness.

Food politics

The food industry and federal governments make for strange bedfellows, recognizing that in most countries the latter are responsible for simultaneously promoting and regulating the former. As stated earlier, world agriculture produces sufficient calories to feed the world. In the USA that figure is roughly 3,800 calories per capita (Nestle 2007). As profit maximizing firms, food companies are motivated to continually increase consumption of their food products; a fact that places their interests in opposition to those of public health (Box 4.3). Governments have proven to be equally culpable in

BOX 4.3 INFANT FORMULA: WHEN PROFITS AND PUBLIC WELFARE ARE AT ODDS

There is no disputing the superiority of mother's milk over that from cows, goats, or a formula container. Particularly in low-income countries, mother's milk provides a child with immunities that can make the difference between life and death in light of the unsanitary conditions they can be exposed to. So how do you market and sell a commodity, namely infant formula, which is vastly inferior to the product it is meant to replace? An enormous controversy erupted in the 1970s over infant formula and the targeted marketing campaign waged by multinational firms who were seeking to expand their market by selling it in developing countries.

Infant formula clearly has some utility. For example, during a devastating cholera epidemic in Chad in 1972, when many mothers died, infant formula was used successfully by international agencies to nourish, and save, a number of infants. Yet it is too often the case that families in poorer countries cannot afford to buy sufficient quantities of the product. Also, many lack access to refrigeration, clean water, and adequate sanitation and may not be supplied with adequate information when it comes to mixing the formula properly. An article in a 1981 issue of the *New York Times* covering the infant formula "controversy" discusses a Jamaican mother who had been instructed to not breastfeed her two youngest children (Solomon 1981). Their diets since birth had been formula. With a family income of roughly US$7 a week, the mother had to dilute the costly formula to make it last longer. It was described how one tin of formula should feed one child for 2–3 days. The mother said she made that one tin last 2 weeks to feed both children.

Formula companies have a lot at stake as they seek to expand globally. And they view developing countries as an important next step as they grow their brands. Infant formula is also becoming a status symbol for many in the developing world (Gerber and Short 1986). This, coupled with its growing use globally (it is now an US$8 billion market), suggests that these campaigns are working. Yet, we must ask, at what and whose expense?

all of this, as often the very governmental officials and organizations responsible for promoting public health through food consumption out of one side of their mouth are, out of the other, cheerleading for the food industry and their sometimes less-than-wholesome products.

Much has also been written about the politics behind the USA's dietary guidelines (Nestle 2007). For example, the statement "decrease consumption of meat," found in the first edition of the US Dietary Goals in 1977, led to the meat industry demanding an immediate revision and the removal of the word "decrease." When the second edition came out in 1977, the statement read "choose meats [. . .] which will reduce saturated fat intake," which was changed again in 1980 to read "choose lean meat." By the 1990 edition, the phrase changed again, to "have two or three servings of meat [. . .] with a daily total of about 6 ounces." Note how the government's phrasing evolved, from explicitly telling consumers to eat less meat in 1977 to explicitly telling them to eat meat – two or three servings in fact – by 1990.

Similar changes have been recorded in Australia, in part because US nutritional policies have been looked to as a resource among Australian nutritionists when crafting national dietary policy (Duff 2004; Lawrence and Germov 2004). There, national nutritional guidelines changed from telling consumers to "avoid too much" meat in 1981 to "choose a diet low in" meat in 1992. As a result of heavy lobbying from agricultural interest groups, the words "avoid," "less," and "too much" disappeared entirely from Australian nutritional guidelines by the 1990s, replaced by words like "care" (Duff 2004). Table 4.2 documents this change, showing dietary guidelines for 1981, 1992, and 2003. Note also the growing inclusion of nondietary items in the dietary guidelines, like the emphasis on physical activity and caring for one's food by way of proper storage and preparation.

There is some debate about whether such nutritional guidelines are justified. For example, a recent article in the *American Journal of Preventative Medicine* raises the fundamental question of whether public health guidelines are appropriate altogether, controversially arguing "the notion that the government should tell people what and how much to eat is inherently paternalistic" (Marantz *et al.* 2008: 234). It is argued that such guidelines are unwillingly forced upon a public that did not ask for them in the first place. Moreover, due to the risk of unintended consequences, some believe guidelines are best avoided until it is unequivocally proved that they will do more good than harm. However, others argue that the food consuming public not only desires this information but have a right to it. How are they ever to become knowledgeable consumers without it? Furthermore, as Woolf and Nestle ask in direct response to the aforementioned argument, in an article published in the same issue of *American Journal of Preventative Medicine*:

> If it was paternalistic for the government to advise people how to eat, as Marantz *et al.* suggest, was it equally paternalistic for Surgeon General Luther

TABLE 4.2 Dietary guidelines for Australian adults for 1982, 1992, and 2003

1981	*1992*	*2003*
Promote breastfeeding	Enjoy a wide variety of nutritious foods	Enjoy a wide variety of nutritious foods:
Choose a nutritious diet from a variety of foods	Eat plenty of breads and cereals (preferably wholegrain), vegetables (including legumes) and fruits	Eat plenty of vegetables, legumes, and fruits
Control your weight	Eat a diet low in fat, and in particular, low in saturated fat	Eat plenty of cereals (including breads, rice, pasta, and noodles), preferably wholegrain
Avoid eating too much fat	Maintain a healthy bodyweight by balancing physical activity and food intake	Include lean meat, fish, poultry, and/or alternatives
Avoid eating too much sugar	If you drink alcohol, limit your intake	Include milks, yogurts, cheeses, and/or alternatives
Eat more bread and cereals (preferably wholegrain) and vegetables and fruit	Choose low fat foods and use salt sparingly	Reduced-fat varieties should be chosen, where possible
Limit alcohol intake	Encourage and support breastfeeding	Drink plenty of water
Use less salt		And take care to:
		Limit saturated fat and moderate total fat intake
		Choose foods low in salt
		Limit your alcohol intake if you choose to drink
		Consume only moderate amounts of sugars and foods containing added sugars
		Prevent weight gain: be physically active and eat according to your energy needs
		Care for your food: prepare and store it safely
		Encourage and support breastfeeding

Based on Duff (2004).

> Terry to alert the public about the hazards of tobacco use and to recommend in 1964 that smokers give up cigarette smoking? Raising concerns about the hazards of riding a motorcycle without a safety helmet, the need to secure children in car seats, and the dangers of hydrocarbon emissions, for example, are, in our view, less exercises in paternalism than the fulfillment of government responsibility to share what is known about health risks and suggested ways to reduce them.
>
> *Woolf and Nestle (2008: 263)*

From an empirical standpoint, some legitimate criticisms can be directed at dietary guidelines. For example, guidelines *in themselves* do little to change actual dietary patterns. This is especially true among the most disadvantaged in society, recognizing that eating healthy often costs more than a diet that does not meet the recommended guidelines (a point discussed earlier in this chapter) (Baghurst 2003). There is also a concern among social scientists that guidelines keep consumers from seeing the food-system-forest for the nutrient-trees. In other words, reducing "food" to its constituent nutrients (or nutritionism), which is a consequence of national dietary recommendations and nutritional standards, makes the larger structures of food production, processing, and distribution more difficult to see. This also makes it easier for food companies to project the image of being "good" corporate citizens, thus absolving them of responsibity for poor public and individual health due to malnutrition and over-consumption (Duff 2004). Making transparent the nutritional value of foods also allows food companies to play the "individual choice" card and thus steer debate away from the food system itself. Under this ideological rubric, obesity, poor diet, and malnutrition all become issues attributable to the individual, via such alleged personal faults as poor self-control, lack of individual dietary knowledge, or just plain laziness. As sociologists, however, we ought to see such "explanations" for what they are: terribly unimaginative answers to questions that deserve being treated to a sociological imagination.

Beyond helping to write (through intense lobbying) nutritional guidelines, agribusiness can influence public debate around food and agriculture in still other ways: namely, by making it illegal to say bad things about them and their food products. Thirteen states in the USA have instituted food libel laws. South Dakota, for example, has a law that specifically prevents people from saying that conventional agricultural practices (like using pesticides) might make food unsafe. Other times – like the famous McLibel case between McDonald's and two citizens of the UK – companies have actually taken people to court in retaliation for negative public claims made against them (Box 4.4).

Transition . . .

Issues of food, as by now should be clear, impact peoples far and wide, within and beyond the farm gate. Yet we've still only scratched the surface when it comes to understanding how current food policies and practices impact (often negatively)

BOX 4.4 McLIBEL

On October 16, 1985, London Greenpeace launched the International Day of Action Against McDonald's (and has been held on that date every year since). As part of this event, the group produced a factsheet "What's Wrong With McDonald's? Everything They Don't Want You to Know." The handout accused the firm of such things as promoting obesity, exploitive labor practices, being responsible for horrendous environmental degradation, and animal cruelty. In response, McDonald's hired private investigators and instructed them to infiltrate the group (at some London Greenpeace meetings there were as many spies as there were legitimate activists). In 1990, McDonald's had enough information to serve libel writs on five Greenpeace activists. They were told to either publically retract the allegations in the leaflets and apologize or go to court (there is no legal aid in the UK for libel cases other than 2 hours of free legal advice). Surprising everyone, two activists decided to challenge McDonald's rather than apologize.

The trial began in June 1994. Three years later (and after 314 days in court), the court ruled that McDonald's marketing "pretended to a positive nutritional benefit which their food (high in fat and salt, etc.) did not match"; that the firm did "exploit children" with their advertising; that they were "culpably responsible for cruel practices regarding the rearing and slaughter of animals"; and that they "pay low wages, thereby helping to depress wages for workers in the catering trade." Nevertheless, the judge did rule that on the other points – that the firm destroys rainforests, causes starvation in poor countries, and knowingly sells contaminated food – the activists did libel McDonald's and were ordered to pay McDonald's £40,000 (they refused but McDonald's hasn't pursued the matter further). In 2005, the European Court of Human Rights ruled that the two activists were denied basic human rights by being refused legal aid. The court awarded a judgment of £57,000 against the UK government. The McLibel Case is universally viewed as a public relations disaster for McDonald's as it brought to light many questionable practices linked to the company (Norton 2005).

human communities. The following chapter (and the new section it kicks off) pulls the veil back still further, exposing additional costs associated with our food to community well-being and public health.

Discussion questions

1 While famines may be rare in democracies (as argued by Sen) hunger, and hidden hunger especially, is a persistent problem, even in the world's most affluent countries. Why is this?

2 Why does obesity continue to be looked upon in many societies as signifying individual fault rather than an issue caused by larger sociological phenomena?
3 Do dietary guidelines and nutritional labeling make it easier for us to make more informed dietary choices? Can consumers ever have too much information when it comes to food labeling and product displays at supermarkets?

Suggested reading: Introductory level

Bello, Walden. 2008. How to manufacture a food crisis. *Development* 51(4): 450–5.

Lockie, Stewart and Susan Williams. 2010. Public health and moral panic: Sociological perspectives on the "epidemic of obesity." In *Food Security, Nutrition and Sustainability*, edited by Geoffrey Lawrence, Kristen Lyons, and Tabatha Wallington. London: Earthscan, pp. 145–61.

Nestle, M. Food Politics Blog, www.foodpolitics.com, last accessed October 26, 2011.

Suggested reading: Advanced level

Friedmann, Harriet. 1992. Distance and durability: Shaky foundations of the world food economy. *Third World Quarterly* 13(2): 371–83.

Guthman, Julie and Melanie DuPuis. 2006. Embodying neoliberalism: Economy, culture, and the politics of fat. *Environment and Planning D: Society and Space* 24(3): 427–48.

Scrinis, Gyorgy. 2008. On the ideology of nutritionism. *Gastronomica* 8(1): 38–48.

Zerbe, Noah. 2009. Setting the global dinner table: Exploring the limits of the marketization of food security. In *The Global Food Crisis: Governance Challenges and Opportunities*, edited by Jennifer Clapp and Marc Cohen. Waterloo, Canada: Wilfrid University Press, pp. 161–77.

References

Baghurst, K. 2003. Appendix B: Social status, nutrition and the cost of healthy eating. In *National Health and Medical Research Council, Dietary Guidelines for Australian Adults*. Canberra: AusInfo, pp. 92–109.

Bello, Walden. 2008. How to manufacture a food crisis. *Development* 51(4): 450–5.

Caballero, Benjamin. 2002. Global patterns of child health: The role of nutrition. *Annuals of Nutrition and Metabolism* 46: 3–7.

Campos, P., A. Saguy, P. Ernberger, E. Oliver and G. Gaesser. 2006. The epidemiology of overweight and obesity: Public health crisis or moral panic? *International Journal of Epidemiology* 35: 55–60.

Carolan, Michael. 2005. The conspicuous body: Capitalism, consumerism, class, and consumption. *Worldviews: Global Religions, Culture and Ecology* 9(1): 82–111.

Carolan, M. 2011. *The Real Cost of Cheap Food*. London: Earthscan.

Cordain, Loren. 1999. Cereal grains: Humanity's double-edged sword. *World Review of Nutrition and Dietetics* 84: 19–23.

Cullather, Nick. 2007. The foreign policy of the calorie. *American Historical Review* 112(2): 337–64.

Dixon, Jane and Dorothy Broom. 2007. *The Seven Deadly Sins of Obesity: How the Modern World is Making us Fat*. Sydney, Australia: UNSW Press.

Dolnick, Sam. 2010. The obesity–hunger paradox. *New York Times,* March 12, http://www.nytimes.com/2010/03/14/nyregion/14hunger.html?src=me, last accessed April 6, 2010.

Drewnowski, Adam and S.E. Specter. 2004. Poverty and obesity: The role of energy density and energy costs. *American Journal of Clinical Nutrition* 79(1–6): 6–16.

Duff, John. 2004. Setting the menu: Dietary guidelines, corporate interests, and nutrition policy. In *A Sociology of Food and Nutrition*, edited by J. Germov and L. Williams. New York: Oxford University Press, pp. 148–68.

Dugger, Celia. 2005. Africa food for Africa's starving is road blocked in Congress. *New York Times,* October 12, 2005. http://www.nytimes.com/2005/10/12/international/africa/12memo.html?ex=1286769600&en=0de1afa6dd7990e7&ei=5090&partner=rssuserland&emc=rss, last accessed February 19, 2010.

FAO (Food and Agriculture Organization). 2002. *Reducing Poverty and Hunger: The Critical Role of Financing for Food, Agriculture, and Rural Development.* Food and Agriculture Organization, International Fund for Agricultural Development, World Food Program, http://www.fao.org/docrep/003/Y6265e/y6265e00.htm, last accessed December, 2010.

Fisher, William. 1943. The Bengal famine: 50,000 Indians weekly succumb to disease and starvation in spreading catastrophe. *Life* 15(21): 16, 19–20.

Freebairn, Donald. 1995. Did the green revolution concentrate incomes? A quantiative study of research reports. *World Development* 23(2): 265–79.

Friedmann, Harriet. 1990. The origins of third world food dependence. In *The Food Question: Profits versus People*, edited by H. Bernstein, B. Crow, M. Mackintosh, and C. Martin. New York: Monthly Review Press, pp. 13–31.

Friedmann, Harriet. 1992. Distance and durability: Shaky foundations of the world food economy. *Third World Quarterly* 13(2): 371–83.

Friel, Sharon and Wieslaw Lichacz. 2010. Unequal food systems, unhealthy diets. In *Food Security, Nutrition and Sustainability*, edited by Geoffrey Lawrence, Kristen Lyons, and Tabatha Wallington. London: Earthscan, pp. 115–29.

Gerber, J. and J. Short. 1986. Publicity and the control of corporate behavior: The case of infant formula. *Deviant Behavior* 7(3): 195–216.

Gupta, Raj and Ashok Seth. 2007. A review of resource conserving technologies for sustainable management of the wheat-cropping systems of the Indo-Gangetic Plains (IGP). *Crop Production* 26(3): 436–47.

Guthman, J. 2011. *Weighing In: Obesity, Food Justice and the Limits of Capitalism.* Los Angeles, CA: University of California Press.

Guthman, Julie and Melanie DuPuis. 2006. Embodying neoliberalism: Economy, culture, and the politics of fat. *Environment and Planning D: Society and Space* 24(3): 427–48.

Hanrahan, Charles and Carol Canda. 2005. International Food Aid: US and other Donor Contributions. Washington, DC: CRS (Congressional Research Service) Report for Congress, US Library of Congress, http://www.au.af.mil/au/awc/awcgate/crs/rs21279.pdf, last accessed July 14, 2010.

Hoddinott, John and Bill Kinsey. 2001. Child growth in the time of drought. *Oxford Bulletin of Economics and Statistics* 63(4): 409–36.

IATP (Institute for Agriculture and Trade Policy). 2005. Agricultural export dumping booms during WTO's first decade, Press release. Minneapolis, MN, February 9, http://www.iatp.org/iatp/press.cfm?refid=89731, last accessed January 12, 2011.

Kelly, T., W. Wang, C.S. Chen, K. Reynolds, and J. He. 2008. Global burden of obesity in 2005 and projections to 2030. *International Journal of Obesity* 32: 1431–7.

Kloppenburg, Jack. 1988. *First the Seed.* Cambridge: Cambridge University Press.

Laraia, Barbara, Anna Maria Siega-Riz, Jay Kaufman, and Sonya J. Jones. 2004. Proximity of supermarkets is positively associated with diet quality index for pregnancy. *Preventative Medicine* 39: 869–75.

Lawrence, Mark and John Germov. 2004. Future food: The politics of functional food and health claim. In *A Sociology of Food and Nutrition*, edited by J. Germov and L. Williams. New York: Oxford University Press, pp. 119–46.

Lockie, Stewart and Susan Williams. 2010. Public health and moral panic: Sociological perspectives on the "epidemic of obesity." In *Food Security, Nutrition and Sustainability*, edited by Geoffrey Lawrence, Kristen Lyons, and Tabatha Wallington. London: Earthscan, pp. 145–61.

Maillot, M., N. Darmon, F .Vieux, and A. Drewnowski. 2007. Low energy density and high nutritional quality are each associated with higher diet costs in French adults. *American Journal of Clinical Nutrition* 86(3): 690–6.

Major, Patrick. 1997. *The Death of KPD: Communism and Anti-Communism in West Germany, 1945– 1956*. New York: Oxford University Press.

Marantz, P., E. Bird, and M. Alderman. 2008. A call for higher standards of evidence for dietary guidelines. *American Journal of Preventative Medicine* 34(3): 234–40.

Martin, Susan. 2010. Restrictions on US food aid waste time and money. *Tampa Bay Times*, February 8, http://www.tampabay.com/news/world/restrictions-on-us-food-aid-waste-time-and-money/1070813, last accessed July 13, 2010.

Monsivais, Pablo, Julia Mclain, and Adam Drewnowski. 2010. The rising disparity in the price of healthful foods: 2004–2008. *Food Policy* 35(6): 514–20.

Morland, Kimberly, Ana Diez Roux, and Steve Wing. 2006. Supermarkets, other food stores and obesity: the atherosclerosis risk in communities of study. *American Journal of Preventative Medicine* 30(4): 333–39.

Morland, Kimberly, Steve Wing, Ana Diez-Roux, and Charles Poole. 2002. Neighborhood characteristics associated with the location of food stores and food service places. *American Journal of Preventative Medicine* 22: 23–9.

Nestle, Marion. 2007 (2002). *Food Politics: How the Food Industry Influences Nutrition and Health*. Berkeley, CA: University of California Press.

Norton, Michael. 2005. *365 Ways to Change the World: How To Make a Difference One Day at a Time*. New York: Free Press.

Peterson, E. Wesley. 2009. *A Billion Dollars a Day: The Economics and Politics of Agricultural Subsides*. Malden, MA: Wiley-Blackwell.

Pimentel, D., L. Hurd, A. Bellotti, M. Forster, I. Oka, O. Sholes, and R. Whitman. 1973. Food production and the energy crisis. *Science* 182(4111): 443–49.

Putman, Judy, Jane Allshouse, and Linda Kantor. 2002. US per capita food supply trends. *Food Review* 25(3): 2–15, http://www.ers.usda.gov/publications/FoodReview/DEC2002/frvol25i3a.pdf

Quigley, Bill. 2008. The US role in Haiti's food riots. *Counterpunch*, April 21, http://www.counterpunch.org/quigley04212008.html, last accessed December 20, 2010.

Romero, Simon and Sara Shahriari. 2011. A food's global success creates a quandary at home. *New York Times*, March 20, http://www.nytimes.com/2011/03/20/world/americas/20bolivia.html, last accessed March 22, 2011.

Rose, Donald and Rickelle Richards. 2004. Food store access and household fruit and vegetable use among participants in the US food stamp program. *Public Health and Nutrition* 7: 1081–8.

Sanghvi, T.G. 1996. *Economic Rationale for Investing in Micronutrient Programs: A Policy Brief Based on New Analyses*. Washington, DC: Office of Nutrition, Bureau for Research and Development, United States Agency for International Development.

Scrinis, Gyorgy. 2008. On the ideology of nutritionism. *Gastronomica* 8(1): 38–48.

Sen, Amartya. 1981. *Poverty and Famines: An Essay on Entitlement and Deprivation*. New York: Oxford University Press.

Sen, Amaryta. 2004. *Rationality and Freedom*. Cambridge, MA: Harvard University Press.

Seshadri, Subadra. 2001. Prevalence of micronutrient deficiency particularly of iron, zinc and folic acid in pregnant women in South East Asia. *British Journal of Nutrition* 85: S87–92.

Sloane, David, Allison Diamant, LaVonna Lewis, Antronette K. Yancey, Gwendolyn Flynn, Lori Nascimento, William J. McCarthy, Joyce Guinyard, and Michael R. Cousineau. 2003. Improving the nutritional resource environment for healthy living through community-based participatory research. *Journal of General Internal Medicine* 18: 568–75.

Solomon, S. 1981. The controversy over infant formula. *New York Times*, December 6, http://www.nytimes.com/1981/12/06/magazine/the-controversy-over-infant-formula.html?pagewanted=1, last accessed March 22, 2011.

Stein, Alexander. 2006. *Micronutrient Malnutrition and the Impact of Modern Plant Breeding on Public Health in India: How Cost-Effective is Biofortification?* Göttingen: Cuvillier Verlag.

UNICEF, WTO, and WFP. 2007. Preventing and Controlling Micronutrient Deficiencies in Populations Affected by an Emergency, Joint Statement. Geneva, Switzerland: http://www.helid.desastres.net/en/d/Js13449e/1.html, last accessed August 9, 2010.

Weis, Tony. 2007. *The Global Food Economy: The Battle for the Future of Farming*. New York: Zed.

Welsh, Ross and Robin Graham. 1999. A new paradigm for world agriculture. *Field Crops Research* 60: 1–10.

Welsh, Ross and Robin Graham. 2002. Breeding crops for enhanced micronutrient content. *Plant and Soil* 245(1): 205–14.

Woolf, S. and M. Nestle. 2008. Do dietary guidelines explain the obesity epidemic? *American Journal of Preventative Medicine* 34(3): 263–5.

World Bank. (2002) *Global Economic Prospects and the Developing Countries: Making Trade Work for the World's Poor, 2002*. Washington, DC: World Bank.

Xizhe, Peng. 1987. Demographic consequences of the Great Leap Forward in China's provinces. *Population and Development Review* 13(4): 639–70.

Zenk, Shannon, Amy Schulz, Barbara Israel, Sherman James, Shuming Bao, and Mark Wilson. 2005. Neighborhood racial composition, neighborhood poverty, and the spatial accessibility of supermarkets in metropolitan Detroit. *American Journal of Public Health* 95: 66–7.

Zerbe, Noah. 2009. Setting the global dinner table: Exploring the limits of the marketization of food security. In *The Global Food Crisis: Governance Challenges and Opportunities*, edited by Jennifer Clapp and Marc Cohen. Waterloo, Canada: Wilfrid University Press, pp. 161–177.

PART II

Community, culture, and knowledge

5

COMMUNITY, LABOR, AND PEASANTRIES

KEY TOPICS

- Decades of empirical research support the conclusion that large-scale, industrial farms have, overall, a negative impact on surrounding rural communities.
- There is considerable evidence of labor exploitation in all sectors of the food system, from pay inequities to hazardous working conditions in the livestock industry and pesticide exposure among field laborers.
- The global food system has negatively impacted peasant communities in less affluent countries; whether through the global trade of agricultural commodities (the USA–Mexico trade relationship is examined) or biotechnology (Monsanto's Smallholder Program is discussed).

Key words

- Acute respiratory symptoms
- CAFO
- Gloves-off economy
- Goldschmidt thesis
- Monsanto Smallholder Program
- NAFTA
- Peasant agriculture
- Pesticide toxicity
- Private standards
- Race to the bottom
- Remittances
- Urban bias
- Washington Consensus
- Yield drag

In this chapter, human impacts associated with the aforementioned changes to the structure of agriculture and the broader food system command the spotlight. The chapter begins with a discussion of what is known as the Goldschmidt thesis, where the impacts of industrial agriculture on rural communities are explored. Attention then turns to the subject of labor, noting various injustices faced by those who feed us. The chapter concludes by examining some of the impacts that agricultural practices and policies have had on peasant communities.

Goldschmidt thesis: Community effects of industrial farming

One research tradition that stands out for its lengthy pedigree looks at community impacts linked to the changing structure of agriculture. In Chapter 1, when briefly discussing the evolution of the subdiscipline, a number of studies were mentioned from the first half of the twentieth century that examined the relationship between a community's structure and the characteristics of its surrounding farms. The most famous of these studies was conducted almost three-quarters of a century ago. Overseen by Walter Goldschmidt, an anthropologist at the US Department of Agriculture (USDA), and funded by the USDA, the study looks at two California communities in the early 1940s: Arvin, where large, absentee-owned, nonfamily operated farms were more numerous; and Dinuba, where locally owned, family operated farms were the norm (the names of both towns are pseudonyms). Goldschmidt concluded that industrial agriculture has, overall, a negative impact on a variety of community quality of life indicators (Goldschmidt 1978a). He noted that, for example, relative to Dinuba, Arvin's population had a smaller middle class, a higher proportion of hired workers, lower mean family incomes, higher rates of poverty, poorer quality schools, and fewer churches, civic organizations, and retail establishments. Residents of Arvin also had less local control over public decisions due to disproportional political influence by outside agribusiness interests.

Goldschmidt's research went virtually unmentioned in the literature until the 1970s (for a superb review of this literature see Lobao and Stofferahn 2008). Rodefeld (1974) is often credited with its "rediscovery" and subsequent popularization in rural sociological circles, especially after the publication of *Change in Rural America* (Rodefeld *et al.* 1978). Regardless of the reasons behind its resurgence, a steady stream of Goldschmidt-type research has been conducted since the 1970s (for an overview of this literature see Table 5.1). The tradition is especially strong in the USA, where the so-called traditional rural community has been perceived as coming under threat by broader structural changes both within and outside of agriculture (see Table 5.1, under column labeled Region).

Most research since the 1970s has given varied levels of support to the Goldschmidt thesis (see Table 5.1, under column labeled Results). Neo-Goldschmidt studies have also expanded upon the original methodology. For example, Goldschmidt's study was a highly qualitative, comparative anthropological study.

TABLE 5.1 Summary of 51 studies examining the effects of industrialized farming on community well-being

Study	*Methodology*	*Region*	*Measures of industrialized farming*	*Community well-being indicators*	*Results*
Goldschmidt (1968, 1978a)	Comparative case study: two communities	California	Scale/organization	Socio-economic and social fabric (class structure, local services and organizations, politics, retail trade)	Detrimental: variety of community indicators
Tetreau (1938, 1940)	Survey design study: 2,700 households	Arizona	Scale/organization	Socio-economic and social fabric (class structure)	Detrimental: increased class inequality, rise in number of poor farm workers
Heffernan (1972)	Survey design study: 138 broiler producers, contract farming	Louisiana	Organization	Social fabric (social psychological indicators, community involvement)	Detrimental: poorer social psychological well-being, less community involvement
Heady and Sonka (1974)	Regional economic impact: 150 producing areas	Continental USA	Scale	Socio-economic	Mixed: large farms generate less total community income but also lower food costs
Rodefeld (1974)	Survey design study: 180 producers from 100 farms	Wisconsin	Scale/organization	Socio-economic and social fabric (class structure, services, population size)	Detrimental: variety of community indicators
Martinson *et al.* (1976)	Survey design study: 180 producers	Wisconsin	Organization	Social fabric (social psychological indicators)	Detrimental: community isolation of farm workers
Fujimoto (1977)	Macro-social accounting: 130 towns	California	Scale	Social fabric (community services)	Detrimental: fewer and poorer quality services
Flora *et al.* (1977)	Macro-social accounting: 105 counties	Kansas	Scale/organization	Socio-economic and social fabric (class structure, retail sales, crime)	Mixed: industrialized farming related to income inequality and crime but also to higher income; other relationships less consistent

TABLE 5.1 *(cont'd)*

Study	*Methodology*	*Region*	*Measures of industrialized farming*	*Community well-being indicators*	*Results*
Small Farm Viability Project (1977)	Comparative case study: reanalysis of Arvin and Dinuba	California	Scale/organization	Socio-economic and social fabric (class structure, services)	Detrimental: variety of community indicators
Goldschmidt (1978b)	Macro-social accounting: states	Entire USA except Alaska	Scale	Social fabric (agrarian class structure)	Detrimental: poorer class structure
Heffernan and Lasley (1978)	Survey design study: 36 grape producers	Missouri	Organization	Social fabric (community social and economic involvement)	Mixed: industrialized farmers less involved in community socially but not more involved in economic control
Wheelock (1979)	Macro-social accounting: 61 counties	Alabama	Scale	Socio-economic and social fabric (class structure, population size)	Mixed: rapid increases in farm scale related to decline in income, population, and white collar workers, but scale also positively related to income in a cross-time model
Marousek (1979)	Regional economic impact: one community	Idaho	Scale	Socio-economic	Mixed: large farms generate less community employment but also greater income
Buttel and Larson (1979)	Macro-social accounting: state-level data	Entire USA	Scale/organization	Environment (energy usage)	Detrimental: industrialized farming conserves less energy
Heaton and Brown (1982)	Macro-social accounting: county-level data	Continental USA	Scale/organization	Environment (energy usage)	No detrimental: industrialized farming conserves more energy

Swanson (1980)	Macro-social accounting: 27 counties	Nebraska	Scale	Socio-economic and social fabric (population size)	Detrimental: variety of community indicators
Poole (1981)	Survey design study: 78 producers	Maryland	Scale	Social fabric (involvement in community organizations)	Detrimental: large farms related to less community involvement
Harris and Gilbert (1982)	Macro-social accounting: state-level data	Continental USA	Scale/organization	Socio-economic and social fabric (class structure)	Mixed: industrialized farming produces poorer community class structure but also greater local income
Swanson (1988)	Macro-social accounting: 520 communities	Pennsylvania	Scale	Social fabric (population size)	No detrimental: farm size has little effect on change in population size
Green (1985)	Macro-social accounting: 109 counties	Missouri	Scale/organization	Socio-economic and social fabric (services, population size)	No detrimental: farm size/ organization have little effect on community indicators
Skees and Swanson (1988)	Macro-social accounting: 706 counties	Southern USA, excluding Florida, Texas	Scale/organization	Socio-economic	Mixed: large farms related to higher unemployment and also to poorer conditions over time, but cross-sectional models show some positive effects on income
MacCannell (1988)	Macro-social accounting: 98 counties	Arizona, California, Florida, Texas	Scale/organization/ capital intensity	Socio-economic and social fabric (population size, local trade, local government)	Detrimental: variety of community indicators

TABLE 5.1 *(cont'd)*

Study	*Methodology*	*Region*	*Measures of industrialized farming*	*Community well-being indicators*	*Results*
Flora and Flora (1988)	Macro-social accounting: 234 counties	Great Plains and West	Scale	Socio-economic and social fabric (local retail trade, population size)	Mixed: large farms related to lower retail sales and population decline but not related to poverty or income
Buttel *et al.* (1988)	Macro-social accounting: 105 counties	Northeast	Organization	Socio-economic and social fabric (housing, retail trade, property taxes)	No detrimental: farm organization has little effect on community indicators
van Es *et al.* (1988)	Macro-social accounting: 331 counties	Corn belt	Scale/organization	Socio-economic and social fabric (population size)	No detrimental: farm scale/organization have little effect; in a few areas, large farms related to higher income
Gilles and Dalecki (1988)	Macro-social accounting: 346 counties	Corn belt and Central Plains	Scale/organization	Socio-economic	Mixed: farm organization (hired labor) related to poorer conditions but larger scale related to better conditions
Lobao (1990)	Macro-social accounting: 3,037 counties	Continental USA	Scale/organization	Socio-economic and social fabric (teenage fertility, infant mortality)	Mixed: industrialized farming related to higher income inequality and births to teenagers, and over time to higher poverty and lower income, but other relationships not significant

Lobao and Schulman (1991)	Macro-social accounting: 2,349 rural counties	USA and four regions	Scale/organization	Socio-economic	No detrimental: industrial farming has little relationship to poverty
Barnes and Blevins (1992)	Macro-social accounting: 2,000 rural counties	USA	Scale	Socio-economic	No detrimental: larger farms related to higher income and lower poverty, but controls for hired labor show detrimental impacts
Durrenberger and Thu (1996)	Macro-social accounting, 99 counties	Iowa	Scale, hog farms	Social fabric (food stamp need)	Detrimental: large farms related to greater use of food stamps
Seipel *et al.* (1998)	Hedonic price analysis: one county	Missouri	CAFOs	Sales prices of farmland parcels with and without homes	Detrimental: reduction in property prices of $144 per hectare within 3.2 km of a hog CAFO
Schiffman *et al.* (1998)	Quasi-experimental design: 88 matched individuals who vary by residence near CAFOs	North Carolina	CAFOs	Social fabric (social-psychological distress)	Detrimental: residents living near hog CAFOs are more depressed due to psychological and physical effects of odors
Wing and Wolf (2000)	Survey design study: 155 residents, three communities	North Carolina	CAFOs	Social fabric (health status, quality of life)	Detrimental: residents of hog CAFO community report greater respiratory and gastrointestinal problems and eye irritations, poorer quality of life

TABLE 5.1 *(cont'd)*

Study	*Methodology*	*Region*	*Measures of industrialized farming*	*Community well-being indicators*	*Results*
NCRCRD (1999)	Comparative case study: 14 farm dependent counties, one which recruited a hog CAFO	Oklahoma	CAFOs	Socio-economic, social fabric (population size, retail sales, school drop-out rates, crime, social conflict, property values, and other well-being indicators), and environment	Mixed: detrimental on most social fabric and environment indicators; no appreciable gains in per capita income and jobs; beneficial effects for a few indicators (increases in population size, retail sales, and property values)
Irwin *et al.* (1999)	Macro-social accounting: 3,024 counties	Continental USA	Organization	Social fabric (residential stability)	No detrimental: industrialized farming has little relationship to non-migration
Gomez and Zhang (2000)	Regional economic impact models: (1,106 towns and cities)	Illinois	Scale, focus on hog farms	Social fabric (retail spending)	Detrimental: larger farms related to less retail spending, weaker economic growth
Lyson *et al.* (2001)	Macro-social accounting: 433 counties	USA all agriculturally dependent counties	Scale/organization	Socio-economic and social fabric (civically engaged middle class, crime, low birth weight babies)	Detrimental: industrialized farming related to a less civically engaged middle class, low birth weight babies, and greater crime; the civically engaged middle class also mediates other effects of industrialized farming

Wright *et al.* (2001)	Case study: six counties with CAFOs	Minnesota	CAFOs/scale	Social fabric: (quality of life, community interaction, social capital)	Mixed: for quality of life, negative effects for neighbors, younger and mid-sized producers; positive effects for those who expanded operations; no effects for those who are not neighbors or not expanding. Community social capital and interaction quality declines
Foltz *et al.* (2002)	Regional economic impact models: 100 dairy farms in three communities	Wisconsin	Scale	Social fabric (farm input purchases made locally)	Detrimental: larger farms related to less input purchases made locally
Peters (2002)	Macro-social accounting: all agriculturally dependent counties	Iowa, Kansas & Missouri	Organization	Social fabric (children-at-risk, composite index of health, education, and general welfare)	Detrimental: industrialized farming related to higher children-at-risk scores
Wilson *et al.* (2002)	Macro-social accounting: census blocks in rural counties with swine CAFOs	Mississippi	CAFOs	Social fabric (environmental injustice)	Detrimental: CAFOs more likely to be located census block with poor African-Americans
Deller (2003)	Regional economic impact: 2249 nonmetro counties	Nonmetro US counties	Scale	Socio-economic	Detrimental: large farms related to slower growth in per capita income

TABLE 5.1 *(cont'd)*

Study	*Methodology*	*Region*	*Measures of industrialized farming*	*Community well-being indicators*	*Results*
Reisner *et al.* (2004)	Survey design study: 109 stakeholders in 52 counties with swine CAFOs	Illinois	CAFOs	Social fabric (perceptions of community problems caused by CAFOs)	Detrimental: residents reported greater dissatisfaction with CAFOs, odors, loss of values of homes, and water quality problems
Crowley and Roscigno (2004)	Macro-social accounting: 1,054 counties	North Central States	Scale/organization	Socio-economic	Detrimental: industrialized farming related to higher poverty and income inequality
Smithers *et al.* (2004)	Survey design study: 120 farmers in two townships	North Huron County, Ontario	Scale	Social fabric (community involvement, purchasing behavior, perceptions of community)	Detrimental: farmers expanding in scale participated less in community activities and organizations and were less committed to sourcing locally
Lyson and Welsh (2005)	Macro-social accounting: 433 agriculturally dependent counties	US all agriculturally dependent counties	Scale/organization	Socio-economic	Detrimental: industrialized farming related to greater poverty and unemployment, with corporate farming laws mediating these effects. Counties in states with weak or no anticorporate farming laws have poorer conditions

Constance and Tuinstra (2005)	Case study design: three clusters of rural communities with poultry CAFOs	East Texas	CAFOs	Social fabric (general quality of life, stress, odor, water quality, health, property values)	Detrimental: deterioration of quality of life along a variety of indicators experienced by those living closer to CAFOs
Whittington and Warner (2006)	Case study design: two communities with large-scale dairies	Ohio	Scale	Social fabric: (perceptions of local capacity to manage risks of large-scale dairies)	Detrimental: residents see weak capacity of local institutions, feelings of hopeless to address problems
Jackson-Smith and Gillespie (2005)	Survey design study: 836 residents from nine dairy-dependent rural areas in seven states	New York, New Mexico, Texas, Minnesota, Utah Wisconsin, and Idaho	Scale	Social fabric (farmers' and neighbors' relationships, community involvement, neighbors' complaints about odor, flies, and noise)	Mixed: dairy farm size has little relationship to most variables, but it is the strongest predictor of neighbors' complaints
Foltz and Zueli (2005)	Survey design study: 141 dairy farmers in three dairy dependent Wisconsin towns	Wisconsin dairy dependent towns	Scale	Social fabric (farm input purchases made locally)	No detrimental: little evidence that large farms are less likely to buy locally. Purchasing patterns are commodity specific and not determined by farm size
McMillan and Schulman (2003)	Case study: two CAFO counties, four focus groups	North Carolina	CAFOs	Social fabric (relations with neighbors, health and environmental concerns, enjoyment of property, local democratic participation, community cohesiveness)	Detrimental: variety of community indicators

CAFO, concentrated animal feeding operations.
Source: Based on Lobao and Stofferahn (2008).

Much of the research today, conversely, is highly quantitative, often involving large data sets and/or surveys (see Table 5.1, under column labeled Methodology). Since the 1970s, researchers have also developed quite elaborate indicators of community well-being, looking at, among other things, socio-economic, psychological, and even environmental variables (see Table 5.1, under column labeled Community well-being indicators; also see Box 5.1).

BOX 5.1 SOCIAL SCIENTISTS REPORT THE FOLLOWING NEGATIVE IMPACTS THAT INDUSTRIALIZED FARMS HAVE UPON COMMUNITY WELL-BEING

Socio-economic well-being

1 Greater income inequality and/or higher rates of poverty (Tetreau 1940; Goldschmidt 1978a; Heady and Sonka 1974; Rodefeld 1974; Flora *et al.* 1977; Wheelock 1979; Lobao 1990; Crowley 1999, Deller, 2003; Crowly and Roscigno 2004: Peters 2002; Welsch and Lyson 2001; Durrenberger and Thu 1996)
2 Higher rates of unemployment (Skees and Swanson 1988; Welsch and Lyson 2001)
3 Reduced employment opportunities (Marousek 1979; Thompson and Haskins 1999)

Social fabric

1 Decline in local population (Goldschmidt 1978a; Heady and Sonka 1974; Rodefeld 1974; Wheelock 1979; Swanson 1980)
2 Social class structure becomes poorer (due to, for instance, increases in hired labor) (Gilles and Dalecki 1988; Goldschmidt 1978a; Harris and Gilbert 1982)
3 Social disruption

- Increases in crime rates and civil suits (North Central Regional Center for Rural Development 1999)
- General increase in social conflict (Seipel *et al.* 1999)
- Greater childbearing among teenagers (Lobao 1990)
- Increased stress and social–psychological problems (Martinson *et al.* 1976; Schiffman *et al.* 1998)
- Swine CAFOs located in census blocks with high poverty and minority populations (Wilson *et al.* 2002)
- Deterioration of relationships between farming neighbors (Jackson-Smith and Gillespie 2005; McMillan and Schulman 2003)

- More stressful neighborly relations (Constance and Tuinstra 2004; Smithers *et al.* 2004)

4 Deterioration in community organizations; less involvement in social life (Goldschmidt 1978a; Heffernan and Lasley 1978; Poole 1981; Rodefeld 1974; Lyson *et al.* 2004; Smithers *et al.* 2004)
5 Decrease in local level political decision-making (as outside interests gain influence) (Tetreau 1940; Rodefeld 1974; Goldschmidt 1978a; McMillan and Schulman 2003)
6 Reduction in the quality of public services (Tetreau 1940; Fujimoto 1977; Goldschmidt 1978a; Swanson 1980)
7 Decreased retail trade and fewer, less diverse retail firms (Goldschmidt 1978a; Heady and Sonka 1974; Rodefeld 1974; Fujimoto 1977; Marousek 1979; Swanson 1980; Skees and Swanson 1988; Foltz *et al.* 2002; Foltz and Zueli 2005; Smithers *et al.* 2004; Gomez and Zhang 2000)
8 Reduced enjoyment of outdoor experience (especially when living near CAFO) (Schiffman *et al.* 1998; Wing and Wolf 1999, 2000; Constance and Tuinstra 2005; Reisner *et al.* 2004; Wright *et al.* 2001; Kleiner 2003; McMillan and Schulman 2003)
9 Neighbors of hog CAFOs report upper respiratory, digestive tract disorder, and eye problems (Wing and Wolf 1999, 2000; Constance and Tuinstra 2005; Reisner *et al.* 2004; Wright *et al.* 2001; Kleiner 2003)
10 Residences closest to hog CAFOs experience declining property values relative to those more distant (North Central Regional Center for Rural Development 1999; Seipel *et al.* 1998; Constance and Tuinstra 2005; Reisner *et al.* 2004; Wright *et al.* 2001)

Environment

1 Depletion of water and other energy resources (Tetreau 1940; Buttel and Larson 1979; North Central Regional Center for Rural Development 1999)
2 Environmental consequences of CAFOs (increase in Safe Drinking Water Act violations, air quality problems, and increased risks of nutrient overload in soils) (North Central Regional Center for Rural Development 1999)

(Based on Stofferahn 2006)

It is important to clarify that "industrial agriculture" in this tradition often goes beyond just scale, looking also at indicators of farm organization (see Table 5.1, under column labeled Measures of industrialized farming). Scale is a measure of an operation's sales or acreage. Yet scale alone does not seem to capture the organizational characteristics of industrial agriculture. According to the research, features like absenteeism (where land is leased and the owners live outside the community), contract farming, dependency on hired labor, and operation by farm managers (as opposed to owner-operator situations) seem far more likely to place communities at risk than if surrounding farms (even large ones) lack these characteristics. According to the research, large-scale family owned farms tend to still purchase their inputs, farm equipment, and services (e.g. custom baling) locally, volunteer their time to local organizations (church groups, school fundraisers, etc.), and regularly interact with their neighbors.

As noted in both Box 5.1 and Table 5.1, Goldschmidt-type research has also looked at the impact of confined animal feeding operations (CAFOs) on community well-being and public health. Workers are not the only ones whose health is negatively impacted by CAFOs. Neighbors of CAFOs have been shown to have higher levels of respiratory and digestive disturbances (Radon *et al.* 2007; Wing and Wolf 2000). They also have abnormally high rates of psychological disorders, such as anxiety, depression, and sleep disturbances (Carolan 2008; Schiffman *et al.* 1995). Children living on and near hog farms have abnormally high rates of asthma (Chrischilles *et al.* 2004). These rates are known to increase proportionally with the size of the operation (Donham *et al.* 2007). School administrators in North Carolina whose schools were within 3 miles of one or more large hog feeding facilities report rates of asthma among the student population at levels well above the state average (Mirabelli *et al.* 2006).

Post-traumatic stress disorder (PTSD) has also been reported among residents living near CAFOs (Donham *et al.* 2007: 318). This can result from such things as a reduction in one's quality of life and/or property values – for example, one study calculated an average reduction in property prices of US$144 per hectare within 3.2 km of a CAFO (Seipel *et al.* 1998). Researchers have also documented that a significant level of social tension – specifically between producers and their neighbors – can emerge when CAFOs are sighted within a community (Constance and Tuinstra 2005; McMillian and Schulman 2003).

When taken as a whole, this research would seem to support recent calls for smaller farms and more local food systems as inherently more sustainable, compassionate, and socially just models of food production. We must be careful not to extrapolate such a general conclusion from these findings. I address the problems of such an assumption in Chapter 12, when discussing what is known as the local trap. Without getting ahead of myself, I will merely point out that Goldschmidt-type research says nothing about one model of food production being *inherently* "better" (e.g. sustainable, just) than another. Indeed, reading the Goldschmidt literature closely you'll find that it is not really industrial agriculture *per se* that negatively impacts community but the *socio-organizational forms* that tend to be associated with this model of food production. Do people trust each other? Do they interact regularly? Are levels of income inequality in the community low? Are social institutions (e.g. education, church groups, bowling leagues) robust? Communities that can answer "Yes" to these questions tend to be considerably socially and economically healthier than those answering "No." And the Goldschmidt research tells us is that communities surrounded by "industrial" farms are more likely to fall into the latter category.

Labor

Wage inequities

In Chapter 1 I talk briefly about how a public sociology of food and agriculture is about consciousness raising by making the hidden visible. Few aspects of our food system require more unmasking than labor. The environmental consequences of our food system are routinely discussed, as evidenced by the growing popularity of organic food and talk about food miles. The subject of labor, however, seems to get lost in the discussion. Occasionally it enters into the public's consciousness – Eric Schlosberg (2001), for example, did an excellent job of discussing labor beyond the farm gate in his book, *Fast Food Nation*. Yet usually the issue of labor takes a back seat to topics like environmental sustainability, organic agriculture, and local food, which is an egregious omission when you consider just how populated the food system is (Box 5.2).

Looking at labor in the food system reveals, for instance, that race matters. Figure 5.1 breaks up the food sector in the USA into four sectors, looking at minority involvement in each. People of color are clearly over-represented in some sectors of the food system.

Figure 5.2 illustrates the disparities in median hourly and annual wages between racial/ethnic groups across the various sectors of the food system in the USA (compared with the wage gap between "whites" and "people of color" for the country as a whole). In each of the four sectors represented, whites are paid

BOX 5.2 SOME OF THE LABOR THAT FEEDS US

Farming and fishing occupations: Farmers/ranchers; field laborers; and graders and sorters; fishermen/women

Food production occupations: Butchers; meat, poultry and fish processing workers; food cooking machine operators; food and tobacco roasting, baking, and drying operators

Food transportation occupations: Drivers/sales workers; hand packers and packagers; laborers and freight, stock, and material movers

Food preparation and serving occupations: Bartenders; cooks; dishwashers; fast food employees; and waiters

Food retail: Grocer stockers; checkout personal; and baggers

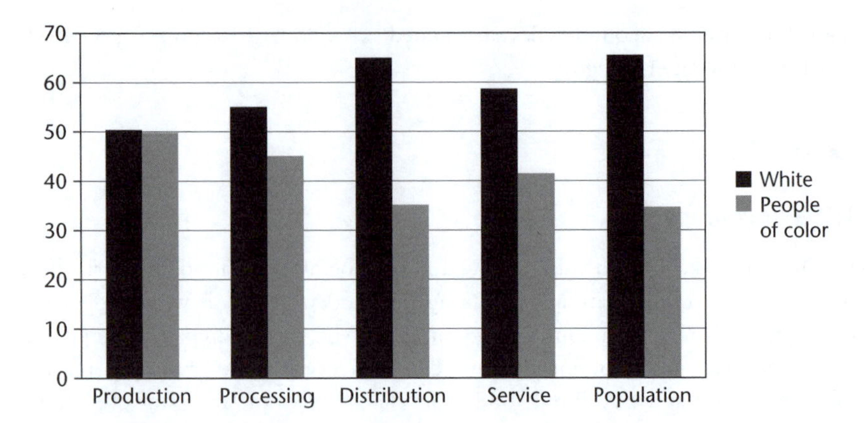

FIGURE 5.1 Employment in food sectors by "white" and "people of color" designation. Based on Liu and Apollon (2011; only those employed, working over 25 hours a week, and earning an income from employment were selected for analysis)

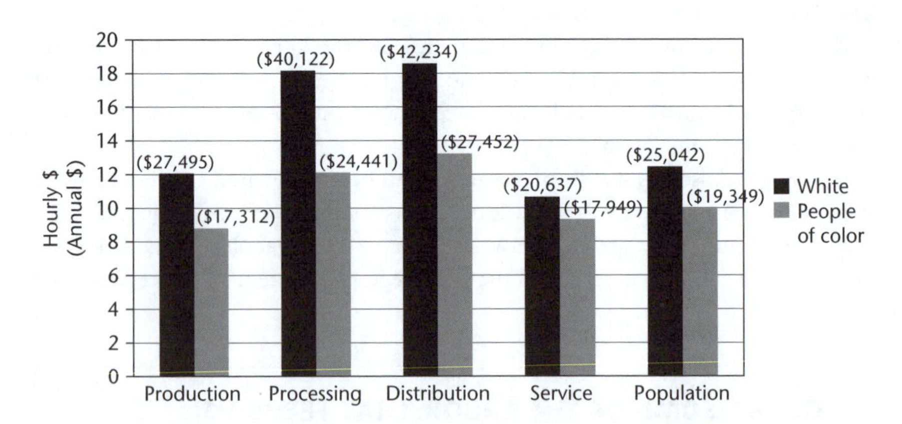

FIGURE 5.2 Hourly wage in food sectors by "white" and "people of color" designation (annual income in parentheses). Based on Liu and Apollon (2011; only those employed, working over 25 hours a week, and earning an income from employment were selected for analysis)

notably more than non-whites. This gap varies considerably, however, as some sectors show greater income equity (e.g. service) than others (e.g. processing).

Let's now add gender into the mix. As detailed in Figure 5.3, we find that white men earn the highest wages in the food system when accounting for both gender and ethnicity. When reviewing the figure, keep in mind that over 70 percent of workers who grade and sort recently harvested crops – a particularly low paying job – are Latino. Moreover, at least 6 out of every 10 farm workers are undocumented immigrants (who are also overwhelmingly Latino).

FIGURE 5.3 Median average salary by ethnicity and gender (to each dollar earned by white male). Based on Liu and Apollon (2011; only those employed, working over 25 hours a week, and earning an income from employment were selected for analysis)

Thinking sociologically about low wages

Competition forces firms to seek out ways to reduce their operating costs, as they strive to outperform (and undercut) their competitors. This has historically been achieved in a number of ways. One is to replace humans with machines, which is a popular strategy in developed countries where the costs of labor have been rising faster than technology. Firms can also seek out cheaper labor. One relatively easy way to do this is to move operations overseas. Another option, which is often employed in labor-intensive sectors of agriculture (e.g. vegetable production), where it's hard to move one's farm to another country, is to employ immigrant labor, who, for various reasons, can be more easily exploited.

Let's think a moment about the pressures this creates on the wage structure for those sectors impacted by these practices. A popular term to explain the effect of such pressures "is 'race to the bottom'" (Tonelson 2002). The term refers to an international dynamic between (typically poor) countries competing with each other to attract foreign investment, racing to see who can come closest to offering firms what amounts to a free ride. In actuality, however, there is nothing "free" about it as the ride comes at great expense to the environment, worker rights, and tax revenue, for example (Carolan 2011). A similar phenomenon has been documented between firms *within* a given country, resulting in a "gloves-off economy" (Bernhardt *et al.* 2008). This refers to employer strategies and practices, even within well-governed nations, that evade or outright violate laws and standards meant to govern job quality and safety. Doing this creates pressure for even reputable employers to take their "gloves off" too, as unsavory employers gain unfair market advantages by violating labor laws and standards (Box 5.3).

Another cost reducing strategy is to externalize costs onto the environment – or, to put it another way, pollute. In truth, however, it is not just the environment that

BOX 5.3 THE US FOOD SYSTEM: A GLOVES-OFF ECONOMY?

The following examples come from the USA, where workplace regulations, while far from perfect, are still stricter than in many other countries. Is this evidence of a gloves-off economy?

- Though low paying, field work is very dangerous: 32.5 of every 100,000 farm workers die for reasons linked to their employment.
- Twenty-five percent of *all* grocery workers in the USA experience minimum wage violations.
- Real wages for meatpacking workers have *fallen* from US$20 per hour in 1977 to US$10.50 per hour in 2001.
- Wages in the poultry industry are 24 percent *lower* than they were in 1977.
- Medium wages among food system employees in the USA are consistently below what's considered an adequate living wage.

(Institute for Food and Development Policy 2009; Liu and Apollon 2011; Roberts 2008)

pays these costs. Future generations incur them, as do farm laborers (as I document later) and society as a whole (e.g. as taxpayers, residents living near CAFOs).

Finally, do low wages really make things cheaper for consumers? On the one hand, cheap labor might help lower the *retail price* of goods. But, on the other, low wages also trigger one's eligibility for social safety nets. So while low wages might reduce the retail price of a good, consumers are still paying for that cheap labor as taxpayers. *Savings* for consumers that result in *expenses* for taxpayers (after all, these are the same people!) is a questionable accounting practice when the end goal is affordable food. But it's happening all around us.

A study published by UC Berkeley's Labor Center concludes that California taxpayers are spending US$86 million annually providing health care and other public assistance (such as food stamps and subsidized housing) to the state's 44,000 Walmart employees (Dube and Jacobs 2004). Two factors account for much of this expense: the company's low wages and the fact that 23 percent fewer Walmart workers are covered by employer-sponsored health insurance than those of other large retail firms. The study estimates that if competing retailers were to adopt Walmart's wage and benefit levels, California's taxpayers would have to pay an additional US$410 million a year in public assistance. US Representative George Miller (2004) oversaw a similar study, with similar conclusions. The report estimates that a Walmart store with 200 workers costs federal taxpayers US$420,750 a year. The following are some of the listed expenses for families who, because of their low wages at Walmart, qualify for government assistance programs:

- US$36,000 a year for free and reduced lunches
- US$42,000 a year for Section 8 housing assistance
- US$125,000 a year for federal tax credits and deductions
- US$100,000 a year for the additional Title I (educational) expenses
- US$108,000 a year for the additional federal health care costs of moving into state children's health insurance programs (S-CHIP).

So, the next time you see an advertisement for cheap food (or cheap anything) ask yourself this one question: At *what* and *whose* expense is it being made cheap?

Health impacts

A considerable amount of research has been conducted over the last 10 years, by health professionals, epidemiologists, and sociologists, examining the impacts of CAFOs upon worker health (Figure 5.4). A highly cited review on the health effects of CAFOs published in 2007 counted 70 peer-reviewed studies documenting the adverse health effects of the confinement environment on swine producers (Donham *et al.* 2007: 317). Over 25 percent of CAFO workers develop respiratory diseases, ranging from bronchitis to mucus membrane irritation, asthma, and acute respiratory distress syndrome (Donham *et al.* 2007: 318). Exposure to high

FIGURE 5.4 Chicken confined animal feeding operation (CAFO). Photo courtesy of USDA

concentrations of bioaerosols (biological in origin airborne particles) has been linked to organic dust toxic syndrome, which affects more than 30 percent of workers in the swine industry (Donham *et al.* 2007: 318).

Acute respiratory symptoms have been documented early in the work history of some individuals sufficiently severe to warrant immediate and permanent removal from the workplace (Dosman *et al.* 2004: 298). The greatest risk, however, is among those working full-time for extended periods (Mitloehner and Calvo 2008: 167). Roughly 60 percent of swine confinement workers who have worked for over 5 years in a CAFO environment experience, on average, at least one respiratory symptom (Donham 2010: 107). Public health scientists have placed the maximum recommended exposure of dust in livestock buildings at 2.5 mg/m^3. This figure is well below the 15 mg/m^3 rate set by, for example, the US Occupational Health and Safety Administration (Donham 2010: 107). With this in mind, recognize also that it is not unusual to find CAFO dust concentrations levels of 10–15 mg/m^3 in winter months (when facilities are more likely to be closed up) or when animals are being moved (an activity that can kick up considerable dust) (Donham 2010: 107; Mitloehner and Calvo 2008: 175).

CAFOs are not the only hazardous work environment in the food system. The field has proven an equally dangerous place to work. For example, 81 percent of pineapple farmers and 43 percent of vegetable farmers in the West African country of Benin reported "considerable" negative health effects from pesticide exposure (Williamson 2005: 171–2). It has been estimated that approximately 7.5 percent of agricultural workers in Sri Lanka experience occupational pesticide poisoning annually (van der Hoek *et al.* 1998). In Costa Rica and Nicaragua, those figures are

4.5 percent and 6.3 percent, respectively (Wesseling *et al.* 1993; Garming and Waibel 2009). Annually, worldwide, more than 26 million individuals suffer from nonfatal pesticide poisoning, while an additional 220,000 poisonings result in death (Richter 2002: 3). Pesticide exposure in the Brazilian state of Paraná cost taxpayers approximately US$443 million annually due to costs incurred through the Brazilian public health system (Soares and Porto 2009). The total number of pesticide poisonings annually in the USA is believed to be around 300,000 (Pimentel 2005: 230). When broken down according to ethnicity, poisoning rates in the USA disproportionately affect Latinos (Box 5.4) (Liu and Apollon 2011). A survey of epidemiological literature indicates that the incidence of cancer in the US population due to pesticides ranges from about 10,000 to 15,000 cases per year (Pimentel 2005: 231).

Pesticide toxicity figures based upon government statistics tend to understate the problem. For example, a study of Indonesian farmers found 21 percent of those interviewed reported three or more neurobehaviorial, respiratory, and/or

BOX 5.4 UNITED FARMWORKERS OF AMERICA (UFA) CANCER REGISTRY STUDY

A seminal study examining the pesticide–cancer link in farm laborers was published in 2001. Supported in part by grants from the US Centers for Disease Control and Prevention and the California Cancer Research Program, the study analyzed cancer incidence among California Latino farmworkers who had been members of the United Farmworkers of America (UFW) union (Mills and Kwong 2001). Just over 1,000 – out of roughly 140,000 – farmworkers had been diagnosed with cancer between 1973 and 1997. Compared with the general Latino population, the study found that UFW farmworkers were 69 percent more likely to acquire stomach cancer. Moreover, the studied farmworkers had a 68, 63, and 59 percent increased likelihood of developing uterine, cervical, and certain types of leukemia, respectively, over the general Latino population.

The lead investigators of the study attributed this cancer rate disparity to the farmworkers regular exposure to pesticides. Farmworkers came into contact with these chemicals routinely: while mixing and applying them to crops; during planting, weeding, thinning, irrigating, pruning, and harvesting crops; and at home, as a result of living near treated fields, eating pesticide contaminated food, and bringing the chemicals home on their clothes. The study also notes that farmworkers tend to be diagnosed at a later stage than the general Latino population, indicating a lack of health care for most laborers. Furthermore, since many pesticide poisonings go unreported – those affected might move away, never seek treatment, or seek treatment in Mexico (Murray 1994: 47) – the data used in this study likely *underestimate* the actual trauma rate.

intestinal symptoms indicative of pesticide toxicity. Yet only 9 percent of the farmers in the study said they had reported being poisoned in a government survey (Kishi *et al.* 1995). There are numerous reasons for the under-reporting of pesticide poisonings. Many of the affected simply ignored the symptoms, as they can't afford to leave their job and/or medical attention is beyond their financial means. Another potential explanation for under-reporting has to with *who* epidemiological studies have tended to examine – namely, the field workers who apply the pesticides. This population is overwhelming male. Yet women also spend long hours working in freshly sprayed fields. Moreover, exposure can occur at home, from, say, the husband's contaminated clothes, which are often mixed and washed with other laundry (Kishi *et al.* 2005: 26).

It is also unfair to blame the victims of pesticide poisonings by attributing exposure to ignorance. Most of those exposed know the dangers of pesticides. Yet other reasons often override farmers' and farm laborers' concerns about safety when applying these chemicals: necessary protective equipment might be too expensive or too uncomfortable when working in oppressive heat and humidity; washing facilities are often inconveniently located relative to fields; and too often agricultural communities use pesticide-laden irrigation channels for bathing and washing purposes (Kishi *et al.* 2005; Mayer *et al.* 2010; Murray and Taylor 2000).

Impacts on peasant communities

Before turning to conventional agriculture's impact on peasant communities, it might help the reader to have an understanding of what "peasant" means, at least for sociologists of food and agriculture. This point is addressed further in Chapter 11, when discussing peasant-based social movements. For the moment just remember that "peasant," in sociological parlance, is not meant as a pejorative term – it does not imply backwards, traditional, or anything like that. It is a title that brings great pride and a shared identity, uniting hundreds of millions of small-scale landholders throughout the developing world. Unfortunately, throughout much of the twentieth century, food and international development policies where clearly biased against this population. The so called Washington Consensus – those widely adopted neoliberal policies that have been dictating developmental (and food) policy for much of the last 30 years – viewed peasants, to be blunt, as wasted potential urban labor. A central goal of the Consensus has been to free peasants "from the shackles of unremitting toil on the land" (Ellis 2005: 144).

A significant part of the problem is that traditional farming is viewed as too labor intensive. Leave food production to capital intensive enterprises that have the capacity to produce far more with far less labor, or so the logic goes. Doing this would free up literally hundreds of millions to work in factories – there are roughly 1.2 billion peasants in the world today (van der Ploeg 2008) – where the real wealth is generated. At the same time, as peasant farmers are dispossessed and replaced by fewer modern enterprises, those farmers that remain need to produce food more efficiently with the help of technology, inputs, and advancements in seed breeding. This will help not only feed a growing non-farming population but, as overproduction begins driving down prices, avoids rising food prices. This cheap food will serve multiple developmental ends – from helping to offset low wages in cities to redistributing wealth from the countryside to urban areas as the wages of urban dwellers go toward buying more than just food (Box 5.5).

In practice, however, the non-farming sectors of developing economies rarely grow fast enough to absorb the surplus labor freed from the "shackles" of farming. For example, while Mexico gained jobs in the manufacturing sector thanks to the North American Free Trade Agreement (NAFTA), employment gains have been outpaced by losses in agriculture, which has seen its total employment numbers drop from 8.1 million in the 1990s to 5.8 million in the second quarter of 2008 (Zepeda *et al.* 2008). There is therefore growing resistance to the Washington Consensus and its rather pejorative view towards peasant agriculture.

An estimated 2.5 billion people worldwide farm and three out of every four people on the planet live in a rural area (World Bank 2007). Critics of the Washington Consensus argue that it makes sense to come up with a developmental strategy that is pro-peasant, as peasants continue to represent a sizable portion of

BOX 5.5 RATIONALE SUPPORTING THE ELIMINATION OF PEASANT AGRICULTURE

Labor efficiency

- A countryside populated by smallholders is an inefficient distribution of labor.

Production efficiency

- Small farmers can't produce food as cheaply in part because of their heavy reliance on labor (their own), whereas large farms, substituting capital for labor, can afford to make less per unit because they are producing many more units.

Redistribution of wealth from rural areas (and agriculture) to urban areas (and industry)

- A surplus of cheap grain not only feeds the growing urban populations but also helps ensure that some of the wages earned there can be used to buy more than just food.

Source: Carolan (2011)

the population in developing economies. The research seems to back up their argument, showing consistently that public investments in agriculture in poor countries, in terms of research and extension, yield higher societal returns than expenditures in other productive sectors (Bezemer and Headey 2008; Fan *et al.* 2000). Nevertheless, agricultural research and development in developing nations is on the decline.

The neglect of peasant agriculture and the rural poor in general in developing countries has been referred to as the "urban bias" in international policy. Popularized by Michael Lupton (1977), the term refers to a tendency in developmental and international agricultural policy circles, dating back to World War II, to under-allocate resources to, and extract surplus from, the rural class of poor countries. The urban bias, in other words, is just as it suggests: a developmental bias toward urban areas and the sectors of the economy that tend to be located there. So why have these policies been allowed to continue even though they undermine the livelihoods of billions?

One reason lies in the fact that poor rural populations are arguably the world's most politically disenfranchised group (Grindle 2004). This helps explain why, for example, in India – the world's largest democratic state – 80 percent of the population live in rural areas but 80 percent of government spending goes into urban areas (Patel 2009: 241). Throughout much of Africa and Central and South America, rural populations are geographically isolated not only from centers of power but also from each other, making collective mobilization difficult (Bezemer and Headey 2008: 1348). The story in Asia is slightly different. Many Asian countries had and still have a high rural density, which makes the organization of rural pressure groups easier. The threat of a rural-based Communist insurgency – in countries like South Korea, Taiwan, Malaysia, and Indonesia – also made the political elite historically more sensitive to the interests of their country's small farmers (Bezemer and Headey 2008: 1348). And in those countries, small farmers are notably better off.

An alternative paradigm is therefore emerging; one that views development and food security as something that occurs *because* of the peasantry rather than in spite of it. An example of this comes from a recent report released by the United Nations Special Rapporteur on the Right to Food (United Nations 2010). The report, *Agro-ecology and the Right of Food*, notes that small-scale sustainable farming has the potential to double food production within 5–10 years in places that are the most food insecure. In a press release, Olivier De Schutter, United Nations Special Rapporteur on the Right to Food, is quoted as saying, "We won't solve hunger and stop climate change with industrial farming on large plantations." Instead, "the solution lies in supporting small-scale farmers' knowledge and experimentation, and in raising incomes of smallholders so as to contribute to rural development."[1]

Peasant agriculture might be more labor-intensive but by all accounts it works (I discuss this point further in Chapter 9 when discussing agroecology).

Remember also, as most peasants do have the resources to practice labor-intensive agriculture, labor – namely, their own – is the one thing they can afford. Of 57 poor countries surveyed in the aforementioned UN report, yields increased by an average of almost 80 percent when farmers used methods like placing weed-eating ducks in rice paddies in Bangladesh or planting desmodium (an insect repellant) in Kenyan cornfields. The labor-intensity of the practices highlighted in the report would also bring jobs to the rural poor in less affluent economies, thus helping to reverse the bias referred to above.

Let us turn to a specific case and look at more closely how food policy and international trade in general has tended to harm the peasant class. The case: the North American Free Trade Agreement (NAFTA).

NAFTA's impact on smallholder farmers in Mexico

Implementation of NAFTA began on January 1, 1994. The agreement removed many barriers to trade and investment between the USA, Canada, and Mexico. NAFTA created one of the world's largest free trade areas, linking 444 million people producing US$17 trillion worth of goods and services. NAFTA, it was promised, would be a boom for Mexican farmers, giving them access to the world's largest consumer market located just north of their border in the USA.

Roughly 18 years later, we can see that NAFTA has not been good for small-scale Mexican farmers. For one thing, they are having a difficult time competing against their capital intensive, highly subsidized neighbors to the north: US farmers. Evidence of this can be seen in that fact that the total value of Mexican agricultural imports from the USA increased 280 percent from 1992 to 2008 (Wise 2009: 4). The Mexican government agreed to lower its support for agriculture while in the USA government support has never been higher. In the late-1990s, the USA extended Mexico roughly US$3 billion in export credits, which the latter then used to buy still more corn from the former (Bello 2009: 45). It is not surprising, then, that between 1993 and 2004 the price of corn and soybeans in Mexico fell by approximately 50 percent (Perez *et al.* 2008: 8).

It is not just peasant grain farmers in Mexico that have been hit hard by NAFTA. Mexican cattle producers too have found it hard to compete against their US competitors. It has been estimated that as many as 80 percent of the cattle in Mexico are pasture fed; as opposed to beef cattle in USA, which spend the last 12 months of this life being "finished" with a corn-based diet. The aforementioned cheap corn has therefore not helped Mexican beef producers to the same extent as beef producers in the USA. One estimate places losses to the Mexican beef industry between 1997 and 2005 at US$1.6 billion (Wise 2009: 26).

NAFTA is also, though rarely discussed in these terms, a de facto immigration policy. Though the borders between the USA and Mexico have been highly fluid ever since these two countries came into existence, the pace of immigration actually *increased* during the NAFTA years, from about 350,000 per year before NAFTA to approximately 500,000 per year by the early 2000s (Wise 2009: 13). It seems NAFTA is driving people north, across the border into the USA.

There is also evidence that, were it not for immigration in the USA, small-scale farmers in Mexico would be in even worse shape. The Mexican peasant is being aided by remittances – the money migrant workers in the USA send home to families in Mexico. Remittances helps explain why 3 million farms continued to grow corn in Mexico, in spite of their tremendous comparative disadvantage relative to US growers. Laura Carlsen (2003), Director of the Mexico City-based Americas Program of the International Relations Center (IRC), claims remittances serve two roles:

> First, the money sustains agricultural activities that have been deemed non-viable by the international market but that serve multiple purposes: family consumption, cultural survival, ecological conservation, supplemental income, etc. Second, by sending money home, migrants in the US seek not only to assure a decent standard of living for their Mexican families but also to maintain the campesino identity and community belonging that continue to define them in economic exile.
>
> *Carlsen (2003)*

In the end, NAFTA, as a product of the Washington Consensus, has dealt with peasant farmers by displacing them. It has been estimated that NAFTA has displaced roughly 15 million Mexican peasants (Bello 2009). "Freed" from subsistence agriculture, small-scale Mexican farmers have been encouraged to grow nontraditional commodities for export, like certain fruits, vegetables, and cut flowers. But these recommendations lack sufficient understanding of sociological conditions on the ground. First, there is the issue of finance: farms need access to credit if they are to switch over and grow an entirely different commodity (not to mention a different skill set, expertise, etc.). While corn requires an investment of approximately US$210 per hectare, nontraditional commodities typically come with a considerably higher per hectare investment: melons, US$500–700; cauliflower, US$971; broccoli, US$1,096; and snow peas, US$3,145 (Bello 2009: 49). Yet, as part of trade liberalization policies, government credit in developing countries has evaporated. This makes the financing of such ventures impossible for the rural poor.

Small-scale farmers are also having a difficult time complying with private standards, which are often dictated by supermarket firms. These standards can spell out any number of things, such as what methods of production to use, storage requirements, and even how something ought to look. Most of the certifying organizations that oversee compliance to these standards are located in affluent nations. Primus Labs in the USA, for instance, certifies for 68 percent of the fruit and vegetable firms in northwest Mexico (Narrod *et al.* 2008: 361). The costs of certification can therefore present insurmountable barriers for small-scale growers looking to raise nontraditional commodities for export (certification is discussed in greater length in Chapter 12). According to one estimate, costs of certification can be as high as US$850 per hour (Narrod *et al.* 2008: 361).

Peasant farmers often therefore find themselves in an unenviable position. Unable to afford the costs that go with switching over to higher value export commodities (and with no access to credit), while simultaneously unable to compete with growers in countries like the USA when it comes to producing traditional commodities like corn, soybeans, and beef. Faced with such harsh realities, small-scale farmers in Mexico (like many peasant farmers) are left with few options.

Biotechnology: Pro-poor, pro-peasant strategy?

Under the heading "Produce More, Conserve More, Improve Farmers' Lives," the Monsanto website reads as follows:

> Grasp tomorrow's challenge. By 2050, say United Nations' experts, our planet must double food production to feed an anticipated population of 9.3 billion people. (That figure is 40 percent higher than today's 6.6 billion.)[2]

Later in the webpage, Monsanto pledges to help the world by, among other things:

- Developing improved seeds that help farmers double yields from 2,000 levels for corn, soybeans, cotton, and spring-planted canola, with a US$10 million grant pledged to improve wheat and rice yields.
- Helping improve the lives of all farmers who use our products, including an additional 5 million people in resource-poor farm families by 2020.

It is fair for the social scientists to look at how well these statements stack up to empirical reality. First, let's take the point made by Monsanto, as it is made endlessly by biotechnology proponents, that we have essentially reached the limits of conventional breeding methods and that future yield increases will only be attained through genetic engineering. The problem with this claim is that a trait directly responsible for yield increase has *never* been inserted into any transgenic crop. Thus, if the past is any guide, the promise of "double yields from 2,000 levels" is a puffery at best. For example, the widely popular Roundup Ready varieties have been found to produce *lower yields* than conventional seed – a phenomena known as "yield drag." (To put it very simply: when a plant is engineered to do something – like be resistant to a particular herbicide – energy and nutrients that otherwise would have gone toward yield are often redirected elsewhere.) Thus, the promise for higher yields with some of their new products – like Roundup Ready 2 Yield needs contextualizing. This new "high yield" soybean line claims 7–10 percent yield increases compared with the original Roundup Ready, which will likely make its yields *comparable* to conventional seeds.

Let's now look at why Monsanto has gone to such length to promote itself as a "pro-poor" firm who is ideally positioned to make the world more food secure. What does it have to gain from such an identity? And, importantly, do its commodities improve the lives of the world's peasants? To answer these questions we'll need to go back a number of years and put its current actions in the context of those from the past.

Massive investments made by Monsanto over the last couple of decades in the area of biotechnology needed to be justified, not only to shareholders but also to the food consuming public. Monsanto spent approximately US$9 billion in the 1990s buying up biotech and seed companies (as evidenced by earlier mentioned dominance in the industry; see Chapter 3). For many, Monsanto looked to be attempting to monopolize food itself (see Shiva 2000: 92). Monsanto sought to counter this image and provide an alternative narrative for why this massive capital buildup into the food industry was necessary. They argued that international rural development and food security could most aggressively be tackled by biotechnology and the massive R&D expenditures such techniques require. Monsanto executives concluded that being viewed as a company striving to eliminate hunger while improving the lives of disadvantaged farmers around the world would be better than the less flattering narrative pushed by their detractors.

An example of just such a pro-poor strategy was the Monsanto Smallholder Program (SHP); an initiative undertaken between 1999 and 2002. According to the company's website, the SHP claimed to be aimed at smallholder farms to provide them with "existing commercial technologies, including improving seeds, biotechnology traits where approved and applicable, conservation tillage practices, crop protection products and other inputs, as well as training and technical assistance."[3] The SHP was of strategic importance to Monsanto. Long known in the developing world as an herbicide company, Monsanto needed to build brand recognition as a seed company in the late 1990s. In addition, while accustomed to dealing with large-scale commercial farmers, especially in North America, Monsanto's global expansion plans required greater dealings with the smallholder segment of the market in developing countries (Glover 2007). By 2001, the SHP was funding 21 projects in 13 countries, purportedly enrolling more than 320,000 smallholder farmers.

Glover (2007) spent time in India studying the methods and assumptions of SHP. His conclusion was that the SHP closely followed the "old" unidirectional model of technology transfer, where the flow of information is essentially one-way (discussed further in Chapter 7). Assuming to know what the wants and needs of the farmer are, this model is about getting information and technology to the farmers (rather than learning from them). Glover also noted the "commodity approach" of the SHP, at least in India, where corn and cotton production was pushed heavily and involving, not surprisingly, transgenic crops licensed by Monsanto. This was a decision, Glover (2007: 68) speculates, that "coincided with the global parent company's strategic decision to withdraw from the rice sector and concentrate on maize, soybean, cotton, and wheat."

It is also important to point out that the smallholders working with Monsanto were not really "small" by Indian landholding standards. A St. Louis-based marketing executive told Glover during an interview that the company, when deciding how small is too small, took into account the "long-term potential" and "objectives of the farmer" when deciding whether to enroll them through the SHP (Glover 2007: 69). Based upon what he saw and heard from those interviewed, Glover concludes that "larger and more prosperous farmers were given special attention" (Glover 2007: 69). Workers employed to help carry out these projects also explained that they felt pressure, as projects progressed, to focus more on "boosting sales" of Monsanto products (Glover 2007: 70).

Though the SHP initiative is no longer in effect, talk of helping small holder farmers is rife in Monsanto publications (see, e.g., Monsanto 2009: 5, 8, 18, 19, 30, 31, and 32). Moreover, under the guise of "philanthropy," the company continues to direct monies to smallholders – for example, US$162,000 was given over 3 years to 20 rural districts in Brazil (Monsanto 2009: 32). Yet, like earlier smallholder initiatives, there is no evidence to suggest that the main goals of these awards are anything but commercial. They are not only positive PR opportunities but also evidence that Monsanto remains committed to the "old" top-down model of technology transfer, which, they hope, will translate into greater future sales.

Transition . . .

Having touched on how the dominant food system impacts rural communities, labor, public health, and peasant agriculture I'd like to slightly shift focus in the next chapter by examining more closely the subject of culture. Structural changes in agriculture have altered social relationships in a variety of ways, some of which were discussed in this (and previous) chapters. As itself a product of social relationships – with, for example, other individuals, animals, and material things – culture has not been immune to those structural changes either. I describe in the following chapter some of those cultural impacts as well as a broader cultural turn that occurred in rural studies and agrifood scholarship in the 1990s.

Discussion questions

1 Do you think the Goldschmidt thesis romanticizes the family farm?
2 Why do you think we often forget about labor issues (and abuses/injustices) when thinking about our food, especially compared to, say, environmental sustainability? How often does the issue of labor enter into your mind when you think about food?
3 Think further through the earlier-mentioned point that profit-maximizing firms are not interested in food security *per se* but in making a profit. How might the goals and interests of seed companies like Monsanto not align with the goals and interests of the world's billions of peasants? Can we rely on market forces alone to bring food security to the world's poor?

Suggested reading: Introductory level

Charlotte Observer. 2008. The cruelest cuts: The human cost of bringing poultry to your table, 6-day special report, February 10–15, http://www.charlotteobserver.com/poultry/

Constance, Douglas and Reny Tuinstra. 2005. Corporate chickens and community conflict in East Texas: Growers' and neighbors' views on the impacts of the industrial broiler production. *Culture and Agriculture* 27(1): 45–60.

Suggested reading: Advanced level

Brown, S. and C. Gertz. 2008. Privatizing farm worker justice: Regulating labor through voluntary certification and labeling. *Geoforum* 39(3): 1184–96.

Lyson, T., R. Torres, and R. Welsh. 2001. Scale of agricultural production, civic engagement, and community welfare. *Social Forces* 80(1): 311–27.

van der Ploeg, J. 2010. The peasantries of the twenty-first century: The commoditization debate revisited. *Journal of Peasant Studies* 37(1): 1–30.

Notes

1 http://uk.ibtimes.com/articles/20110315/farmers-in-developing-nations-can-double-food-output.htm
2 http://www.monsanto.com/responsibility/sustainable-ag/default.asp
3 http://www.monsantoafrica.com/layout/about_us/default.asp

References

Barnes, D. and A. Blevins. 1992. Farm structure and the economic well-being of non-metropolitan counties. *Rural Sociology* 57: 333–46.

Bello, Walden. 2009. *The Food Wars*. New York: Verso.

Bernhardt, Annette, Heather Boushey, Laura Dresser, and Chris Tilly. 2008. *The Gloves-Off Economy: Workplace Standards at the Bottom of American's Labor Market*. Champaign, IL: Labor and Employment Relations Association.

Bezemer, Dirk and Derek Headey. 2008. Agriculture, development and urban bias. *World Development* 36(8): 1342–64.

Buttel, Frederick, Mark Lancelle, and David Lee. 1988. Farm structure and rural communities in the northeast. In *Agriculture and Community Change in the US: The Congressional Research Reports*, edited by Louis Swanson. Boulder, CO: Westview Press, pp. 181–257.

Buttel, Frederick and Oscar Larson. 1979. Farm size, structure, and energy intensity: An ecological analysis of US agriculture. *Rural Sociology* 44: 471–88.

Carlsen, Laura. 2003. The Mexican Farmers's Movement: Exposing the Myths of Free Trade. International Forum on Globalization, Americas Policy Report, February 25, http://www.ifg.org/analysis/wto/cancun/mythtrade.htm, last accessed July 25, 2010.

Carolan, Michael 2008. When good smells go bad: A sociohistorical understanding of agricultural odor pollution. *Environment and Planning A* 40(5): 1235–49.

Carolan, Michael. 2011. *The Real Cost of Cheap Food. London*: Earthscan.

Chrischilles, Elizabeth, Richard Ahrens, Angela Kuehl, Kevin Kelly, Peter Thorne, Leon Burmeister, and James Merchant. 2004. Asthma prevalence and morbidity among rural Iowa schoolchildren. *Journal of Allergy and Clinical Immunology* 113(1): 66–71.

Constance, Douglas and Reny Tuinstra. 2005. Corporate chickens and community conflict in east texas: Growers' and neighbors' views on the impacts of the industrial broiler production. *Culture and Agriculture* 27(1): 45–60.

Crowley, Martha. 1999. The impact of farm sector concentration on poverty and inequality: An analysis of north central US counties. Master's thesis, Department of Sociology, The Ohio State University, Columbus, OH.

Crowley, Martha and Vincent Roscigno. 2004. Farm concentration, political economic process and stratification: The case of the north central US. *Journal of Political and Military Sociology* 31: 133–55.

Deller, Steven. 2003. Agriculture and rural economic growth. *Journal of Agricultural and Applied Economics* 35: 517–27.

Donham, Kelly. 2010. Community and occupational health concerns in pork production: A review. *Journal of Animal Science* 88: 102–11.

Donham, Kelley, Steven Wing, David Osterberg, Jan Flora, Carol Hodne, Kendall Thu, and Peter Thorne. 2007. Community health and socioeconomic issues surrounding concentrated animal feeding operations. *Environmental Health Perspectives* 115(2): 317–20.

Dosman, J., J. Lawson, S. Kirychuk, Y. Cormier, J. Biem, and N. Koehncke 2004. Occupational asthma in newly employed workers in intensive swine confinement facilities. *European Respiratory Journal* 24(4): 698–702.

Dube, Arindrajit and Ken Jacobs. 2004. Hidden cost of Wal-Mart Jobs: Use of safety net programs by Wal-Mart workers in California. UC Berkeley Labor Center, Briefing Paper Series, August 2, UC Berkeley, Berkeley, CA, http://laborcenter.berkeley.edu/retail/walmart.pdf, last accessed September 27, 2010.

Durrenberg, Paul and Kendall Thu. 1996. The expansion of large scale hog farming in Iowa: The applicability of Goldschmidt's findings fifty years later. *Human Organization* 55: 409–15.

Ellis, Frank. 2005. Small farms, livelihood diversification, and rural-urban transitions: Strategic issues in Sub-Sahara Africa. In The Future of Small Farms: Proceedings of a Research Workshop, Organized by International Food Policy Research Institute and Overseas Development Institute, pp. 135–49. Imperial College, London, http://citeseerx.ist.psu.edu/viewdoc/download?doi=10.1.1.139.3719&rep=rep1&type=pdf#page=142, last accessed March 25, 2011.

Fan, Shenggen, P. Hazell, and S. Sukhadeo Thorat. 2000. Government spending, growth and poverty in rural India. *American Journal of Agricultural Economics* 82(4): 1038–51.

Flora, Cornelia and Jan Flora. 1988. Public policy, farm size, and community well-being in farming dependent counties of the Plains. In *Agriculture and Community Change in the US: The Congressional Research Reports*, edited by Louis E. Swanson. Boulder, CO: Westview Press, pp. 76–129.

Flora, Jan, Ivan Brown, and Judith Conby. 1977. Impact of type of agriculture on class structure, social well-being, and inequalities. Paper presented at the annual meetings of the Rural Sociological Society, Burlington, VT, August.

Foltz, Jeremy, Douglas Jackson-Smith, and Lucy Chen. 2002. Do purchasing patterns differ between large and small dairy farms? Economic evidence from three Wisconsin communities. *Agricultural and Resource Economics* 31: 28–32.

Foltz, Jeremy and Kimberly Zueli. 2005. The role of community and farm characteristics in farm input purchasing patterns. *Review of Agricultural Economics* 27: 508–25.

Fujimoto, Isao. 1977. The communities of the San Joaquin Valley: The relation between scale of farming, water use, and quality of life. In *US Congress, House of Representatives, Obstacles to Strengthening the Family Farm System. Hearings Before the Subcommittee on Family Farms, Rural Development, and Special Studies of the Committee on Agriculture, 95th Congress, first session.* Washington, DC: US Government Printing Office, pp. 480–500.

Garming, H. and H. Waibel. 2009. Pesticides and farmer health in Nicaragua: A willingness-to-pay approach to evaluation. *European Journal of Health Economics* 10(2): 125–33.

Gilles, Jere Lee and Dalecki, Michael. 1988. Rural well-being and agricultural change in two farming regions. *Rural Sociology* 53: 40–55.

Glover, D. 2007. Beyond corporate social responsibility? Business, poverty and social justice. *Third World Quarterly* 28(4): 851–67.

Goldschmidt, Walter. 1968. Small business and the community: A study in the central valley of California on effects of scale of farm operations. In *US Congress, Senate, Corporation Farming, Hearings Before the Subcommittee on Monopoly of the Select Committee on Small Business, US Senate, 90th congress, 2nd session, May and July 1968*. Washington, DC: US Government Printing Office, pp. 303–433.

Goldschmidt, Walter. 1978a. *As You Sow: Three Studies in the Social Consequences of Agribusiness*. Montclair, NJ: Allanheld, Osmun and Company.

Goldschmidt, Walter. 1978b. Large-scale farming and the rural social structure. *Rural Sociology* 43: 362–6.

Gomez, Miguel and Lying Zhang. 2000. Impacts of concentration in hog production on economic growth in rural Illinois: An econometric analysis. Paper presented at the annual meeting of the American Agricultural Economics Association, Tampa, FL.

Green, Gary 1985. Large-scale farming and the quality of life in rural communities: Further specification of the Goldschmidt hypothesis. *Rural Sociology* 50: 262–73.

Grindle, Merilee. 2004. Good enough governance: Poverty reduction and reform in developing countries. *Governance* 17(4): 525–48.

Harris, Craig and Jess Gilbert. 1982. Large-scale farming, rural income, and Goldschmidt's agrarian thesis. *Rural Sociology* 47: 449–58.

Heady, Earl and Steven Sonka. 1974. Farm size, rural community income, and consumer welfare. *American Journal of Agricultural Economics* 56: 534–42.

Heaton, Tim and David Brown. 1982. Farm structure and energy intensity: Another look. *Rural Sociology* 47: 17–31.

Heffernan, William. 1972. Sociological dimensions of agricultural structures in the United States. *Sociologia Ruralis* 12(3–4): 481–99.

Heffernan, William and Paul Lasley. 1978. Agricultural structure and interaction in the local community: A case study. *Rural Sociology* 43: 348–61.

Institute for Food and Development Policy. 2009. Labor in the food system. Institute for Food and Development Policy, Oakland, CA, November 6, http://www.foodfirst.org/en/node/2620, last accessed March 9, 2011.

Irwin, M., C. Tolbert, and T. Lyson. 1999. There's no place like home: Non-migration and civic engagement. *Environment and Planning* A 31: 2223–38.

Jackson-Smith, Douglas and Gilbert Gillespie. 2005. Impacts of farm structural change on farmers' social ties. *Society and Natural Resources* 18: 215–40.

Kishi, Misa, Norbert Hirschhorn, Marlinda Djajadisastra, Latifa Satterlee, Shelley Strowman, and Russell Dilts. 1995. Relationship of pesticide spraying to signs and symptoms in Indonesian farmers. *Scandinavian Journal of Work, Environment and Health* 21: 124–33.

Kleiner, Anna. 2003. Goldschmidt revisited: An extension of Lobao's work on units of analysis and quality of life. Paper presented at the annual meeting of the Rural Sociological Society. Montreal, Quebec, Canada, July.

Liu, Yvonne and Dominique Apollon. 2011. *The Color of Food*. New York: Applied Research Center.

Lobao, Linda M. 1990. *Locality and Inequality: Farm and Industry Structure and Socioeconomic Condition*. Albany, NY: State University of New York Press.

Lobao, Linda and Michael Schulman. 1991. Farming patterns, rural restructuring, and poverty: A comparative regional analysis. *Rural Sociology* 56: 565–602.

Lobao, Linda and Curtis Stofferahn. 2008. The community effects of industrialized farming: social science research and challenges to corporate farming laws. *Agriculture and Human Values* 25(2): 219–40.

Lupton, Michael. 1977. *Why Poor People Stay Poor: A Study of Urban Bias in World Development.* London: Temple Smith.

Lyson, Thomas, Robert Torres, and Rick Welsh. 2001. Scale of agricultural production, civic enagagement, and community welfare. *Social Forces* 80: 311–27.

Lyson, Thomas and Rick Welsh. 2005. Agricultural Industrialization, anticorporate farming laws, and rural community welfare. *Environment and Planning* A 37: 1479–91.

MacCannell, Dean. 1988. Industrial agriculture and rural community degredation. In *Agriculture and Community Change in the US: The Congressional Research Reports*, edited by Louis Swanson. Boulder, CO: Westview Press, pp. 15–75.

McMillian, MaryBe and Michael Schulman. 2003. Hogs and citizens: A report from the North Carolina front. In *Communities of Work: Rural Restructuring in Local and Global Contexts*, edited by William Falk, Michael Schulman, and Ann Tickamyer. Dayton, OH: University of Ohio Press, pp. 219–39.

Marousek, Gerald. 1979. Farm size and rural communities: Some economic relationships. *Southern Journal of Agricultural Economics* 11: 57–61.

Martinson, Oscar, Eugene Wilkening, and Richard Rodefeld. 1976. Feelings of powerlessness and social isolation among "large-scale" farm personnel. *Rural Sociology* 41: 452–72.

Mayer, Brian, Joan Flocks, and Paul Monaghan. 2010. The role of employers and supervisors in promoting pesticide safety behavior among florida farmworkers. *American Journal of Industrial Medicine* 53(8): 814–24.

Miller, George. 2004. A Report by the Democratic Staff of the Committee on Education and the Workforce, US House of Representatives, February 16, http://wakeupwalmart.com/facts/miller-report.pdf, last accessed September 30, 2010.

Mills, Paul and Sandy Kwong. 2001. Cancer Incidence in the United Farmworkers of America (UFW), 1987–1997. *American Journal of Industrial Medicine* 40(5): 596–603.

Mirabelli M., S. Wing, S. Marshall, T. Wilcosky. 2006. Asthma symptoms among adolescents who attend public schools that are located near confined swine feeding operations. *Pediatrics* 118(1): 66–75.

Mitloehner, F. and M. Calvo. 2008. Worker health and safety in concentrated animal feeding operations. *Journal of Agricultural Safety and Health* 14(2): 163–87.

Monsanto. 2009. Grown for the Future: Monsanto Company, 2008–2009: Corporate Responsibility and Sustainability Report, Monsanto Corporation, St. Louis, MO, http://www.monsanto.com/pdf/responsibility/2008-2009_mon_csr_report.pdf, last accessed June 23, 2011.

Murray, Douglas. 1994. *Cultivating Crisis: The Human Cost of Pesticides in Latin America.* Austin, TX: University of Texas Press.

Murray, Douglas and Peter Taylor. 2000. Claim no easy victories: evaluating the pesticide industry's global safe use campaign. *World Development* 28(10): 1735–49.

Narrod, Clare, Devesh Roy, Belen Avendano, and Julius Okello. 2008. Standards on smallholdlers: Evidence from three cases. In *The Transformation of Agri-Food Systems: Globalization, Supply Chains, and Small Farmers*, edited by Ellen McCullough, Prabhu Pingali, and Kostas Stamoulis. London: Earthscan, pp. 355–72.

North Central Regional Center for Rural Development (NCRCRD). 1999. The impact of recruiting vertically integrated hog production in agriculturally-based counties of

Oklahoma. Report to the Kerr Center for Sustainable Agriculture. Ames, IA: Iowa State University.

Patel, Raj. 2009. *Stuffed and Starved: The Hidden Battle for the World Food System*. Brooklyn, NY: Melville House Publishing.

Perez, Mamerto, Sergio Schlesinger, and Timothy Wise. 2008. The Promise and Perils of Agricultural Trade Liberalization: Lessons from Latin America. White Paper, Washington Office on Latin America, Washington, DC.

Peters, David J. 2002. Revisiting the Goldschmidt hypothesis: The effect of economic structure on socioeconomic conditions in the rural Midwest. Technical Paper P-0702-1, Missouri Department of Economic Development, Missouri Economic Research and Information Center, Jefferson City, MO.

Pimentel, David. 2005. Environmental and economic costs of the application of pesticides primarily in the United States. *Environment, Development and Sustainability* 7: 229–52.

Poole, Dennis. 1981. Farm scale, family life, and community participation. *Rural Sociology* 46: 112–27.

Radon, K., A. Schulze, V. Ehrenstein, R. van Strien, G. Praml, and D. Nowak. 2007. Environmental exposure to confined animal feeding operations and respiratory health of neighboring residents. *Epidemiology* 18: 300–8.

Reisner, Ann, Dawn Coppin and Pig in Print Group. 2004. But What Do the Neighbors Think? Community Considerations and Legal Issues Paper, Swine Odor Management Papers number 5, March, 17. University of Illinois at Urbana-Champaign.

Richter, E.D. 2002. Acute human pesticide poisonings. In *Encyclopedia of Pest Management*, Vol. 1, edited by David Pimentel. New York: Dekker, pp. 3–6.

Roberts, Paul. 2008. *The End of Food*. New York: Houghton Mifflin Company.

Rodefeld, Richard. 1974. The changing organizational and occupational structure of farming and the implications for farm work force individuals, families, and communities. PhD dissertation, University of Wisconsin.

Rodefeld, Richard, Jan Flora, Donald Voth, Isao Fujimoto, and Jim Converse (eds.) 1978. *Change in Rural America: Causes, Consequences and Alternatives*, St. Louis, MO: C.V. Mosby Co.

Schiffman, Susan, Elizabeth Miller, Mark Suggs, Brevick Graham. 1995. The effect of environmental odors emanating from commercial swine operations on the mood of nearby residents. *Brain Research Bulletin* 37: 369–75.

Schiffman, Susan, Elizabeth Slatterly-Miller, Mark Suggs, and Brevick Graham. 1998. Mood changes experienced by persons living near commercial swine operation. In *Pigs, Profits, and Rural Communities*, edited by Kendall M. Thu and E. Paul Durrenberger, Albany, NY: The State University of New York Press, pp. 84–102.

Schlosberg, Eric. 2001. *Fast Food Nation*. New York: Houghton Mifflin.

Seipel, Michael, Katie Dallam, Anna Kleiner, and Sanford Rikoon. 1999. Rural residents' attitudes toward increased regulation of large-scale swine production. Paper presented at the Annual Meetings of the Rural Sociological Society, August.

Seipel, Michael, Mubarak Hamed, J. Sanford Rikoon, and Anna M. Kleiner. 1998. The impact of large-scale hog confinement facility sightings on rural property values. Conference Proceedings: Agricultural Systems and the Environment, Des Moines, Iowa, July, pp. 415–18.

Shiva, V. 2000. *Stolen Harvest*. Cambridge, MA: South End Press.

Skees, Jerry and Louis Swanson. 1988. Farm structure and rural well-being in the south. In *Agriculture and Community Change in the US: The Congressional Research Reports*, edited by Louis Swanson. Boulder, CO: Westview Press, pp. 238–321.

Small Farm Viability Project. 1977. The Family Farm in California: Report of the Small Farm Viability Project, Sacramento, California. Employment Development, the Governor's Office of Planning and Research, the Department of Food and Agriculture in the Department of Housing and Community Development.

Smithers, John, Paul Johnson, and Alun Joseph. 2004. The dynamics of family farming in north Huron County, Ontario. Part II: Farm–community interactions. *Canadian Geographer* 48: 209–24.

Soares, Wagner and Marcelo Porto. 2009. Estimating the social cost of pesticide use: An assessment from acute poisoning in Brazil. *Ecological Economics* 68: 2721–8.

Stofferahn, Curtis. 2006. Industrialized Farming and Its Relationship to Community Well-Being: An Update of a 2000 Report by Linda Lobao. Prepared for the State of North Dakota, Office of the Attorney General, September.

Swanson, Larry. 1980. A study in socioeconomic development: changing farm structure and rural community decline in the context of the technological transformation of American agriculture. PhD dissertation, University of Nebraska, Lincoln.

Swanson, Louis E. 1988. *Agriculture and Community Change in the US: The Congressional Research Reports*. Boulder, CO: Westview.

Tetreau, E.D. 1938. The people of Arizona's irrigated areas. *Rural Sociology* 3: 177–87.

Tetreau, E.D. 1940. Social organization in Arizona's irrigated areas. *Rural Sociology* 5: 192–205.

Thompson, N. and Haskins, L. 1999. *Searching for 'Sound Science: A Critique of Three University Studies on the Economic Impacts of Large-Scale Hog Operations*. Walthill, NE: Center for Rural Affairs.

Tonelson, Alan. 2002. *Race to the Bottom: Why a Worldwide Worker Surplus and Uncontrolled Free Trade are Sinking American Living Standards*. Boulder, CO: Westview Press.

United Nations. 2010. Agro-ecology and the Right to Food, United Nations Special Rapporteur on the Right to Food, December 20, http://www.srfood.org/images/stories/pdf/officialreports/20110308_a-hrc-16-49_agroecology_en.pdf, last accessed March 10, 2011.

van der Hoek, W., F. Konradsen, K. Athukorala, and T. Wanigadewa. 1998. Pesticide poisoning: a major health problem in Sri Lanka. *Social Science and Medicine* 46(4): 495–504.

van der Ploeg, Jan. 2008. *The New Peasantries: Struggles for Autonomy and Sustainability in an Era of Empire and Globalization*. London: Earthscan.

van Es, John, David Chicoine, and Mark Flotow. 1988. Agricultural technologies, farm structure and rural communities in the Corn Belt: Policies and implications for 2000. In *Agriculture and Community Change in the US: The Congressional Research Reports*, edited by Louis Swanson. Boulder, CO: Westview Press, pp. 130–80.

Welsh, Rick and Thomas A. Lyson. 2001. Anti-corporate Farming Laws, the 'Goldschmidt Hypothesis' and Rural Community Welfare. Paper presented at the annual meeting of the Rural Sociological Society, Albuquerque, NM, August.

Wesseling, C., L. Castillo and C. Elinder. 1993. Pesticide poisonings in Costa Rica. *Scandinavian Journal of Work, Environment and Health* 19(4): 227–35.

Wheelock, Gerald. 1979. Farm size, community structure and growth: specification of a structural equation model. Paper presented at the annual meetings of the Rural Sociological Society, Burlington, VT, August.

Whittington, Susie and Kellie Warner. 2006. Large scale dairies and their neighbors: A case study of perceived risk. *Journal of Extension* 44, Article No. 1FEA4 http://www.joe.org/joe/2006february/a4.shtml

Williamson, Stephanie. 2005. Breaking the barriers to IPM in Africa: Evidence from Benin, Ethiopia, Ghana, and Senegal. In *The Pesticide Detox*, edited by Jules Petty. London: Earthscan, pp. 165–80.

Wilson, Sacoby, Frank Howell, Steve Wing, and Mark Sobsey. 2002. Environmental injustice and the Mississippi hog industry. *Environmental Health Perspectives* 110(2): 195–201.

Wing, Steve and Susanne Wolf. 1999. Intensive Livestock Operations, Health, and Quality of Life Among Eastern North Carolina Residents. Report to the North Carolina Department of Health and Human Services. Chapel Hill, NC: Department of Epidemiology, University of North Carolina.

Wing, Steve and Susanne Wolf. 2000. Intensive livestock operations, health and quality of life among eastern North Carolina residents. *Environmental Health Perspectives* 108(3): 233–8.

Wise, Timothy. 2009. Agricultural Dumping Under NAFTA: Estimating the Costs of US Agricultural Policies to Mexican Producers, GDAE (Global Development and Environment Institute at Tufts University) Working Paper No. 09-08, Medford, MA, http://www.ase.tufts.edu/gdae/Pubs/wp/09-08AgricDumping.pdf, last accessed July 25, 2010.

World Bank. 2007. *World Development Report 2008: Agriculture for Development.* Washington, DC: World Bank.

Wright, Wynne, Cornelia Flora, Kathy Kremer, Willis Goudy, Clare Hinrichs, Paul Lasley, Ardith Maney, Margaret Kronma, Hamilton Brown, Kenneth Pigg, Beverly Duncan, Jean Coleman, and Debra Morse. 2001. Technical Work Paper on Social and Community Impacts. Prepared for the Generic Environmental Impact Statement on Animal Agriculture and the Minnesota Environmental Quality Board.

Zepeda, Eduardo, Timothy Wise, and Kevin Gallagher 2008. Rethinking Trade Policy for Development: Lessons from Mexico Under NAFTA, Policy Outlook, Carnegie Endowment for International Peace, Washington, DC, www.carnegieendowment.org/files/nafta_trade_development.pdf, last accessed July 25, 2010.

6

FOOD AND CULTURE

KEY TOPICS

- Taste and memory as sociological concepts are explored, building up to a discussion of how these phenomena are shaped by (and in turn help shape) food systems.
- As the structure of agriculture has changed so too has changed our relationship to nonhuman animals; a fact of considerable sociological consequence.
- The literature expands noticeably in the 1990s – taking what has come to be known as a "cultural turn" – to include issues related to gender, masculinity, and ethnicity.

Key words

- Agnotology
- Agricultural masculinities
- Ark of Taste
- Cultural capital
- Cultural turn
- Epistemic distance
- Geographical Indications of Origin (GI)
- Invisibility (of gender and whiteness)
- Masculine in the rural
- Modernization theory
- Moral boundaries (in-/out of-place)
- Protected Designation of Origin (PDO)
- Protected Geographic Indication (PGI)
- Risk society thesis
- Rural in the masculine
- Taste
- The third shift
- Women and Development
- Women in Development
- Women, Environment and Sustainable Development

Food is clearly more than just a material artifact; more than a mere vessel of macro and micro nutrients. Food blurs boundaries, between internal/external, self/other, and nature/culture (Gething 2010; Lupton 1996). As one noted scholar recently put it, "You are what you eat – literally – but also *how*, *when*, *where* and *why* you eat" (Goodman 2011: 250). This chapter explores in some detail food's fluid identity, as both a thought and a thing. The meanings and practices attached to food have changed considerably over the last century, for a variety of reasons. Some of the changes can be linked fairly convincingly to modifications to the structure of agriculture. These changes, and the "cultural turn" they evoked in the subdiscipline, are the subject of this chapter.

Thinking about taste sociologically

Western philosophy long ago established a hierarchy of the senses. Assumed within this hierarchy is the belief that distance – between knower and known – has a cognitive, moral, and aesthetic advantage, while a bodily sense like taste brings one dangerously close to the object of perception (Korsmeyer 1999). Plato, in his famous discussion of the cave in *Republic* VII, uses a wealth of metaphors to convey the intellectual power of sight: shadows, light, the sun, the darkness of a cave, and Forms. The belief, then as now, has been that vision, though open to creating illusions, is freer from the pull of emotions and appetite than other more corporeal forms of knowledge, like taste. For Plato, the philosophic life required the denial of these lower senses. In his *Symposium*, for example, friends gathered for a banquet. But as they philosophize about love they make sure not to eat or drink so as to keep their intellect sharp. Similarly, the denial of the more fleshy "lower" senses has been a constant theme in theological scholarship for over two millennia.

Not until the twentieth century did scholars began to think sociologically about taste. Norbert Elias (2000) devoted most of his career to uncovering the sociological underpinnings of this phenomenon. For Elias, our feelings about what constitutes good and bad manners, proper and improper behaviors, and civil and uncivil tastes and dispositions are ultimately social in nature. Yet their social nature does not make them any less objective and real. Over time, through repetitive practices, these social constructions become taken for granted and their origins less transparent – in a word they become "natural."

Elias provides considerable insight into this process, specifically in regards to those feelings and tastes that pertain to manners and our understandings of civility. For Elias, tastes do not occur randomly. Taste is strategic, as evidenced by the fact that these dispositions have been used to maintain, and when possible expand, class divisions in society. Socially desirable tastes often take time, and frequently money (e.g. taking etiquette lessons), to acquire. This point has also been famously articulated by French social theorist Pierre Bourdieu (1984: 13), who too spent much of his career talking about the strategic role of taste (Box 6.1). For Bourdieu, taste can take the form of a type of capital – what he called "cultural capital." According to both Elias and Bourdieu, taste orders society. It organizes people into groups and is therefore central to class and individual identities.

BOX 6.1 BOURDIEU ON CULTURE, CLASS, AND CONSUMPTION

Pierre Bourdieu, most notably through his book *Distinction* (1984), has had a significant impact on sociological studies on the intersection between food consumption, class, and culture. Studying the consumption practices of the French in the 1960s, Bourdieu points to empirical evidence suggesting that class-based "tastes" – in art, literature, films, and food – are an important means by which class differences are produced and maintained. On the subject of food, Bourdieu maintained that while the French working class has "a taste for the heavy, the fat, and the coarse," the palette among the upper class centered on foods that are "light," "refined," and "delicate" (Bourdieu: 1984: 185). But taste does more than just mark one's place in the social hierarchy; it reproduces those very class distinctions, acting like a barrier for those hoping to move up the socio-economic ladder.

The term "cultural capital" is important in Bourdieu's argument. What you like – your tastes – can help (or hinder) your ability to access other forms of capital, like social capital (social networks) and financial capital (money and credit). For Bourdieu, all the various forms of capital are inter-related. Tastes associated with the upper class (so-called "highbrow" culture) are often associated with artifacts that are expensive – think of caviar or an old bottle of a dry red wine. Moreover, often the tastes themselves take time to acquire, which is precisely why they are valued. If the tastes were easily learned they would not signal as strongly one's position in the class structure.

The ability to distinguish the subtle differences between wines – in terms of, say, their "notes" and "bouquet" – is often cited for its function as a class marker (Charters 2006). The acquisition of this knowledge requires time (to practice drinking wine), money (to purchase the wine and perhaps sessions at a wine tasting class), and access to particular social networks (not knowing anyone with this knowledge would make it considerably more difficult to acquire). Remember also that we are not just talking about the acquisition of knowledge but also physical routines, such as learning to "properly" swirl the glass of wine so as to release the aromas and open it up, slurping the wine, and swirling it about in one's month before either swallowing or spitting it out. These routines too take time, money, and social networks to develop before the wine tasting performance can be successfully pulled off. And why are these tastes, knowledges, and performances like a type of capital? Because, according to Bourdieu, knowing them makes it easier to socially maneuver amongst the upper echelons of society. While, conversely, not knowing them can be detrimental to one's hopes of ever acquiring access these social networks.

It is not unusual for food writers and activists (e.g. Pollan 2008; Waters 2008) to talk about how, say, local or organic food is better tasting or fresher than that provided by conventional means. Claims such as these must be treated with caution (a point explained further in Chapter 12). We know, for example, that "freshness" in the context of food is dependent on a host of carefully coordinated practices and technologies, like plastic salad bags employing modified atmosphere packaging (MAP) technology (which involves a special concoction of gases designed the retard the growth of micro-organisms) (Freidberg 2009). We also know that designates like "better tasting," "fresh," and "quality" can vary considerably based upon one's social networks and personal connections to food (Carolan 2011). Then there is the issue of affordability. Local, organic, and artisan foods are notoriously more expensive than conventional fare (though, in some cases this price disparity is narrowing). This led Guthman (2003) to at one point classify organic food as "yuppie chow," as opposed to the counter-cuisine organic food that emerged out of the counter-culture movement of the 1960s, which brings us back to Elias' and Bourdieu's point about taste as a potential marker of social status.

Food and memory

Slow Food is taken up further in Chapter 11, where I talk more about it as a social movement. Seeing as how this movement concerns itself with food culture, specifically its conservation, I feel I must, at least briefly, address it here. I'll leave a detailed description for Chapter 11. For the moment, the reader needs only to know that the movement attempts to save, among other things, food diversity, as this diversity is culturally consequential.

I was lecturing recently about Slow Food, noting that while society places importance on saving biodiversity, language, and material cultural artifacts like paintings, pottery, and historical buildings, we have yet to think about food in a similar way (Box 6.2). I could tell my students still were not buying the argument, so I decided to give them a personal example. I explained to them how I felt more closely connected with my Czechoslovakian heritage than to either my Irish or German roots. This connection, I told them, has nothing to do with my familiarity with authentic Czech music, clothing, or language. Rather, the connection I feel to this heritage is almost entirely felt through and because of food. Even my limited grasp of the Czechoslovakian language is mostly confined to either foods (such as the delicious *kolache* [a fruit-filled bread]) or food-related artifacts (such as *kuchenka* [cooker/stove]); knowledge that, coincidently, was most likely acquired while sitting around the table eating. This is what I mean when I say food diversity is culturally consequential.

A leader of the Slow Food movement describes a lesson that occurred in an Australian primary school (Hayes-Conroy and Hayes-Conroy 2008: 467). The students were given a blind taste test of homemade and commercial strawberry jams. Before the experiment they were asked which jam they thought they would like most. Most answered the homemade jams. Their taste buds, however, betrayed their expectations. The vast majority of students reported actually preferring the

BOX 6.2 THE ARK OF TASTE

You can't save polar bears by eating them but consuming rare foods can lead to their preservation. This is a core premise of Slow Food: want to save rare foods – get people excited about growing, preparing, and consuming them. Slow Food organizers therefore created an ark, like the Biblical character Noah, to help save food – the Ark of Taste. Slow Food activists recognize that every set of genes on this ark encodes for not only biological traits but also cultural practices. Launched in 1996, the Ark of Taste selects, catalogues, and promotes endangered foods, and the practices needed to keep those foods alive, on all continents, save for Antarctica. Yet, identification and promotion alone cannot save all foods that are rare. Consequently, in 1999 Slow Food created *presida*, an arm of the Ark of Taste designed specifically to assist artisans and producers of items admitted to the ark by helping them purchase equipment, create necessary infrastructure, and market their products (Allen and Albala 2007).

To qualify for the Ark, food products must be:

1. Outstanding in terms of taste (as defined in the context of local traditions and uses)
2. At risk biologically or as culinary traditions
3. Sustainably produced
4. Culturally or historically linked to a specific region, locality, ethnicity, or traditional production practice, and
5. Produced in limited quantities (by farms or by small-scale processing companies).

Product categories include (but are not limited to) beverages, cereals, cheeses, fish, fruits, honey, legumes, livestock, nuts, shellfish, spices, syrups, vegetables, vinegars, and wild game. There are currently over 800 products (from more than 50 countries) on this list.

jam they grew up eating – commercial jam. After describing this story the Slow Food member remarked: "It illustrates the need to educate their tastes [...] or they won't be able to appreciate other tastes and they will always go for the industrial products" (as quoted in Hayes-Conroy and Hayes-Conroy 2008: 467). Admittedly, this suggestion is not without its problems. I can't help but wonder: *who* does such educating and *what* tastes ought to be appreciated (and, finally, who determines this)? But it does point back to the interrelationship between food and memory. If we learn as a society to prefer industrial, highly-processed food products over, say, whole (or at least less processed) foods, how will that shape future food choices and food system trajectories? Similarly, if I lose the taste for (and fail to acquire the knowledge that would allow me to make) those *kolaches* that tie me to my Czechoslovakian heritage I would undoubtedly lose a piece of who I am. For these reasons, are not foods (and tastes) worth saving?

Food is also important in building national and regional identities in addition to remembering past events that further feed into that larger collective identity. Caldwell (2009), for example, writes about how food in the former Soviet Union emerged as a commemorative medium for past victories. Even farm implements, livestock, peasant farming traditions, and other agricultural themes had long been part of the communist mythology utilized by the state to create a collective "Soviet" identity.

All nations utilize food to build a connection to their collective national past and present. Quoting an elderly Korean woman who, since emigrating to Japan, is unable to eat the "Korean food" she knew as a child, S. Lee (2000) demonstrates just how powerful the connection is between our sense of self and the foods we eat: "My stomach and heart ache together from even just smelling Korean food, because it brings back all the hardship I have suffered in my life. Even if I wanted to forget, I cannot. My body has absorbed the past like a sponge. Forgetting is an impossibility" (S. Lee 2000: 216).

Knowing food through images and stories

I've already discussed at length the story about the changing structure of agriculture. Yet to be discussed, however, are the equally momentous technological advances in food preservation. Any move to free people from agriculture, so they can go to work in the city, presupposes the simultaneous development of techniques to keep the food from spoiling before it reaches its final destination. Thanks to food preservation techniques, people could be separated by great distances from the land, farms, and farmers that fed them.

Arguably the first modern food preserving innovation (recognizing that people had been salting, smoking, fermenting, and drying food for millennia) occurred in 1809, when Nicolas Appert won 12,000 francs from Napoleon for inventing a technique for preserving food in bottles that was then quickly adopted by the military during the Napoleonic Wars. In 1819, William Underwood applied the same principles in the USA with tin canisters – hence the term "canned food." Then Louis Pasteur's pasteurization process was invented in 1862. And the pieces for a food revolution began falling into place.

People initially viewed industrialized food with a degree of suspicion. Previous to this, food was largely something people produced themselves, or it was produced by someone familiar. Understanding of freshness and quality was consequently wrapped up in a particular set of relations, which usually involved people being in intimate contact with what they ate. So, understandably, when food began coming in a can or box most were initially suspicious. For example, W.K. Kellogg initially believed consumers needed symbols they could identify with that would give them confidence about the quality of the food they were buying. He therefore began in 1906 to print his signature on every box of corn flakes in an attempt to build trust in the Kellogg brand. In 1907, Kellogg placed an actual face on their product. Called the "Sweetheart of Corn," the illustrated image was of a young girl holding an ear of corn in each hand. Between 1905 and 1940 Kellogg spent $100 million on advertising in an attempt to further assure consumers of the quality of their product. The industry as a whole followed suit and by 1940 Americans consumed nearly as much processed foods as fresh food (Blay-Palmer 2008).

Given the stance of medical and health professionals today towards processed foods it might surprise readers to learn that this early push towards industrial food was helped along by health professionals at the turn of twentieth century. Their argument – echoed by such industrial food giants as W.K. Kellogg and C.W. Post – was that the early pioneer diet was too rich for the sedentary, urban lifestyle (Blay-Palmer 2008). What the "modern" body needed, so it was argued, was access to standardized, safe, and healthy processed foods. This argument was used quite effectively and led, for example, to the flaked cereal revolution at the beginning of the twentieth century.

Companies knew early on the importance of savvy marketing for gaining consumer trust. Consumers, for their part, sought out this knowledge too. Looking to fill in these knowledge gaps brought on by an increasingly industrialized food system, consumers turned to labels and advertisements to make sense of food. Early food manufacturers constructed a "founder" and a "founding story" through which they communicated their product. For example, with the passage of the 1870 federal trademarks law in the USA, firms could protect images and narratives designed to provide a proxy of sorts for the first-hand experience previous generation had of their food. The Quaker Oats Man was "born" in 1877, followed by Aunt Jemima (1905), the Sweetheart of Corn (1907), the Morton Salt Girl (1911), Betty Crocker (1921), and the Jolly Green Giant (1926). All sought to place a face – real or imagined – on food which had been removed by the industrialization process.

Most early food packages also contained text; indeed, by today's standards a remarkable amount of text. Its purpose, as best as I can tell, was to provide consumers additional "information" about the food they were thinking about purchasing and assuage any fears about its distant origins. From today's perspective – where the 5-second sound-bite and two or three word catchphrase dominate food adverts – it is difficult to understand why firms a century ago would have bothered offering so much detail within their ads. One must assume consumers read them. Otherwise this strategy would not have been so widely used; though I firmly doubt many consumers today would take the time to read, say, an advert containing an entire page with text about a particular brand of honey (Figure 6.1). Perhaps this gives some insight into the level of suspicion people at the time had of food raised without a face. This suspicion was so great that consumers actually took the time to read the stories contained in the ads in the hope of learning more about their food (Carolan 2011).

We also know food through the stories we tell, which in part explains its power to produce and reproduce culture and identity. In the past, these stories were

What a Stray Swarm of Bees Started Fifty-Odd Years Ago

ONE August day, in 1865, a swarm of bees passed over the jewelry shop of a young man in the little town of Medina, Ohio. Jokingly, the young man told one of his assistants he would give a dollar for the bees. The man rushed out and in a few minutes returned with the bees all snugly housed in a light grocery box. The young man paid the dollar, took the bees and, being a lover of nature, at once became intensely interested in his new friends. It was the beginning. From that day this man and his family have devoted the best part of their lives to the study of bees, their lives, varieties, habits, culture and housing and to the betterment of their wonderful product, honey, through improvement of bee colonies and their surroundings. For his well-proved theory always has been that if you improve the worker and the worker's living conditions you're bound to improve the product.

The name of that man is A. I. Root, who stands today as the head and father of the bee industry in America. Recognized and quoted as an authority on bees and honey in that great work the Encyclopedia Britannica. In their long article on bees, reference after reference is made to A. I. Root, the A. I. Root Company and to Mr. Root's book on bees entitled "A. B. C. of Bee Culture." These references and illustrations are to be found in Volume 3, Eleventh Edition of the Encyclopedia; pages 630-632-633-634-636 and on page 638, where his name comes under the heading, "authorities."

Today, the house founded by A. I. Root and his stray swarm of bees is acknowledged to be the biggest producer of and dealer in bees, bee products and bee-keeper's supplies in the world. It is this firm, Mr. Root's pride and pleasure — a house maintaining firmly the ideals of its founder and operated entirely by his family—that offers you and guarantees to you the quality and purity of Airline Honey.

Though honey has always been used by many housewives in preference to sugar, the present situation has done much to teach vast numbers of others the superiority of honey.

The war has raised the price of sugar until it is an item well worth considering.

Honey makes you partially independent of sugar. Not only is it cheaper than sugar, but it is better for many things you're in the habit of using sugar for. It's more wholesome, it is more delicious and it gives better results.

Honey makes the finest, clearest preserves you ever tasted, and preserves the fruit better.

There never was such delicious cake, or cake that stays moist so long as that made with honey — nearly all kinds of cakes and cookies.

And candy! Just try some candy made with honey and you'll always use honey in making candy thereafter.

Spread honey on your batter cakes, waffles, bread and biscuits — there's no flavor in the world like it.

Use Airline Honey for many things in the way of cooking and sweetening and you'll rejoice in your discovery of honey's superior goodness — you'll benefit by its superior wholesomeness and profit by its economy.

You Need This Cook Book

It contains over 100 recipes for things in which Airline Honey is used instead of sugar. Also general directions giving the amount of honey to be used in replacing sugar as a sweetening, for one uses less honey than sugar. Send us your grocer's name and the book will go to you, free.

Send for Our Trial Jar

of Airline Honey. A generous quantity of the extracted honey in an individual jar. Send 10 cents for this treat—it will make you a honey enthusiast.

We Have Some Candy for You

—Delicious honey candy—several kinds—neatly and substantially packed. Ten cents in stamps to pay packing and postage will get you this candy. Send for it.

You can buy Airline Comb Honey in air-tight packages or extracted in glass jars (several sizes) with patent easily removable tops, at good grocers. Served in individual packages on most all railroad dining cars, at leading hotels and restaurants. This in itself is a striking endorsement of Airline quality — these people seek only the best and purest.

THE A. I. ROOT CO. Medina, Ohio

"The Home of the Honey Bees"

FIGURE 6.1 1917 advertisement for Airline Honey

often "entangled" (Goodman 2011; R. Lee 2006) in the physical, social, and symbolic acts of raising, harvesting, preserving/storing, and consuming food. Food today still comes with stories. Yet as our physical relationship to food has changed so too have these narratives. Many people, after all, still go to the grocery store looking for a good story. Perhaps it is a tale of how many miles a food has traveled before reaching one's dinner table. Or the story of the organic label, telling consumers about how a particular commodity was raised and processed. Or, following the a-picture-is-worth-a-thousand-words logic, some consumers may merely be looking for a friendly face – an Uncle Ben or a Betty Crocker – that they've known since childhood.

Speaking of a good story . . .

Geographical indications: GIs, PDOs, and PGIs

Food labels have changed considerably since the days that gave birth to such icon figures as the Morton Salt Girl and the Jolly Green Giant. Labels from a century ago seemed to mask specifically where food was coming from. There were no images of, or descriptions about, say, the specific slaughterhouse or farm from whence the food came (and for good reason, if Upton Sinclair's book *The Jungle* has any truth to it, which described the working conditions of the former in nightmarish terms). Today, however, some labels attempt to do exactly this, while also seeking to promote rural development, preserve local artisan techniques, and thus conserve, in a word, culture.

Geographical Indication of Origin (GI) is an example of just such a label. While there is no universally agreed upon definition of GI, the following captures its general spirit:

> A Geographical Indication identifies a good as originating in a delimited territory or region where a noted quality, reputation or other characteristic of the good is essentially attributable to its geographical origin and/or the human or natural factors there.
>
> *Giovannucci* et al. *(2009: 2)*

GIs serve multiple functions. They can promote rural development while simultaneously preserving local foods and artisan practices, culture, and tradition (Barham 2007; Stevenson and Born 2007). They have been called an example of "glocalization" (Giovannucci *et al.* 2009: 3), in that they require participation in global markets while and at the same time supporting local culture and economies. GIs help products distinguish themselves from the otherwise undifferentiated mass that is bought and sold based largely on the commodity's price. This gives GI designated commodities a competitive advantage – and the region supplying the commodity a monopoly – precisely because no one else in the world can produce that particular branded good.

Worldwide, there are over 10,000 protected GIs with an estimated trade value over US$50 billion (Giovannucci *et al.* 2009). The more well-known GIs include Darjeeling tea (which must come from the Darjeeling region in West Bengal,

India), Bordeaux wine (any wine produced in the Bordeaux region of France), Parmigiano-Reggiano cheese (named after the producing areas near Parma, Reggio Emilia, Modena, Bologna, Mantova), and Idaho potatoes (which can only be grown in the US state of Idaho). These material objects can still be produced anywhere in the world, like champagne, though it's usually called something like "sparkling wine," as "champagne" is a protected GI. That said, producers can get away with designations like American (or Australian, New Zealand, etc.) produced "camembert," "champagne," "feta," "gruyere," and "sangria," even though these titles are all GI protected. Roughly 9 out of 10 of all GIs are located in the 34 Organization for Economic Co-operation and Development (OECD) countries. This should not come as a surprise, as establishing GIs often come at some expense, in terms of organizational and institutional structures, infrastructural requirements, and ongoing operational requirements (like marketing and legal enforcement) (Josling 2006). As a rural development tool for less affluent nations, then, GIs might have limited impact.

The scope of protections offered by a GI is extensive and not limited only to a specific geographic name, like Darjeeling tea. A GI can also be a picture, like that of a Cornish pasty for genuine Cornish pasties, the outline of a geographic area, such as the outline of the state of Florida for Florida oranges, or almost anything else that can identify the source of the commodity. GIs are secured through multiple layers of protection. This can include, but is not limited to, trademark laws, consumer fraud protection laws, and laws that explicitly recognize (and thus protect) individual GIs.

Two examples of GIs are Protected Designation of Origin (PDO) and Protected Geographic Indication (PGI). These legal frameworks are specific to the European Union (EU). PDOs and PGIs differ most significantly in the extent of their link to a specific geographic region. A PDO product must be composed of materials originating from the area, have characteristics attributable to the locale (associated with, say, climate or soil type), and be produced and processed in the defined GI region. A PGI, in contrast, only need to have at least one of the production or processing stages take place in the defined area. In most other respects PDOs and PGIs are the same, from the application process to control systems and consumer guarantees (Giovannucci *et al.* 2009).

Sociological significance of the human–animal relationship

Talking about animals in a chapter on culture might seem, to the uninitiated, a little odd. Its placement reflects the fact that when sociologists talk about animals they are not talking about nonhuman animal societies *per se* – after all, the field is usually called the "sociology of animals" never "animal sociology" (although socio-biologists occasionally evoke the latter term). When social scientists study animals what they are really interested in is our (ever-changing) associations with them, namely, the human–animal *relationship*. And as this relationship is arguably a cultural artifact – as cultures clearly vary in how they relate to, think about, and treat animals – it deserves inclusion here.

One strand of this literature draws directly on the work of twentieth century social and cultural anthropologist Mary Douglas and her research into societal definitions of "pollution." According to Douglas, these definitions are culturally variable and reflect societal views of defilement or disorder. This led to her definition of pollution as "matter out of place" (Douglas 1966: 35). Public understandings of pollution thus reflect socially defined moral boundaries – that is, of what is (and is not) natural and thus right. These moral boundaries are defined spatially, in terms of phenomena being "in place" (e.g. the good/the right) or "out of place" (e.g. defilement/the bad/the unnatural).

This framework has been used to understand our changing relationship and attitudes toward animals by explaining how animals (and in particular livestock) were slowly excluded from city spaces in the late eighteenth and early nineteenth centuries. Initially, this shift in public understandings of animals within city spaces had links to public health discourses and fears of moral degeneracy. In terms of the former, health reformers were looking to cleanse the filthy urban environment, with animal secretions (e.g. odor, manure) being at the top of their list of prime offenders. As for the latter, there was a fear that humans could become morally debased if they were frequently exposed to the unregulated urges of the uncivil beast (Philo 1995).

Thus began the process of extricating animals from the urban environment; a process that has been of sociological consequence to how we understand them and the food they provide. Previous to this, animals and humans regularly lived in close proximity to one another. Not only did humans and animals share urban environments, but animals readily transgressed boundaries between the "in here" of the home and the "out there" found beyond one's doorstep. Centuries ago it was not uncommon to find, for example, chickens, goats, and pigs scurrying about one's home, particularly among the peasant class (and in some parts of the world we still see this). It was not only peasants who allowed such transgressions to occur, kings too were known to favor having animals in their place of residence. The discovery a few years ago of three lion skulls in the Tower of London points to the location of the royal menagerie back to the twelfth century. In addition to lions, the menagerie is known to have held an elephant (although only briefly), ostriches, bears, leopards, and tigers (Owen 2005).

By the early nineteenth century, not only was it becoming unacceptable to allow animals into one's home (save for "pets"), it was also becoming increasingly objectionable for animals to be located near large populations of people – hence, their expulsion to the countryside. In both cases, within the "in here" of one's home and the "in here" of the city, animals had become "out of place." The countryside, in many respects, now represents the last bastion of space – except for zoos, stockyards, dog parks, and the like – where animals are still be considered "in place."

Yet this is also changing. Recent trends threaten to make animals "out of place" even in the countryside. The cause of this most recent movement can be traced, at least in part, to structural changes within agriculture and the parallel movement of nonfarmers into the countryside.

Take the consolidation of the hog industry in the USA, which (with poultry) has been at the forefront of livestock consolidation trends. While the number of hogs in the USA has changed little over the last century, the number of farms raising hogs has decreased precipitously. The number of farms producing hogs in the USA has declined from almost 700,000 in 1980 to approximately 75,000 in 2007. At the same time, the percentage of hogs raised by the largest sized farms (greater than 5,000 head inventory) rose from 20 percent of total US production in 1992 to approximately 60 percent in 2007. As livestock become concentrated in fewer and fewer areas, many other areas are becoming sanitized as animals are disappearing from them.

Contrast this to the experience of our ancestors, who, at least among those residing in the countryside, lived in very close proximity to animals. This is becoming less the case today. For example, poultry has gone from being raised on approximately 78 percent of all farms in the USA to just less than 4 percent during the latter half of the twentieth century. For hogs, that figure, during the same time period, went from 56 percent to less than 4 percent of all farms. Dairy cows likewise went from being raised on 68 percent of all farms in mid-century to less than 4 percent of all farms 50 years later. Similar trends have been recorded throughout the developed world, from, for example, Canada, to New Zealand, Australia, and England. These trends have created a conception of rural life remarkably devoid of livestock; a conception of rural life where even livestock are increasingly becoming "out of place" (Carolan 2008).

These trends also lie at the center of a number of arguments pertaining to how we relate to and thus treat animals. Philosopher Hub Zwart (1997) argues that the conventional Western view of animals, where they are reduced to amoral machines, has helped legitimize their objectification and sustained the view that they are resources for human use. Elsewhere I have written on what I call "epistemic distance" (Carolan 2006). I coined this term to speak specifically about how many of the benefits of sustainable agriculture cannot be immediately experienced – for instance, improvements in soil fertility and tilth take time to occur after one adopts more sustainable farm management practices. Conventional agriculture, conversely, has witnessed such success in part because many of its benefits are immediate in comparison to its costs, like applying fertilizer at high application rates and witnessing increases in yields. But the term has wider applications. When taken as a whole, food production has become epistemologically distant from the average consumer. This distance benefits the food system status quo because it creates gaps and distortions in terms of the information available to consumers (Carolan 2011). This distance, one could argue, also has an impact of our ethical orientation toward distant phenomena, as it makes it easier for us to act with moral indifference toward those faraway people, places and things that provide us to with what we eat.

Beck (1992) has written at length about how nonknowledge is an inevitable product of the expansion of capital and techno-scientific rationality. Following Beck's thesis – what is famously known as the "risk society thesis" – this unawareness allows for the global flow of information and technologies. According to Beck, risks today are an ever-present and largely unavoidable by-product of modernity with its associated technological and organizational forms. Take our global food system. While production efficiencies and scales of economy make food today remarkably inexpensive, they also presuppose an organizational form that is inherently risky. As slaughter facilities, for example, continue to increase their speed and improve their efficiency mistakes will inevitably happen. Increasing production line speeds increases the risk of cutting open internal organs while the animals are being eviscerated, thus exposing meat to fecal matter. And as the meat from different animals becomes mixed in processing facilities, so increase the chances of cross-contamination as the bacteria of one animal suddenly finds itself spread across thousands.

Another term is helpful here: agnotology. Coined by Proctor (1996), agnotology refers to manufactured ignorance. A prime example of agnotology is the tobacco industry's decades-long public relations assault that involved producing doubt about the cancer risks of tobacco use (Michaels and Monforton 2005; Proctor 1996). But agnotology need not only refer to ignorance induced by lies, deceptions, and distortions. More generally, agnotology serves to alienate us from other ways of knowing that may challenge conventional patterns of social organization.

Let's apply this framework to how we think about "livestock" – a word that itself seems to reduce animals to a mere resource. In the book *Chicken World: Profitable Poultry Production*, published back in 1910, M.G. Kains tells the reader to think of the hen as another piece of machinery. When calculating production costs one must "not only [think of] the cost of the egg as a market commodity, but the cost of making the machine, the hen, which is to manufacture the egg" (Kains 1910: 18, as quoted in Squier 2006: 76). A nonchicken example where the machine trope is used to describe livestock can be found in the book *Old MacDonald's Factory Farm*, which explains that "the breeding sow should be thought of as, and treated as, a valuable piece of machinery whose function is to pump out baby pigs like a sausage machine" (Coats 1989: 32). Is it not reasonable to assume that in thinking of animals as machines we'll eventually start treating them as such?

The gender, masculinity, and ethnicity "turn"

Having already discussed a number of "turns" in the sociology of food and agriculture literature, I would like to introduce a couple more. In the 1990s, the boundaries of agrarian political economy that dominated scholarship in the 1980s opened up allowing for what others have called a "cultural turn" in rural studies scholarship (Buttel 2001: 172). "The thrust of cultural-turn rural studies," explains Fred Buttel (2001: 172), "has been less to resurrect rural sociological voluntarism or contest the notion that power relations are central to understanding the character of rural and agri-food systems than to add new tools to the analysis of rural power relations." You have already been given a taste of this turn in preceding sections. But there is another body of scholarship that deserves its own "turn" status: that which focuses on gender and ethnicity. Like the aforementioned cultural turn, the gender and ethnicity turn is interested in analyzing power in novel ways. And, like the cultural turn, its roots can be traced to the early 1990s (see, e.g., Symes 1991; Whatmore 1991), if not earlier (see, e.g., Sachs 1983).

Studying gender

The earliest research looking at gender in the context of rural sociology focused primarily (if not exclusively) on unequal gender divisions of labor in agricultural production. This research points especially to the importance of farm wives' economic contribution to the rural household and rural economy more generally, even though the labor tended to be undervalued and invisible (when was the last time you heard the term "farm husband?") (see, e.g., Alston 1995; Berlan-Darqué and Gasson 1988; Buttel and Gillespie 1984; Sachs 1983). Changes to the structure of agriculture over the last quarter of a century are making it increasingly difficult in affluent countries for farms to survive without a supplemental source of income. Often, it is the wives who work off-farm to provide this alternative source of income (as well as retirement and health care benefits). This fact seems to have bolstered the position of women in (at least some) farm families. There is some research indicating that the financial independence wives feel when working off-farm positively impacts their psychological well-being (Rogers and DeBoer 2001). Furthermore, O'Hara (1998), studying farm families in Ireland, concludes that off-farm work empowers women to have a larger role in the decision-making process when it comes to household and farm-related matters.

Others are hesitant to describe women's off-farm employment as empowering. They point out that the off-farm work taken by women is often low-paying (Mammen and Paxson 2000). Working off the farm also further marginalizes women's role on the farm, reducing their involvement to tending animals and other tasks deemed menial (Saugeres 2002a). Some go so far as to write "that it is women's off-farm work that keeps farming male" (Shortall 2002: 171). Moreover, women working off the farm continue to be responsible for child care and most household work while also contributing to farm chores and administrative work. In this context, off-farm work, rather than empowering, is yet another job that women are expected to do – what has been called "the third shift" (Gallagher and Delworth 1993). Others note that the historically gendered nature of farming and control over resources greatly mediates (and typically negates) the emancipatory effect of women's off-farm income (Kelly and Shortall 2002; Shortall 2002).

Female-headed farms earn, on average, less than male-headed farms in affluent countries (Leck 1993; Weinberg 2000). There is also evidence that women have trouble establishing themselves in conventional agriculture and its associated male-dominated social networks (Carolan 2005). Female farmers (unlike their male counterparts) are also responsible for child and household care, while often also working off-farm, as female-operated farms are on average considerably smaller than male-operated farms. Not surprisingly, then, female farmers have less leisure time and get less sleep than male farmers (Gallagher and Delworth 1993).

It is worth noting that most of the research into gender in the sociology of food and agriculture literature centers on farming in developed regions, most notably North America, the UK, the EU, and the Antipodes (Australia and New Zealand). In these parts of the world, farming is widely thought of as a historically male occupation. This is a cultural artifact that is not universally held in every country. In fact, women farmers produce more than half of the food grown in the world (and close to 90 percent of all food grown in Africa). Approximately 1.6 billion women depend on agriculture for their livelihoods (Box 6.3). Yet even though the majority of farmers in many developing countries are women, most are unable to benefit from agricultural funding and participate

BOX 6.3 WOMEN IN DEVELOPMENT, WOMEN AND DEVELOPMENT, AND WOMEN, ENVIRONMENT, AND SUSTAINABLE DEVELOPMENT

While arguably not part of the sociology of food and agriculture literature *per se*, there is a rich research tradition that explores the links between gender and agriculture in the field of international development. Three of the most popular frameworks operating within this tradition are Women in Development (WID), Women and Development (WAD), and Women, Environment and Sustainable Development (WED).

The oldest and arguably still most dominant tradition is WID. Its programs are informed by modernization theory, which casts traditional societies as authoritarian and male-dominated and modern societies as democratic and market-driven. WID is therefore sensitive to the oppression faced by women in less affluent nations and works toward the full integration of women into these societies as workers and producers. WID is most interested in opening up the formal economy to women and pays less attention to the informal economy (e.g. inside the home), believing the democratization of the former will naturally result in the democratization of the latter.

WAD emerged as a critique of modernization theory and the WID approach. WAD argues for the need to also focus on the private sphere within the home, highlighting women's role as unpaid domestic laborers. WAD proponents argue that without paying attention to the domestic sphere and valuing the work done there women will not achieve true equity to men.

WED argues women must be taken into consideration when initiating and planning sustainable development programs. The rationale for this is simple, as women have a close working relationship with the natural resources in developing countries. Women do most of the farming and water and wood gathering in these societies for their households. Women must therefore be centrally involved in the process of developing and implementing strategies that produce development that meets the needs of the present generation without compromising future generations' ability to meet theirs.

in the market due to institutional and cultural barriers, including lack of access to land, credit, and education (Motzafi-Haller 2005; Quisumbing and Pandolfelli 2010).

The subject of gender in agriculture is revisited in Chapter 13, specifically to discuss how it remains on the margins of sociology of agriculture (Allen and Sachs 2007). This critique does not apply, however, to the sociology of food and food consumption in particular. There is a rich literature detailing how and why food consumption is gendered – ever noticed how Western diets have tended to view steaks, hamburgers, and beer as masculine foods while salads, yogurt, and fruit have long been associated with what women eat (see, e.g., Sobal 2005)? Decades of research also connects the dots between food consumption and the female body image (see, e.g., Bordo 1993; Lupton 1996). More recently, many of those same scholars have turned to their gaze to men, noting how they too consume foods according to cultural codes that describe how the masculine body ought to look (see, e.g., Bordo 1999). Indeed, gender scholarship on the whole has, in the last 20 years or so, expanded its gaze, as evidenced by the somewhat recent emergence of the field of rural masculinity studies.

Studying masculinities

The beginning of rural masculinity studies has been traced to 1995, with Brandth's pioneering research of masculinities and farming practices in Norway (Coldwell 2009). Another memorable moment for this literature is 2000, when a special issue on rural masculinities in the journal *Rural Sociology* (co-edited by Campbell and Bell) was published. The publication of rural masculinities research increased substantially after this special issue.

Masculinity research should not be viewed as a departure from earlier gender studies but rather an extension of it. Gender research seeks to understand gender relations, specifically unequal distributions of power between men and women, which lead to the latter becoming marginalized. This marginality – this otherness – is relational in nature (Cloke and Little 1997), which means there is nothing "natural" about it. As a sociological artifact, *all* involved parties and practices need to be studied – the subordinate as well as the dominant position. Understanding this otherness of traditional gender studies requires an investigation into how and why some identities have become legitimate while others have not, which inevitably has led researchers to look more closely at men. Tying this insight back to food and agriculture scholarship, we might say that "if some identities get to be legitimate, stable and natural whereas others are marginalised and devalued, then it is likely that farming practices might also be organised in a similar way" (Coldwell 2009: 175).

As Coldwell's (2009) review of rural and agricultural masculinity scholarship details, the literature on the subject is diverse. This empirical and conceptual diversity makes a thorough summary of the literature impossible in the space of a couple of paragraphs (or even a couple of pages). Rather than attempt the impossible, I will try to give readers a taste of this diversity: Table 6.1. Building upon Coldwell's thematic organizational structure, the table breaks the rural and agricultural masculinity research into the following categories: "masculine in the rural," "rural in the masculine," and "agricultural masculinities." The final category, "agricultural masculinities," is then broken down further, into "farmer identity," "farming technologies," "crisis in agriculture," and "sustainable agriculture."

The categories "masculine in the rural" and "rural in the masculine" come from Campbell and Bell (2000). While ultimately an analytic distinction (which means the categories are admittedly artificial), it is a useful reminder that "masculinity" and "rurality" are identities that individuals work at performing. The "masculine in the rural" refers to the different ways in which masculinity is produced, maintained, and contested in those spaces rural social scientists deem "rural." Putting it another way: "masculinity" might mean different things depending on whether you are in a small rural community or in a large urban center. The "rural in the masculine," in contrast, refers to those symbolic codes embedded in our collective understanding of "rural" that also constitute masculinities. There are, in other words, a number of things associated with rurality that are also deemed masculine, like pickup trucks, tractors, guns, cowboy hats, chewing tobacco, and the like. The concept the "rural in the masculine" allows us to talk about how "rural" *itself* has become coded with particular masculine signifiers. Thus, by evoking those codes one can simultaneously perform both "rurality" and "masculinity."

Studying race and ethnicity

The issue of race and ethnicity is no less complex when it comes to the sociology of food and agriculture literature. For example, some arguments surrounding "fat acceptance" assert that representations disparaging large bodies must reject this universalizing (white/privileged) discourse and allow for a diversity of body sizes (and ethnicities, class positions, etc.) (Azzarito 2008). Others look at the how race has been portrayed in food advertisements, cookbooks, and ethnic dishes. Zafar (1999), for instance, wonders if the updating of the image of Aunt Jemima on the pancake box reflects a broader change of the collective American consciousness about African-Americans and their relationship to food. Food also helps feed racial identity. "Eating black" (Slocum 2011: 4) remains a source of connection for some African-Americans as their ethnic and cultural identities become more diverse. Consuming soul food has been found to be, at least for some African-Americans, an expression of racial pride and solidarity as its roots trace back to slavery (Bailey 2007). For others, particularly younger African-Americans, connections to black history through food appears to be on the decline, as soul food is decried by some health professionals as being too high in saturate fats and sodium (Zafar 1999). Elsewhere it has been noted how Black Muslims of the 1970s condemned soul food, as it represented a "the diet of a slave mentality"

TABLE 6.1 Summary of a sample of rural and agriculture masculinity scholarship, categorized by focus

Study	*Research site*	*Findings*
Masculine in the rural		
Kenway *et al.* (2006)	Four rural towns in Australia	Globalization is influencing masculine identities and the lives of young men in rural Australia in uneven ways. Authors find no singular masculine identity or a unitary notion of hegemonic rural masculinity. For example, in one of the rural towns studied, young men remain bound to traditional notions of masculinity and to the heavy industrial work that is slowly disappearing. In other towns studied new industries like tourism are emerging, creating new employment opportunities and with that new acceptable masculinities.
Campbell (2000)	Pubs in rural New Zealand	Clear codes and norms that govern who may drink, where they may drink, and how much they may drink in rural pubs. In addition to gender, localness and labor were shown to be important entry requirements for pub drinking. Men had to display their masculinity by how they talked and how much they could drink without showing the effects. Men who control the public drinking space of the rural pub are also influential in the wider community. Campbell suggests that the act of pub drinking expresses and helps solidify power for certain men in the community.
Rural in the masculine		
Campbell and Kraack (1998)	New Zealand	New Zealand beer advertisements show "the rural" as the opposite of "the urban," in that it is problem-free, friendly, clean, egalitarian, and peaceful. Yet it is also shown as rough and rugged and in need of being brought under human (read: male) control. Beer advertisers thus exploit a perceived inherent masculinity contained within "the rural."
Campbell *et al.* (2006)	USA, New Zealand, and Australia	References politicians in the USA and Australia, noting how leaders (like former US President George Bush) often draw upon the imagery of "the rural" – with things like a cowboy hat and boots, rolled up sleeves, and pickup truck – to demonstrate and reinforce masculine identity.

Agricultural masculinities: farmer identity		
Saugeres (2002b)	France	Many of the men in this study feminized the land they were farming – that is, they likened it to a nurturing mother or a sexualized object that is potentially threatening. While not all the farming men interviewed felt this way, they all defined themselves as legitimate farmers. This was not the case with women; a position reinforced by discourse suggesting that women's bodies, compared to men's, are inadequate for the physical work of farming. The author argues that the subjectivities of men and women on the farms are largely shaped through discourse and socialization processes that reflect hard to change cultural stereotypes.
Kroma (2002)	USA	Analyzed petrochemical advertisements between 1975–1976 and 1995–1996 in the magazines *Wallace's Farmer* and *Successful Farming*. Found advertisements to be filled with masculine symbols. For example, 68 percent of the advertisements had names that metaphorically reflect dominant male imagery. These names for pesticides included "Bicep," "Counter Lock 'n' Load," "Lasso," "Marksman," "Prowl," "Squadron," and "Warrior." In response to changing environmental values, by the 1990s the advertisements also evoked feminine imagery, in an attempt to soften the product's image when it came to issues of environmental impacts.
Agricultural masculinities: farming technologies		
Saugeres (2002a)	Southern France	Machines reinforce men's distance from women while helping to establish their connections to other men. The tractor, in particular, has become an important symbol of masculine power and stands as a boundary between men's and women's work on the farm. The removal of women from the field is viewed by respondents as a consequence of mechanization and not a consequence of men's appropriation of women's work in the fields. Mechanization could have enabled women to do more field work. Instead, the process removed them from the space.
Brandth (1995)	Norway	Farming is moving away from the physical to the technological and businesslike. The author points to the tractor when making this argument. The driver's seat has become more like an office chair, positioning the farmer within easy reach of the instruments. The cabin is enclosed against the wind, noise, and dust. Heating and cooling systems are like those in an office and noise levels are now so low that the driver can easily listen to the radio. The tractor is presented not just as a symbol of conquering and controlling nature, but also as a comfortable indoor workplace. The author suggests that these changes mirror a changing masculinity in farming towards a less physical, more white-collar image.

TABLE 6.1 *(cont'd)*

Study	*Research site*	*Findings*
Agricultural masculinities: crises in agriculture		
Ni Laoire (1999, 2001, 2005)	Ireland	Examines impact of rural restructuring on young farming men's masculine identities and the struggles they face. Many of these men will inherit their farm. But this comes with duty and responsibility: namely, that they stay behind and keep their family farms going while others, typically women and those with better education, leave. Even the language of "staying behind" implies for some of these men a sense of failure and exclusion. While cautious not to draw any distinct lines of causation between high rates of suicide in young men and rural agricultural change, the author notes that agriculture is the foundation upon which male farmers have traditionally based their identities. Thus, when agriculture changes some men will struggle to adjust.
Courtenay (2006)	USA	Links rural men's health outcomes to images of hegemonic masculinity. Rural men often deny their need for health care and deny the existence of stress and psychological illness – even in times of agricultural crisis – because they expect to be rewarded with money and respect in the community and because they believe that's just what "real" men do.
Agricultural masculinities: sustainable agriculture		
Liepins (2000)	Australia and New Zealand	Identifies in rural media a hegemonic discourse around the "tough" male farmer. Farmers, in this context, battle nature, possess physical strength, and display a rugged individualism. The author further identifies the emergence of a new masculine farming identity that emphasizes entrepreneurial over physical skills, that is committed to more sustainable agricultural practices, and which advocates for more egalitarian on-farm gender relations.
Peter *et al.* (2000)	USA	The shift to sustainable and organic farming heralds a new masculinity. Unlike conventional farmers, men practicing sustainable farming techniques are more likely to reveal their mistakes and emotions as well as less intent on controlling and dominating nature. Discourses of individualism, physicality, and toughness are less central to men involved in sustainable farming, distinguishing them from men who farm conventionally, who define their masculinity in terms of doing battle with the land and through the machines that they use. Men practicing sustainable agriculture are also more open to gender equality.

Table assembled with help from Coldwell (2009).

(Wit 1999: 260), while the Black Panther Party celebrated it (Slocum 2011: 4). There is also some evidence that people of color who reject vegetarianism and veganism do so because those lifestyles represent food choices of the privileged (Bailey 2007).

Race and ethnicity also comes into play when discussing the issue of food access. As discussed in Chapter 4, people of color are disproportionately located in food deserts. Research also suggests that people of color – and African-Americans in particular – are less likely to participate in farmers' markets and community supported agriculture (CSA). While sociologists of food and agriculture have a lengthy track record examining issues related to class, race and ethnicity remain underexplored in this literature. The most comprehensive study of farmers' markets to look at race was conducted by the Agricultural Marketing Service of the United States Department of Agriculture (USDA) (Payne 2002). This nationwide study found that on average 74 percent of farmers' market attendees were white, 14 percent African-American, 5 percent Asian, and 6 percent Hispanic; though, as detailed in Table 6.2, racial composition of the markets varied considerably by region. While an equally comprehensive study of CSA does not yet exist, there is considerable anecdotal evidence that the majority of its participants are white (see, e.g., Hinrichs and Kremer 2002; Guthman 2008a; Slocum 2008).

Slocum provides us with some of the most novel research on the inter-relationships between alternative food provisioning spaces (like farmers' markets) and racial identities. In a 2006 article, for example, she discusses how and why Local Food movements have been slow to tackle issues of white privilege (Slocum 2006). She argues that this failing is the result of the invisibility of whiteness; least we forget, "whiteness" too is a racial – and cultural – category. Think about the foods found in these spaces and how their presence reflects ethnic and cultural categories. If, for example, the vast majority of the foods for sale at a given farmers' markets or CSA are fruits, vegetables, and spices that constitute standard fair in European diets (and recipes), and little is available for, say, the preparation soul food, an argument could be made that this space is shaped by an underlying ideology of "whiteness." Slocum also notes that there is resistance within the local movement to talk explicitly about race, out of fear of sounding racist.

Elsewhere, in a study of a Minneapolis farmers' market, Slocum (2008) highlights how a politics of racial identity – and middle-class "whiteness" in particular – is continually played out within this space as definitions about local consumption are shaped predominately by individuals from white, middle-class communities. To put it simply (and greatly oversimplifying the dynamic): white, affluent consumers attend farmers' markets, purchase certain foods, and thus send signals to farmers about what to grow, which in turn slowly restricts the availability of alternative foods for minority groups. Supply in these spaces does not just reflect demand; supply also *reshapes* demand by ignoring certain populations and privileging others. Similar arguments are found in Guthman's (2008b) analysis of food justice projects, where growing, donating, and educating African-American communities about food production and consumption were found to reflect the "white desires"

TABLE 6.2 Customer's race in US farmers' markets, by region

Region	*White (%)*	*Black or African-American (%)*	*Native American (%)*	*Asian (%)*	*Native Hawaiian (%)*	*Other (%)*
Far West (Alaska, California, Hawaii, Nevada, Oregon, and Washington)	69	5	1	10	1	13
Rocky Mountain (Arizona, Colorado, Idaho, New Mexico, Montana, Utah, and Wyoming)	77	7	2	3	Less than 1	10
Southwest (Arkansas, Louisiana, Oklahoma, and Texas)	69	14	1	3	1	13
North Central (Illinois, Indiana, Iowa, Kansas, Michigan, Minnesota, Missouri, Nebraska, North Dakota, Ohio, South Dakota, and Wisconsin)	81	9	1	5	Less than 1	4
Southeast (Alabama, Florida, Georgia, Kentucky, Mississippi, North Carolina, South Carolina, and Tennessee)	75	19	1	2	Less than 1	2
Mid-Atlantic (Delaware, District of Columbia, Maryland, New Jersey, Pennsylvania, Virginia, and West Virginia)	68	25	Less than 1	3	Less than 1	3
Northeast (Connecticut, Maine, Massachusetts, New Hampshire, New York, Rhode Island, and Vermont)	84	9	1	3	Less than 1	3

Source: Based on Payne (2002).

of the volunteers more than the actual needs of the communities they sought to serve.

Transition . . .

I have yet to come across a sufficiently complete definition of "culture," which is why I am reluctant to provide one here. Culture can be expressed in many ways. It involves ways of thinking (such as about things like taste), as well as ways of doing (such as in the case of performing masculinities). We can even see culture, if we look hard enough, in the very foods offered for sale at farmers' markets, such as in foods that reflect eating white (versus, say, eating black). Yet food and agriculture link up with culture in still other ways that I have yet to address. The following chapter continues this exploration into culture by looking more specifically at its links to agrobiodiversity.

Discussion questions

1 How have our tastes as they relate to food changed over the years? Can these changes be explained using the concept "cultural capital"?
2 The majority of the agrifood literature looking at race comes from scholars located in the USA. What might some reasons be for this?
3 Rachel Slocum argues that Local Food movements often forget about issues related to race due to the "invisibility of whiteness." What do you think she means by this?

Suggested reading: Introductory level

Anderson, E. 2005. Me, myself and others: Food as social marker. In *Everybody Eats: Understanding Food and Culture*. New York: New York University Press, pp. 124–39.

Suggested reading: Advanced level

Coldwell, Ian. 2009. Masculinities in the rural and the agricultural: A literature review. *Sociologia Ruralis* 50(2): 171–97.

Mintz, S. and C. DuBois. 2002. The anthropology of food and eating. *Annual Review of Anthropology* 31: 99–119.

Slocum, R. 2011. Race in the study of food. *Progress in Human Geography* 35(3): 303–27.

Wilk, R. 2010. Power at the table: Food fights and happy meals. *Cultural Studies, Critical Methodologies* 10(6): 428–36.

References

Allen, Gary and Ken Albala. 2007. *The Business of Food: Encyclopedia of the Food and Drink Industries*. Westport, CT: Greenwood Publishing.

Allen, P. and C. Sachs. 2007. Women and food chains: The gendered politics of food. *International Journal of Sociology of Food and Agriculture* 15(1): 1–23.

Alston, M. 1995. *Women on the Land: The Hidden Heart of Rural Australia.* Sydney: University of New South Wales Press.

Azzarito, L. 2008. The rise of the corporate curriculum: Fatness, fitness and whiteness. In *Biopolitics and the "Obesity Epidemic": Governing Bodies*, edited by J. Wright and V. Harwood. New York: Routledge, pp. 183–98.

Bailey C. 2007. We are what we eat: Feminist vegetarianism and the reproduction of racial identity. *Hypatia* 22(2): 39–59.

Barham, Elizabeth. 2007. The lamb that roared: Origin-labeled products as place-making strategy in Charlevoix, Quebec. In *Remaking the North American Food System*, edited by C. Hinrich and T. Lyson. Lincoln, NE: University of Nebraska Press, pp. 277–97.

Beck, Ulrich. 1992. *Risk Society: Towards a New Modernity.* London: Sage.

Berlan-Darqué, M. 1988. The division of labour and decision making in farming couples: Power and negotiation. *Sociologia Ruralis* 28(4): 271–92.

Blay-Palmer, Alison. 2008. *Food Fears: From Industrial to Sustainable Food Systems.* Burlington, VT: Ashgate.

Bordo, Susan. 1993. *Unbearable Weight: Feminism, Western Culture, and the Body.* Berkeley, CA: University of California Press.

Bordo, Susan. 1999. *The Male Body: A New Look at Men in Public and in Private.* New York: Farrar, Straus and Giroux.

Bourdieu, Pierre. 1984. *Distinction: A Social Critique of the Judgment of Taste.* Cambridge, MA: Harvard University Press.

Brandth, B. 1995. Rural masculinity in transition: Gender images in tractor advertisements. *Journal of Rural Studies* 11(2): 121–33.

Buttel, Fredrick. 2001. Some reflections on late-twentieth century agrarian political economy. *Sociologia Ruralis* 41(2): 165–81.

Buttel, F. and G. Gillepsie. 1984. The sexual division of farm household labour: An exploratory study of the structure of on-farm and off-farm labour allocation among farm men and women. *Rural Sociology* 49(2): 183–209.

Caldwell, Melissa. 2009. Introduction. In *Food and Everyday Life in the Post-Socialist World*, edited by Melisssa Caldwell. Bloomington, IN: Indiana University Press, pp. 1–28.

Campbell, H. 2000. The glass phallus: Pub(lic) masculinity and drinking in rural New Zealand. *Rural Sociology* 65(4): 562–81.

Campbell, H. and M. Bell. 2000. Special issue on rural masculinities, rural sociology. *Rural Sociology* 65(4): 532–47.

Campbell, H. and A. Kraack. 1998. Beer and double vision: Linking the consumer to the rural through New Zealand television beer advertising. In *Australasian Food and Farming in a Globalised Economy: Recent Perspectives and Future Prospects*, edited by D. Burch, G. Lawrence, R. Rickson, and J. Goss. Monash Publications in Geography no. 50. Melbourne, Australia: Monash University, pp. 159–76.

Campbell, Hugh, Michael Bell, and Margaret Finney. 2006. Masculinity and rural life: An introduction. In *Country Boys: Masculinity and Rural Life*, edited by Campbell, Hugh, Michael Bell, and Margaret Finney. University Park, PA: Pennsylvania State University Press, pp. 1–22.

Carolan, Michael. 2005. Barriers to the adoption of sustainable agriculture on rented land: An examination of contesting social fields. *Rural Sociology* 70: 387–413.

Carolan, Michael. 2006. Do you see what I see? Examining the epistemic barriers to sustainable agriculture. *Rural Sociology* 71(2): 232–60.

Carolan, Michael. 2008. When good smells go bad: A socio-historical understanding of agricultural odor pollution. *Environment and Planning A* 40(5): 1235–49.

Carolan, Michael. 2011. *Embodied Food Politics*. Burlington, VT: Ashgate.

Charters, Stephen. 2006. *Wine and Society: The Social and Cultural Context of a Drink*. Woburn, MA: Butterworth-Heinemann.

Cloke, P. and J. Little. 1997. *Contested Countryside Cultures: Otherness, Marginalisation and Rurality*. London: Routledge.

Coats, C. David. 1989. *Old MacDonald's Factory Farm*. New York: Continuum.

Coldwell, Ian. 2009. Masculinities in the rural and the agricultural: A literature review. *Sociologia Ruralis* 50(2): 171–97.

Courtenay, W. 2006. Rural men's health: Situating risk in the negotiation of masculinity. In *Country Boys: Masculinity and Rural Life*, edited by H. Campbell, M. Bell, and M. Finney. University Park, IL: Penn State University Press, pp. 139–58.

Douglas, Mary. 1966. *Purity and Danger*. London: Routledge and Kegan Paul.

Elias, Norbert. 2000 (1939). *The Civilizing Process*. New York: Wiley.

Freidberg, Susanne. 2009. *Fresh: A Perishable History*. Cambridge, MA: Harvard University Press.

Gallagher, E. and U. Delworth. 1993. The third shift: Juggling employment, family, and the farm. *Journal of Rural Community Psychology* 12(2): 21–36.

Gething, Anna. 2010. Menstrual metamorphosis and "the foreign country of femaleness." In *Rites of Passage: In Postcolonial Women's Writing*, edited by Pauline Dodgon-Katiyo and Gina Wisker. New York: Rodopi, pp. 267–82.

Giovannucci, D., T. Josling, W. Kerr, B. O'Connor, and M. Yeung. 2009. *Guide to Geographical Indications: Linking Products and Their Origins*. Switzerland: International Trade Center, Geneva, http://www.intracen.org/publications/Free-publications/Geographical_Indications.pdf, last accessed March 24, 2011.

Goodman, M. 2011. Towards visceral entanglements: Knowing and growing the economic geographies of food. In *The SAGE Handbook of Economic Geography*, edited by R. Lee, A. Leyshon, L. McDowell, and P. Sunley. London: Sage, pp. 242–57.

Guthman, Julie. 2003. Fast food/organic food: Reflexive tastes and the making of "yuppie chow." *Social and Cultural Geography* 4(1): 45–58.

Guthman, J. 2008a. "If they only knew": Colorblindness and universalism in California alternative food institutions. *Professional Geographer* 60: 387–97.

Guthman J. 2008b. Bringing good food to others: Investigating the subjects of alternative food practice. *Cultural Geographies* 15(4): 431–47.

Hayes-Conroy, Allison and Jessica Hayes-Conroy. 2008. Taking back taste: Feminism, food, and visceral politics. *Gender, Place and Culture* 15(5): 461–73.

Hinrichs, C. and K. Kremer. 2002. Social inclusion in a Midwest local food system. *Journal of Poverty* 6(1): 65–90.

Josling, Tim. 2006. The war on *Terroir*: Geographical indications as a transatlantic trade conflict. *Journal of Agricultural Economics* 57(3): 337–63.

Kains, M.G. 1910. *Chicken World: Profitable Poultry Production*. New York: Orange Judd Company.

Kelly, R. and S. Shortall. 2002. "Farmers' wives": Women who are off-farm breadwinners and the implications for on-farm gender relations. *Journal of Sociology* 38(4): 327–43.

Kenway, J., A. Kraack and A. Hickey-Moody. 2006. *Masculinity Beyond the Metropolis*. Basingstoke: Palgrave Macmillan.

Korsmeyer, Carolyn. 1999. *Making Sense of Taste: Food and Philosophy*. Ithaca, NY: Cornell University Press.

Kroma, M. 2002. Gender and agricultural imagery: Pesticide advertisements in the 21st century agricultural transition. *Culture and Agriculture* 24(1): 2–13.

Leck, G. 1993. Female farmers in Canada and the gender relations of a restructuring agricultural system. *Canadian Geographer* 37(3): 212–30.

Lee, R. 2006. The ordinary economy: Tangled up in values and geography. *Transitions of the Institute of British Geographers* 31(4): 413–32.

Lee, Sandra S.-J. 2000. Dys-appearing tongues and bodily memories: The aging of first-generation resident Koreans in Japan. *Ethos* 28: 198–223.

Liepins, R. 2000. Making men: The construction and representation of agriculture-based masculinities in Australia and New Zealand. *Rural Sociology* 65(4): 605–20.

Lupton, Deborah. 1996. *Food, the Body, and the Self.* Thousand Oaks, CA: Sage.

Mammen, K. and C. Paxson. 2000. Women's work and economic development. *Journal of Economic Perspectives* 14(4): 141–64.

Michaels, David and Celeste Monforton. 2005. Manufacturing uncertainty: Contested science and the protection of the public's health and environment. *American Journal of Public Health* 95: S39–48.

Motzafi-Haller, Pnina (ed.) 2005. *Women in Agriculture in the Middle East.* Burlington, VT: Ashgate.

Ni Laoire, C. 1999. Gender issues in Irish rural outmigration. In *Migration and Gender in the Developed World*, edited by P. Boyle and K. Halfacree. London: Routledge, pp. 223–37.

Ni Laoire, C. 2001. A matter of life and death? Men, masculinities and staying "behind" in rural Ireland. *Sociologia Ruralis* 41(2): 220–36.

Ni Laoire, C. 2005. "You're not a man at all": Masculinity, responsibility and staying on the land in contemporary Ireland. *Irish Journal of Sociology* 14(2): 94–114.

O'Hara, P. 1998. *Partners in Production? Women, Farm and Family in Ireland.* New York: Berghahn Books.

Owen, James. 2005. Medieval lion skulls reveal secrets of Tower of London "Zoo." *National Geographic*, November 3 http://news.nationalgeographic. com/news/ 2005/11/1103_051103_tower_lions.html, last accessed March 1, 2011.

Payne, Tim. 2002. US Farmers' Markets 2000: A Study of Emerging Trends. Washington, DC: USDA, Agricultural Marketing Service, http://agmarketing.extension.psu.edu/ComFarmMkt/PDFs/emerg_trend_frm_mrkt.pdf, last accessed March 28, 2011.

Peter, G., M. Bell, S. Jarnagin, and S. Bauer. 2000. Coming back across the fence: Masculinity and the transition to sustainable agriculture. *Rural Sociology* 65(2): 215–34.

Philo, Chris. 1995. Animals, geography, and the city: Notes on inclusions and exclusions. *Environment and Planning D: Society and Space* 13: 655–81.

Pollan, Michael. 2008. *In Defense of Food: An Eater's Manifesto.* New York: Penguin Group.

Proctor, Robert. 1996. *Cancer Wars: How Politics Shapes What We Know and Don't Know.* New York: Basic Books.

Quisumbing, A. and L. Pandolfelli. 2010. Promising approaches to address the needs of poor female farmers: Resources, constraints, and interventions. *World Development* 38(4): 581–92.

Rogers, S. and D. DeBoer. 2001. Changes in wives' income: Effects on marital happiness, psychological well-being, and the risk of divorce. *Journal of Marriage and Family* 63: 473–9.

Sachs, C. 1983. *The Invisible Farmers: Women in Agricultural Production.* New Jersey: Rowman and Allanheld.

Saugeres, L. 2002a. Of tractors and men: Masculinity, technology, and power in a French farming community. *Sociologia Ruralis* 42(2): 143–59.

Saugeres, L. 2002b. "She's not really a woman, she's half a man": Gendered discourses of embodiment in a French farming community. *Women's Studies International Forum* 25(6): 641–50.

Shortall, S. 2002. Gendered agricultural and rural restructuring: A case study of northern Ireland. *Sociologia Ruralis* 42(2): 160–75.

Slocum, R. 2006. Anti-racist practice and the work of community food organizations. *Antipode* 38(2): 327–49.

Slocum, R. 2008. Thinking race through feminist corporeal theory: Divisions and intimacies at the Minneapolis farmers' market. *Social and Cultural Geography* 9(8): 849–69.

Slocum, R. 2011. Race in the study of food. *Progress in Human Geography* 35(3): 303–27.

Sobal, J. 2005. Men, meat, and marriage: Models of masculinity. *Food and Foodways* 13: 135–58.

Squier, Susan. 2006. Chicken auguries. *Configurations* 14: 69–86.

Stevenson, G. and Holly Born. 2007. The "red label" poultry system in france: Lessons for renewing an agriculture-of-the-middle in the United States. In *Remaking the North American Food System*, edited by C. Hinrich and T. Lyson. Lincoln, NE: University of Nebraska Press, pp. 144–62.

Symes, D. 1991. Changing gender roles in productivist and post-productivist capitalist agriculture. *Journal of Rural Studies* 7(1–2): 85–90.

Waters, Alice. 2008. *Art of Simple Food.* London: Penguin UK.

Weinberg, D. 2000. Earnings by gender: Evidence from census 2000. *Monthly Labor Review* July/August: 26–34.

Whatmore, S. 1991. *Home Farm: Women, Work and Family Enterprise.* London: Macmillan.

Wit, D. 1999. *Black Hunger: Soul Food and America.* Minneapolis, MN: University of Minnesota Press.

Zafar, R. 1999. The signifying dish: Autobiography and history in two black women's cookbooks. *Feminist Studies* 25(2): 449–69.

Zwart, H. 1997. What is an animal? A philosophical reflection on the possibility of a moral relationship with animals. *Environmental Values* 6: 377–92.

7

AGROBIOCULTURAL DIVERSITY AND KNOWLEDGE TRANSFER

KEY TOPICS

- While conventionally viewed as disparate phenomena, cultural and biological diversity are inseparably linked, particularly in the context of food and agriculture.
- Memory banks and heritage seed banks: while gene banks have become wildly popular tools in the conservation of agrobiological diversity, they forget to conserve some really important things, like memories and practical knowledge.
- Biopiracy – when biocultural knowledge is exploited for commercial gain without just compensation to those who originally held it – is a growing problem as seed and food companies scour the planet looking for organisms to patent.
- An extensive literature examines how knowledge is generated and transferred between farmers and specialists, from the classic "extension model," which supports a linear model of knowledge exchange, to more recent models that place the farmer (and their needs) first.

Key words

- Biocultural diversity
- Biopiracy
- Bioprospecting
- Campesino-to-Campesino
- Convention on Biological Diversity
- Crop wild relatives
- Cultural memory
- Expertise (contributory and interactional)
- *Ex situ* conservation
- Extension model
- Farmer Field School
- Farmer first
- *In situ* conservation
- Integrated Pest Management
- Memory banking
- Monoculture within monoculture
- Traditional Knowledge Data Libraries

Wendell Berry once wrote:

> As the local community decays along with local economy, a vast amnesia settles over the countryside. As exposed and disregarded soil departs with the rain, so local knowledge and local memory move away to cities or are forgotten under the influence of homogenized sales talk, entertainment, and education. The loss of local memory and local knowledge – that is, of local culture – has been ignored, written off as one of the cheaper "prices of progress", or made the business of folklorists.
>
> *Berry (1990: 157)*

While speaking of "loss" in rural America, Berry's words are no less true when applied to other countries, poor and affluent alike. There is convincing evidence pointing to a forgetfulness that is settling over us. But this collective amnesia is not only about lost knowledge – what we often attribute to the verb "to forget" – we are forgetting, quite literally, biodiversity itself as well as certain forms of agriculture, as this chapter explains.

In some ways, this chapter is ultimately about knowledge. While knowledge might not seem a terribly sociological topic, social theorists have long been interested in the concept. Knowledge (and especially cultural knowledge) always has an identity; a source of origin. So when we talk about knowledge we ask such questions as "Whose knowledge?" and "Knowledge towards what end?" A sociological understanding of knowledge also forces us to break free from what I call classic understanding of knowledge – the view that knowledge refers only to something we *think* not something we *do* or *practice*. This view of knowledge is not only terribly naïve but, if left uncorrected, may result in the loss of biodiversity while increasing the incidences of biopiracy.

Biocultural diversity

Concerns over the erosion of agrobiodiversity date back to at least the 1930s, when scientists began collecting plant material from around the world to be saved and exchanged for agricultural proposes (Veteto and Skarbo 2009). These concerns were given additional weight with the advent of the green revolution. In its drive to increase yields the green revolution sought to increase the yields of one or two strains of one or two crops – what is known as "monoculture within monoculture." Yet, in promoting just a handful of food crops, the strategy came at the expense of biodiversity (Box 7.1).

Monocultures are more vulnerable, relative to polycultures, to pest and disease infestation. Environmental change (such as climate change) also threatens future food security when that "security" is sought through monocultures. Monocultures are bred to grow (and thrive) within a narrowly defined window of optimal growing conditions. When those growing conditions change, and move outside that window, today's high yielding crops risk becoming tomorrow's low yielding

BOX 7.1 CROP WILD RELATIVES

The gene pool for a crop is composed of:

1 Commercial varieties
2 Landraces (varieties developed by farmers through acts of seed saving rather than through formal plant breeding), and
3 Crop wild relatives.

Crop wild relatives are the wild ancestors of crop plants and other species closely related to crops. Crop wild relatives' conservation has received less attention than either commercial varieties or landraces, even though they are likely to have a significant role in securing future food security. Their tremendous genetic diversity, it is believed, will become an important resource in plant breeding to produce crops that can overcome imminent challenges, from the adverse impacts of climate change to increasing scarcity of water and other inputs. One estimate from 1997 placed the global value of crop varieties bred from crop wild relatives at US$115 billion per year (Hopkins and Maxted 2011). The figure is no doubt considerably higher today as breeders (and, of course, seed companies) continue find more and more value in these wild ancestors.

mistakes. Monocultures also disrupt cultures. The handful of high yielding varieties that now populate most of the world's cultivated land are shallow substitutes for the native and wild relatives they've replace in terms of the cultural significance they had for millions around the world.

Conservation practices tend to be categorized in one of two forms: *ex situ* and *in situ*. *Ex situ* conservation involves the sampling, transferring, and storage of a species in a place other than the original location in which it was found (e.g. zoo or gene bank), while *in situ* conservation involves the management of a species at the location of discovery. The *ex situ* model is what often comes to mind – at least for most in the developed world – when people think of conserving agrobiodiversity. Norway's "doomsday" seed vault (the Svalbard Global Seed Vault) is one such example which has received considerable publicity of late. Officially opened on February 26, 2008, the seed bank – designed to hold 4.5 million seed samples – is constructed 120 meters inside a sandstone mountain on Spitsbergen Island roughly 1,300 kilometers from the North Pole. Its location is said to protect our genetic heritage even in a "doomsday" scenario, hence its namesake (Figure 7.1). But that's just one example. There are approximately 1,400 gene banks (in more than 100 countries) all over the world, such as the International Rice Research Institute (IRRI) in the Philippines, the International Potato Center (CIP) in Peru,

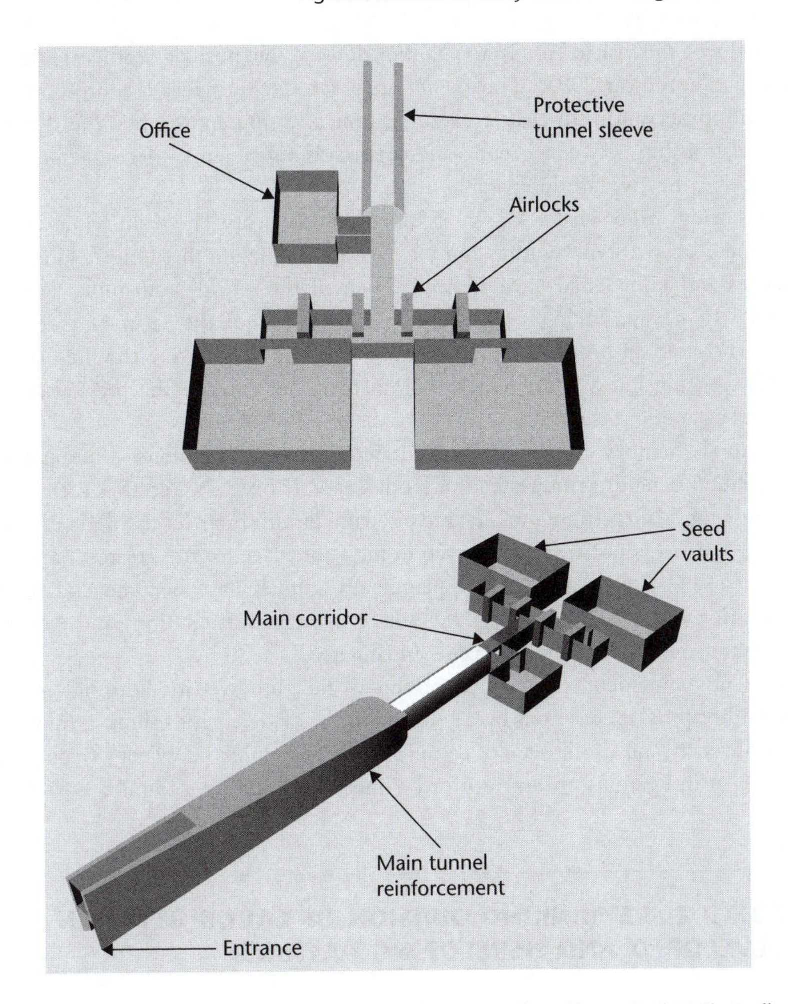

FIGURE 7.1 Svalbard Global Seed Vault. From http://en.wikipedia.org/wiki/File:Svalbard_Global_Seed_Vault.jpg

and the National Center for Genetic Resources Preservation on the campus of Colorado State University in Fort Collins, CO.[1]

The popularity of the *in situ* model, however, has grown considerably in recent decades (Hunter and Heywood 2010). Even after the green revolution, approximately 2,000 plant varieties still feed the continent of Africa. Yet this diversity continues to come under threat. This is why the Food and Agriculture Organization (FAO) of the United Nations now promotes the practice of African farmers breeding a variety of different seed – an example of *in situ* conservation. The policy is viewed as win-win-win: it is pro-poor (as it encourages seed saving and the raising of crops that do not require copious amounts of purchased inputs) and protects biodiversity while also improving food security for the continent (a diverse crop base reduces the risk of total crop failure due to unexpected weather, pests, and disease) (Thompson 2007). The *in situ* approach also provides a more complete form of conservation, in that it captures, unlike *ex situ* approaches, the dynamic inter-relationship between genetic and socio-cultural systems – recognizing, quite rightly, that the two are inseparable.

This brings us to what is known as biocultural diversity. The term acknowledges that biodiversity sustains culture and vice versa, recognizing that long-held cultural practices and knowledge have allowed much of the world's remaining agrobiodiversity to persist. Saving seed, therefore, at the site of the farm or garden by those who hold an intimate knowledge about it will help ensure the preservation of this working knowledge, which will, in turn, help ensure the preservation of the seed's genetic diversity.

Cultural memory is being discussed with increased frequency among social and cultural scientists interested in biodiversity (Box 7.2). Yet this memory is disappearing; a forgetting causally linked with diminishing levels of biodiversity. "Where human populations had stayed in the same place for the greatest duration," Nabhan (1997: 2) explains, "fewer plants and animals have become endangered species; in parts of the country [US] where massive in-migrations and exoduses were taking place, more had become endangered."

It is no coincidence that most of the world's biodiversity hotspots are also cultural hotspots, represented by a concentration of divergent ethnic groups, linguistic diversity, and a multitude of cultural practices and folk knowledge (Figure 7.2). Looking at indigenous groups in North American, some have found a correlation

BOX 7.2 RETHINKING DIVISION OF LABOR BETWEEN DEVELOPED AND DEVELOPING NATIONS

It is often said that developed countries are technology rich but gene poor while developing nations are technology poor but gene rich (e.g. Sedjo 1992). It is true; there are remarkable concentrations of biological diversity in the Global South. But the statement also misses a couple important points; an omission that, in turn, subtly justifies a global division of labor between those with genes (the less affluent South) and those with the means to do something valuable with this biological material (the affluent North) (Nazarea 2005: 57). The genetic material held within seeds, however, is not worth much without corresponding knowledge of the characteristics of the plant (such as its water, soil, nutrient, and climate requirements) and a working knowledge of planting techniques. In many cases this knowledge is held in but one repository: the community from where the genetic material came. Realizing this forces us to re-evaluate the view that the Global South is but a warehouse for the world's genetic resources.

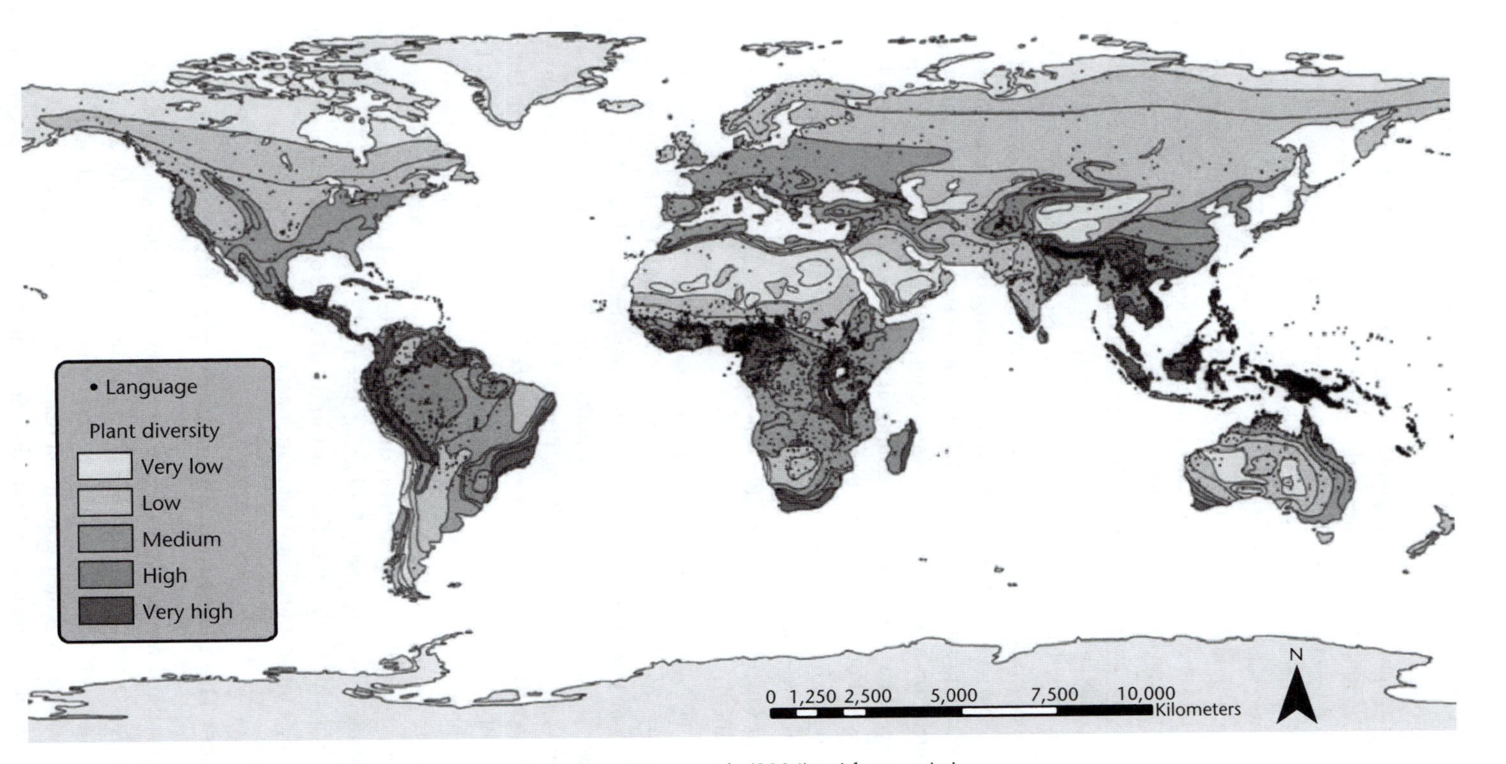

FIGURE 7.2 Plant diversity and language distribution. From Stepp *et al.* (2004) with permission

between measures of biodiversity within a given space and its levels of cultural and linguistic diversity (Smith 2001: 95). Elsewhere, research looking at Native American populations in the southwestern portion of the USA has found a high diversity of folk crop varieties and traditional farming techniques. All of this is maintained by members of these societies actively transmitting cultural knowledge from one generation to the next (Nabhan 1985).

Taste, to evoke a term from the previous chapter, is also invaluable in this quest to maintain biodiversity levels. Nazarea (1998), for example, documents how differences in sweet potato preferences in the Philippines helps sustain the crop's diversity. In Mexico, the great diversity in its corn landraces is expected to survive as long as the country's culinary diversity remains intact and continues to require all those different varieties (Smale *et al.* 2001).

Not to overstate the power of taste in this respect. A country's food traditions can still remain the same while its biodiversity erodes. This occurred in Ethiopia in the 1980s. Its food practices remain unchanged while agrobiodiversity was reduced significantly (Tsegaye and Berg 2007). Traditional wheat varieties were substituted for modern varieties with similar culinary properties to their predecessors. (This, it is worth mentioning, is a somewhat unusual case, as breeding for high yield often changes the texture, taste, and/or cooking properties of food.) Fortunately, when those traditional varieties were reintroduced in the 1990s these shared properties also worked in favor of agrobiodiversity, as the original varieties were quickly reincorporated into the local food–cuisine system (Veteto and Skarbo 2009).

Virginia Nazarea (1998) has written extensively on sweet potato farmers in the Philippines. At one site she studied, they were beginning the processes of commercializing production, while at another they remained firmly at the level of subsistence agriculture. She had hypothesized that commercialization causes a narrowing of genetic and cultural diversity among sweet potatoes raised. Her hypothesis was confirmed. But she also noticed something unexpected. There was a large disparity between the two sites in terms of the number of varieties known or remembered, compared to the biodiversity that actually existed. At the commercial site, farmers had knowledge about a far lower percentage of sweet potato varieties than at the other site. This suggests a faster erosion of cultural knowledge than genetic diversity itself. Reflecting upon this research, Nazarea (2005: 62) writes how this finding signified "that in the context of agricultural development and market integration, knowledge may actually be the first to go."

Practice: The glue of agrobiocultural diversity

MacKenzie and Spinardi (1995) argued a while back that nuclear weapons are becoming "uninvented" due to global nuclear disarmament trends and nuclear test ban treaties (the relevance of this to food and agriculture will quickly become clear, I promise). It is their contention that "if design ceases, and if there is no new generation of designers to whom that tacit knowledge can be passed, than in an important (though qualified) sense nuclear weapons will have been uninvented" (MacKenzie and Spinardi 1995: 44). They are pointing, to put it as simply as possible, to the importance of practice. A lot of knowledge has to be acted out – literally *practiced* – for it to exist and be passed along to others. Merely telling someone about it just won't do. Try teaching someone to ride a bike with words alone – it just doesn't work.

The cultural knowledge tied to traditional crops is very much like this. Many traditional cultures are oral, so you often will not find this folk knowledge written down anywhere. Consequently, once this knowledge is lost there is a good chance it is gone forever. This is what DeLind (2006: 134) was getting at when she wrote, "knowledge that is not used, and information that is not felt, are indistinguishable from ignorance" (Box 7.3).

Less than 3 percent of the 250,000 plant varieties available to agriculture are currently in use (Vernooy and Song 2004: 55). This statistic, however, only tells part of the story. There may be a quarter of a million plant varieties housed in gene banks around the world. Yet how much do we really know about these seeds? In the absence of the cultural knowledge that allowed people for generations (perhaps centuries) to put these seeds to work, their value is regrettably diminished.

BOX 7.3 HOW CUBA ALMOST FORGOT ORGANIC AGRICULTURE

The collapse of the Soviet Union in 1989 not only left Cuba without a major trading partner, but also without a supplier of cheap fertilizers, pesticides, petroleum, and grain. This proved disastrous for not only Cuban farmers, but also for the population as a whole as it left the country hungry and economically vulnerable. This was the first time since the Cuban Revolution (30 years earlier) that the country could not feed itself. While capital-intensive agriculture helped feed the nation it also made it heavily dependent upon (and thus vulnerable to) outside forces that were beyond their control.

Searching for alternatives proved difficult as most farmers and agricultural "specialists" knew only capital-intensive methods of farming – after all, it had been almost 30 years since conventional agriculture took the country by storm. However, a few traditional farmers, and their knowledge, remained. And with their help the process began of integrating scientific and traditional knowledge to make Cuba more food independent. The project has been so successful that in 1999 the *Grupo de Agricultura Organica* (GAO), a Cuban organic farming association which has been at the forefront of the country's transition from industrial to organic agriculture, was awarded the prestigious Right Livelihood Award (also known as the Alternative Nobel Prize) (Wright 2008).

As Pretty *et al.* write, this loss comes at the expense of possible solutions to future global challenges:

> The combined loss of biodiversity and ecological knowledge has implications for human health in the future, as we stand to lose the opportunistic uses and future potential of species, for instance in curing human diseases or feeding growing populations. With this loss of knowledge, we are subsequently losing the adaptive management systems embedded in traditional cultures that have sustained natural resource pools through to the present day and may be a key tool in the future of global biodiversity protection.
>
> *Pretty* et al. *(2008: 9)*

Take the impending agricultural challenges associated with climate change. Crop scientists are looking to traditional crop varieties and planting techniques as they try to find ways to help farmers keep up with climate variability. The tremendous genetic and cultural diversity associated with folk crop varieties, crop wild relatives, and traditional agricultural systems provide far more potential responses to climate change than the monocultures – biological and cultural – that define today's commercial farming systems.

So, if "practice" is the glue linking cultural diversity to biodiversity, what, to keep the analogy going, is the glue bottle? One answer: memory banks.

Memory and heritage seed banks

As opposed to an "extractive" model, memory banking is said to involve "enriching" research (Rhoades and Nazarea 2006: 334). The former benefits only the researchers, whereas the latter seeks to be advantageous to all parties involved. Unlike the basic "passport data" collected for *ex situ* conservation – involving, among other things, recording the sample's species, subspecies and/or variety and the place and date the material was collected – memory banking seeks to conserve an in-depth socio-cultural record of the folk crop variety. The goal is to save more than the seed itself. By keeping record of, for example, life history interviews of those with working knowledge of the plants in question, repositories ensure that further generations will know as much about these plants as those growing them today.

An example of memory banking occurred in the Cotacachi, Ecuador (Rhoades and Nazarea 2006). Cotacachi is losing much of its traditional agrobiodiversity due to such factors as climate change, the green revolution, and the decline of family labor. In this project, children were allocated approximately US$20 to interview their parents, grandparents, or other elders using the memory bank method. This money was sufficient to cover the expenses of school, as it was agreed beforehand that it would be used to pay for participants' education, whether primary, secondary, or college. During the interview process, they asked a series of basic questions: e.g. What varieties did your grandparents grow?; How did farming change over time? In addition to learning how to interview, record information, and store it in the project database, the children also collected culturally significant plants (leaves, seeds, and roots) and prepared them for display and storage. The students also planted and maintained a biodiversity garden and made public exhibitions of the plants and elders' knowledge during special days at the school. Between 1999 and 2004, 15 students received free tuition in exchange for their involvement in the project.

The majority of biocultural diversity studies have been centered on less affluent countries in the Global South (Veteto and Skarbo 2009). Yet developed countries are under no less of a threat when it comes to the erosion of agrobiodiversity. Over 20 years ago, Folwer and Mooney (1990) concluded that as much as 93 percent of folk crop varieties in the USA have been lost. The researchers based their conclusion on an analysis of commercially available seed. Fortunately, just because a seed is not "commercially" available does not mean it is lost.

This is where heritage seed banks – a type of memory bank – come into the picture, with their broad seed saving (and memory banking) agenda. An example of one such bank is the Seed Savers Exchange (SSE), a US-based heritage seed bank in northeastern Iowa (see http://www.seedsavers.org/). Founded in 1975, SSE is a nonprofit organization that both saves and sells heirloom fruit, vegetable, and flower seeds. On this 890-acre farm, which goes by the name of the Heritage Farm, there are 24,000 rare vegetable varieties (including about 4,000 traditional varieties from Eastern Europe and Russia), approximately 700 pre-1900 varieties of apples (which represents nearly every remaining pre-1900 variety left in existence out of the 8,000 that once existed), and a herd of the rare Ancient White Park Cattle (its estimated global population is below 2,000).

SSE is clearly more than a gene bank. Visitors can learn about the seeds being saved – about their history, phenotypic characteristics, best planting practices, recipes, and, if you come at the right time of year, even how their fruit taste. You might even say SSE is more than just a memory bank – as "bank," at least to me, is too passive to explain what goes within this space. As I've detailed elsewhere (Carolan 2011), SSE also practices memory *making*; a point that separates it drastically from the earlier mentioned "doomsday" seed bank in Norway (or any of the other 1,400 conventional seed and/or gene banks in the world today). SSE does not just serve as a repository of seeds and knowledge. Recognizing that some knowledge is inextricably wrapped in cultural practice – in actually *doing* the knowledge in question – SSE provides visitors the opportunity to put this knowledge to work (Box 7.4).

Biopiracy

For evidence of cultural knowledge's value we need only look to how corporations have been attempting to appropriate it for profit. It is estimated, for example, that developing nations would be owed US$5 billion if they received royalties of 2 percent for their contributions in pharmaceutical research and another US$302

BOX 7.4 MEMORY BANKING AT A HERITAGE FARM

The Seed Savers Exchange (SSE) offers an Heirloom Tomato Tasting Workshop. Through this event, participants not only get to taste more than 40 different kinds of tomatoes, but also learn how to save the seeds of their favorite varieties for future planting. The effect of this active learning upon participants is unmistakable, as described by the following visitor to SSE:

> "People often mistake this place as just another seed repository. But they do a lot more than just save seeds here. [. . .] They should change their name to avoid this misconception. This place is also a place where people get real hands on training about things like seed saving; about how to grow these unusual varieties, how and when to pick them, how to eat them, can them, freeze them. That's what we talk about here. It's about so much more than seeds."
>
> "Why is this important?" I asked.
>
> "Because we've forgotten so much. Maybe that's why people think that this place only saves seeds. They've forgotten the rest."
>
> *Quoted in Carolan (2011: 85)*

million for royalties in agricultural products (Anuradha 2001: 28; Ismail and Fakir 2004: 175–7). Biopiracy (or what's called bioprospecting by those doing it) occurs when biocultural knowledge is exploited for commercial gain without just compensation to those who originally held it. It's easy to chalk biopiracy up to powerful multinational corporations exploiting the knowledge and resources of indigenous populations. It's a bit more complicated than that, however.

For one thing, patent law (something already touched on in Chapter 3) makes it exceedingly difficult for indigenous populations to call this cultural knowledge their own. According to patent law, no patent can be issued where prior art exists because of the statutory requirement that patents are to be granted only for new inventions on the basis of novelty and nonobviousness. An invention is generally not regarded as new if it has been patented or described in a printed publication. Indigenous knowledge, conversely, is an oral and embodied effect, acquired through years of literally *doing* that knowledge. (This emphasis on *written* knowledge can be traced back to ancient case law, such as *Clothworkers of Ipswich* [1615] [Mgbeoji 2001: 170–4].) Because it is oral, and not written down, the cultural knowledge of these indigenous societies can be freely plundered for private gain, as evidenced by firms patenting and claiming as their own knowledge that had for centuries been part of the public domain.

Traditional groups are fighting back by publishing this indigenous knowledge in large digital libraries, what are commonly known as Traditional Knowledge Data Libraries (TKDLs). Transforming this age-old embodied and oral knowledge into a written form turns it into something that patent law recognizes and thus prohibits others from patenting it. Many countries – such as South Korea, Thailand, Mongolia, Cambodia, South Africa, Nigeria, Pakistan, Nepal, Sri Lanka, and Bangladesh – are following in the footsteps of India, who developed one of the first TKDLs in the late 1990s in the wake of its successful but expensive legal battle to revoke a US patent for turmeric.

As the former Director General of India's Council of Scientific and Industrial Research, Raghunath Anant Mashelkar, explains, the goal of the TKDLs is to build a "bridge between the knowledge contained in an old Sanskrit Shloka and the computer screen of a patent examiner in Washington" (as quoted in Chander and Sunder 2004: 1358). A study in 2000 by India's National Institute of Science Communication and Information Resources (NISCAIR) found that approximately 80 percent of the 4,896 references to individual plant-based pharmaceutical patents in the US Patents Office related to seven plants from India. According to the director of NISCAIR, this is due largely to that fact that none of the 131 academic journals patent examiners referenced when deciding whether to grant a patent came from developing countries. Thus, TKDLs provide these examiners with an important resource when deciding whether or not grant patent applications.

What's at stake?

What does this mean in practical terms? As the following examples illustrate, it means that others (most notably firms) have been able to claim ownership over (and thus profit from) knowledge, artifacts, and/or practices that have been known to indigenous societies for centuries. What follow are three examples of biopiracy.

Bt *Brinjal*

Bt Brinjal is a genetically modified strain of brinjal (eggplant) created by India's largest seed company, Mahyco, which also happens to be a subsidiary of Monsanto. *Bt* Brinjal has been engineered to be resistant to lepidopteron insects, in particular the brinjal fruit and shoot borer. According to the National Biodiversity Authority (NBA) of India, the development of *Bt* Brinjal is a case of biopiracy, which means Monsanto and Mahyco could face criminal proceedings. The NBA charges the defendants with accessing nine Indian varieties of brinjal to develop their genetically modified eggplant without prior permission from the NBA or relevant national and local boards. This is a violation of the Biological Diversity Act, 2002, which provides for conservation of biological diversity and calls for fair and equitable sharing of the benefits that arise out of the use of biocultural resources. By using the local brinjal varieties without permission, Monsanto and Mahyco weakened India's sovereign control over its resources while denying economic and social benefits to local communities under the benefit-sharing requirements of the Biological Diversity Act (Jebaraj 2011).

Herbal teas

Global food giant Nestlé is facing allegations of biopiracy after applying for patents based on two South African plants without first receiving permission from the South African government to use them. This puts Nestlé in direct violation of South African law as well as the Convention on Biological Diversity (CBD) (Box 7.5). The controversy centers on two South African plants, rooibos and honeybush, both commonly used to make herbal teas with well-known (and long-known) medicinal benefits. A Nestlé subsidiary has filed five international patent applications seeking to claim ownership over some of those medicinal benefits, like using the plants to treat hair and skin conditions as well for anti-inflammatory uses. At issue is benefit-sharing: Nestlé believes it doesn't need to share any future profits that might be derived from these patents. Yet according to the South African Biodiversity Act, firms must obtain a permit from the government if they intend to use the country's genetic resources for research or patenting. And these permits can *only* be obtained in exchange for a benefit-sharing agreement (ICTSD 2010).

Basmati rice

Basmati rice is distinct for both its aroma and flavor (basmati translates to "queen of aroma"). Basmati rice has been raised, it is believed, for thousands of years by Indian and Pakistani growers. Tens of thousands of farmers depend on it for their livelihoods; a livelihood that was almost taken away from them in the late 1990s. In 1997, the Texas-based company RiceTec, Inc., was awarded a US patent on the name "basmati," giving it commercial ownership of the name for rice seed, rice plants, and rice grain. Technically, RiceTec was given a patent on a "new" variety of basmati (remember, Indian and Pakistani growers have been coming up with new varieties of basmati for millennia and have never received a patent for their labors). The patent, however, extended to "functionally equivalent" basmati rice,

BOX 7.5 CONVENTION ON BIOLOGICAL DIVERSITY

The Convention on Biological Diversity (CBD) was signed in 1992 during the UN Conference on Environment and Development (UNCED), also known as the Earth Summit. Currently, over 180 countries are Parties to the agreement (Andorra, Somalia, and the USA have not ratified the agreement). Its objectives are threefold: the conservation of biodiversity; the sustainable use of biodiversity; and the equitable sharing of benefits arising from the use of genetic resources. The CBD represents the very first international legal instrument that recognizes the importance of traditional cultural knowledge.

which meant that other people selling the rice could be restricted by the patent. Ultimately, RiceTec could have had sole right to *all* basmati rice anywhere in the world. The case raised near immediate global outrage, as Indian growers and exporters would have had to pay royalties to RiceTec. Challenges to the patent were quickly launched, culminating in the patent ultimately being revoked in 2001. An additional consequence of the case is that it raised global awareness about biopiracy (Subbiah 2004).

Knowledge generation and transfer in agriculture

On the subject of what to do with that local, cultural knowledge of farmers, there are two general schools of thought: one is to essentially, ignore it; the other is to use it. There is a rich literature examining how knowledge is generated and transferred between farmers and specialists (see, e.g., Chambers *et al.* 1989; Carolan 2008; Pretty 1995; Röling 1988). The classic "extension model" presumes a linear model of knowledge exchange. This model begins with university experiment station "experts," who practice scientific innovation in the laboratory. The knowledge to arise from this activity is then funneled through extension agents to growers, the ultimate consumers of the knowledge and users of the technology (Warner 2008). (If this model sounds familiar the adoption-diffusion model discussed in Chapter 1 views technology transfer in much the same way.) The expertise within this model lies with the scientists alone (Figure 7.3).

This model has come under intense criticism in recent decades. For starters, it problematically attributes failures to adopt new farming technologies to ignorance or cultural conservatism on the part of farmers. Yet, as research over the last three decades has conclusively illustrated, agricultural science succeeds only when the needs and goals of farmers are met and when the conditions on the farm coincide with those of the laboratory (or test field) (Scoones and Thompson 2009). A deeper criticism lies in how the classic extension model (naïvely) views "expertise" as something that only scientists possess. Yet this claim flies in the face of mounting empirical evidence that university scientists do not have a monopoly over expertise when it comes to knowing the production end of the food system (Box 7.6).

Take Wynne's (1996) study on the relationship between scientists and sheep farmers after the radioactive fallout from the Chernobyl disaster contaminated the

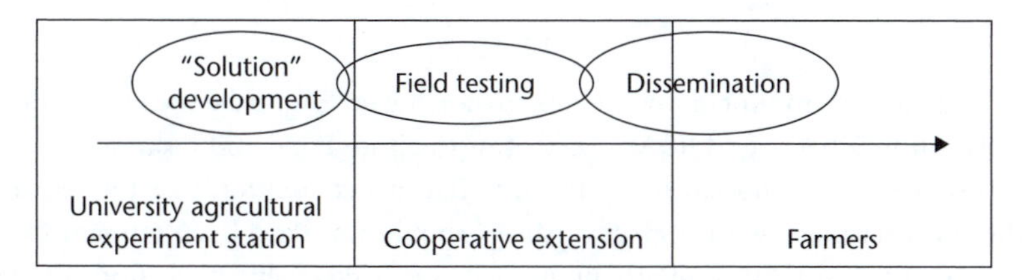

FIGURE 7.3 Classic extension model of knowledge generation/transfer

BOX 7.6 MULTIPLE FORMS OF EXPERTISE IN AGRICULTURE

Building upon Collins and Evans (2002), Carolan (2006) attempts to sketch out different forms of expertise among farmers in Iowa who actively practice sustainable agricultural methods. This research inductively leads him to construct the following typology:

- *No expertise:* A cognitive state that does not fall within either of the following categories.
- *Contributory expertise:* Enough expertise to contribute to the knowledge base of the topic in question, noting, importantly, that such cognitive authority can come in the form of either abstract/generalizable or local/practical knowledge.
- *Interactional expertise:* Having enough expertise to allow for interesting interactions between contributory experts of both abstract/generalizable and local/practical knowledge domains, which allows for an exchange of ideas to occur to the extent that all participants leave the process cognitively changed.

Based upon this research, Carolan argues that it is not enough to merely reverse the classic extension model. Whether scientists or farmers are "first," without interactional expertise, any meaningful communication between university scientists and farmers will likely always be problematic and incomplete. University scientists, of both the "natural" and "social" persuasions, spend most of their careers developing abstract/generalizable contributory expertise. And while they value local knowledge, very little time is actually spent examining how farmers and university scientists – both of whom are arguably in possession of a type of contributory expertise – interact. Carolan (2006: 430–1) concludes arguing that while "the importance of 'local' knowledge has been emphasized, attention must now turn to better understanding how such knowledge is communicated (or not) to certified (contributory) experts and, in doing this, to help nurture institutional arraignments to better utilize the expertise of those on the ground."

Cumbrian fells (a mountainous region in northwest England). Wynne examined the relationships between UK Ministry of Agriculture Food and Fisheries (MAFF) scientists and the Cumbrian sheep farmers. It is his contention that the expertise of the sheep farmers, with respect to sheep, should not have been ignored by the government when they were attempting to assess risks to livestock from the fallout. Wynne documents convincingly that the farmers knew a great deal about the ecology of the sheep, prevailing winds, and the behavior of rainwater on the pasture land that was relevant to discussions of how the sheep should be treated in order to minimize the impact of radioactivity. Nevertheless, the MAFF scientists failed to listen to the farmers because they (the farmers) lacked the proper scientific training and credentials – because, in other words, they were not viewed as "experts." Unfortunately, not understanding the local ecology, weather patterns, and hydrology, government scientists made poor policy recommendations that exacerbated risks to livestock and the farms that depended upon them for their livelihoods.

So what would an alternative to the classic extension model look like? One way to visualize this alternative model is found in Figure 7.4. The model is non-linear, to acknowledge that expertise is not concentrated into the hands of any one group. Note also that university specialists are not insulated from farmers through extension agents. If all parties are to be engaged in social learning, university scientists are going to have to step foot, literally, on the farms of farmers that they will be advising.

If anyone is privileged in this model it is the farmer. By placing farmers first – coincidently, this model has also been called simply "farmer first" (Chambers *et al.* 1989) – the main objective shifts from the linear transfer of knowledge and technology to the empowerment of farmers. The primary research and development location likewise moves from the experiment station or laboratory to the famers' fields. And the "thing" transferred to farmers shifts from a prescribed package of practices to a basket of choices. For, in the end, who knows the "needs" of a farm better than the farmer, who in some cases has spent their entire life on the same plot of land?

Farmer first models are also *empowering*. Versus the passive, sponge-like identity given to farmers under the classic extension model – where growers are expected to soak up what they are told and act accordingly – these alternative approaches

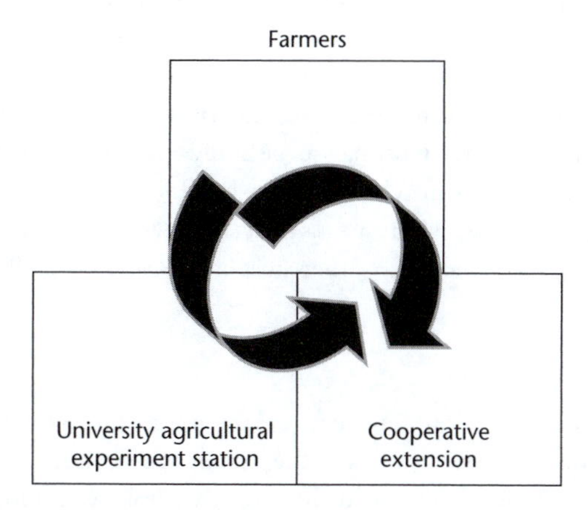

FIGURE 7.4 Model of knowledge generation and transfer based on open understanding of expertise

place farmers (not university or corporate scientists) in the driver's seat. I will conclude by briefly describing two specific active-learning models that have been successfully used to empower and benefit farmers: Farmer Field Schools and Campesino-to-Campesino.

Farmer Field School

Farmer Field School (FFS) is a season-long group training activity that takes place in the field. It has been widely used by a number of governments, nongovernmental organizations (NGOs) and international agencies to promote what is known as Integrated Pest Management (IPM) (a farm management technique that relies on ecological principles rather than petrochemical inputs to control pests). The first FFSs were designed and managed by the UN FAO in Indonesia in 1989. Since then, more than 2 million farmers across Asia have been involved in this type of active learning. The training process is always learner-centered, participatory, and heavily reliant on an experiential learning approach. The basic elements of this approach, as it applies to IPM, include (but are not limited to): the FFS consists of a group of 20–25 farmers; the experience is field-based, lasting for at least one cropping season (from seeding to harvest); farmers conduct a study comparing IPM strategy with common farmers' practice; the FFSs address special topics that are selected by the farmers; and the group is guided by at least one facilitator offering experiential learning opportunities, rather than delivering top-down instruction (IPM DANIDA Project 2005).

Campesino-to-Campesino

Campesino, which translates into peasant, is literally what the name describes, a peasant-centered approach to innovation and knowledge sharing. Campesino-to-Campesino (CAC) is an active-learning model based around farmers who have innovated new solutions to problems common among many farmers or who have recovered or rediscovered older traditional solutions. A central tenet of CAC is that farmers are more likely to believe – and thus more likely to be willing to learn from – a fellow farmer who has successfully used a technique on their own farm, especially when they can witness it first-hand. The CAC model thus involves regular visits to peer farms, where peasants learn new techniques and knowledge, which are then disseminated onto other peasants (Rosset *et al.* 2011).

Transition . . .

This wraps up another section. Yet there remains a lot of terrain still to be covered. The next section, Food Security and the Environment, turns generally around issues pertaining to how we can feed the world without wreaking total havoc on the environment in the process. The section begins by tackling, sociologically, the issue of global food security.

Discussion questions

1 What cultural memories are attached to industrial foods? What are the implications of these memories for biodiversity?
2 To play devil's advocate: might the alleged *global* benefits of biopiracy from "new" (even though they were known to certain groups for perhaps centuries) drugs and foods outweigh the interests of any *one* nation or social group?
3 Why was the classic extension model so popular for as long as it was? Who benefits from it?

Suggested reading: Introductory level

ETC group, http://www.etcgroup.org/en/issues/biopiracy
Nazarea, Virginia. 2005. Chapter One. In *Heirloom Seeds and Their Keepers: Marginality and Memory in the Conservation of Biological Diversity*. Tucson, AZ: University of Arizona Press.
Rhoades, R. 1989. The role of farmers in the creation of agricultural technology. In *Farmer First: Farmer Innovation and Agricultural Research*, edited by R. Chambers, A. Pacey, and L. Thrupp. Sussex, UK: Institute of Development Studies, pp. 3–9.

Suggested reading: Advanced level

DeLind, L. 2006. Of bodies, place and culture: Re-situating Local Food. *Journal of Agricultural and Environmental Ethics* 19(1): 121–46.
Kloppenburg, J. 2010. Impeding dispossession, enabling repossession: Biological open source and the recovery of seed sovereignty. *Journal of Agrarian Change* 10(3): 367–88.
Nazarea, Virginia. 2006. Local knowledge and memory in biodiversity conservation. *Annual Review of Anthropology* 35(1): 317–35.
Wynne, B. 1996. May the sheep safely graze? A reflexive view of the expert–lay knowledge divide. In *Risk, Environment and Modernity: Towards a New Ecology*, edited by S. Lash, B. Szerszynski, and B. Wynne. London: Sage, pp. 44–83.

Note

1 http://www.regjeringen.no/en/dep/lmd/campain/svalbard-global-seed-vault/frequently-asked-questions.html?id=462221, last accessed August 18, 2011.

References

Anuradha, R.V. 2001. IPRs: Implications for biodiversity and local and indigenous communities. *Review of European Community and International Environmental Law* 10(1): 27–36.
Berry, Wendell. 1990. *What Are People For?* New York: North Point Press.
Carolan, M. 2006. Sustainable agriculture, science and the co-production of "expert" knowledge: The value of interactional expertise. *Local Environment* 11(4): 421–31.
Carolan, Michael. 2008. Democratizing knowledge: Sustainable and conventional field days as divergent democratic forms. *Science, Technology and Human Values* 33(4): 508–28.
Carolan, Michael. 2011. *Embodied Food Politics*. Burlington, VT: Ashgate.
Chambers, R., A. Pacey, and L. Thrupp (eds). 1989. *Farmer First: Farmer Innovation and Agricultural Research*. Sussex, UK: Institute of Development Studies.

Chander, Anupam and Madhavi Sunder. 2004. The romance of the public domain. *California Law Review* 92: 1331–73.

Collins, H.M. and R. Evans. 2002. The third wave of science studies: studies of expertise and experience. *Social Studies of Science* 32: 235–96.

DeLind, L. 2006. Of bodies, place and culture: Re-situating Local Food. *Journal of Agricultural and Environmental Ethics* 19(1): 121–46.

Folwer, P. and C. Mooney. 1990. *Shattering: Food, Politics, and the Loss of Genetic Diversity*. Tucson, AZ: University of Arizona Press.

Hopkins, J. and N. Maxted. 2011. Crop wild relatives: Plant conservation for food security. Natural England Research Report NERR037, Natural England, Sheffield, http://www.cropwildrelatives.org/fileadmin/www.cropwildrelatives.org/documents/NaturalEngland ResearchReportNERR037.pdf, last accessed August 18, 2011.

Hunter, D. and V. Heywood (eds.) 2010. *Crop Wild Relatives: A Manual of in situ Conservation*. London: Earthscan.

ICTSD. 2010. Food giant Nestlé accused of biopiracy. *Bridges Trade BioRes: International Center for Trade and Sustainable Development (ICTSD)* 10(10): 3, http://ictsd.org/downloads/biores/biores10-10.pdf, last accessed August 19, 2011.

IPM DANIDA Project. 2005. Farmer Field Schools for IPM: Refresh your memory, strengthening farmers' IPM in pesticide-intensive areas, Bangkok, Thailand, http://thailand.ipm-info.org/documents/Refresh_your_Memory_(English).pdf, last accessed August 20, 2011.

Ismail, Zenobia and Tashil Fakir. 2004. Trademarks or trade barriers?: Indigenous knowledge and the flaws in the global IPR system. *International Journal of Social Economics* 31(1–2): 173–94.

Jebaraj, P. 2011. Development of Bt brinjal: a case of bio-piracy. *The Hindu*, August 10, http://www.thehindu.com/todays-paper/article2341585.ece, last accessed August 19, 2011.

MacKenzie, Donald and Graham Spinardi. 1995. Tacit knowledge, weapons design, and the uninvention of nuclear weapons. *American Journal of Sociology* 101(1): 44–99.

Mgbeoji, Ikechi. 2001. Patents and traditional knowledge of the uses of plants: Is a communal patent regime part of the solution to the scourge of biopiracy? *Indiana Journal of Global Legal Studies* 9(1): 163–86.

Nabhan, Gary. 1985. Native American crop diversity, genetic resource conservation, and the policy of neglect. *Agriculture and Human Values* 11(3): 14–17.

Nabhan, Gary. 1997. *Cultures of Habitat*. Washington, DC: Counterpoint.

Nazarea, Virginia. 1998. *Cultural Memory and Biodiversity*. Tucson, AZ: University of Arizona Press.

Nazarea, Virginia. 2005. *Heirloom Seeds and Their Keepers: Marginality and Memory in the Conservation of Biological Diversity*. Tucson, AZ: University of Arizona Press.

Pretty, J. 1995. Participatory learning for sustainable agriculture. *World Development* 23(8): 1247–63.

Pretty, J., B. Adams, F. Berkes, S. Ferreira de Athayde, N. Dudley, E. Hunn, L. Maffi, K. Milton, D. Rapport, P. Robbins, C. Samson, E. Sterling, S. Stolton, K. Takeuchi, A. Tsing, E. Vintinner, and S. Pilgrim. 2008. How do biodiversity and culture intersect? Plenary paper for Conference Sustaining Cultural and Biological Diversity in a Rapidly Changing World, IUCN-The World Conservation Union/Theme on Culture and Conservation, New York, April 2–5, 2008.

Rhoades, R. and V. Nazarea. 2006. Reconciling local and global agendas in sustainable development. *Journal of Mountain Science* 3(4): 334–46.

Röling, N. 1988. *Extension Science: Information Systems in Agricultural Development*. New York: Cambridge University Press.

Rosset, P., B. Sosa, A. Jaime, and D. Lozano. 2011. The Campesino-to-Campesino agroecology movement of ANAP in Cuba: Social process methodology in the construction of sustainable peasant agriculture and food sovereignty. *Journal of Peasant Studies* 38(1): 161–91.

Scoones, I. and J. Thompson (eds.) 2009. *Farmer First Revisited: Innovation for Agricultural Research and Development*. Rugby, UK: Practical Action Publishing.

Sedjo, R. 1992. Property rights, genetic resources, biotechnological change. *Journal of Law and Economics* 35(1): 199–213.

Smale, M., M. Bellon, and J. Gomez. 2001. Maize diversity, variety attributes, and farmers' choices in southeastern Guanajuato, Mexico. *Economic Development and Cultural Change* 50: 201–25.

Smith, Eric. 2001. On the coevolution of cultural, linguistic and biological diversity. In *Biocultural Diversity: Linking Language, Knowledge and the Environment*, edited by Luisa Maffi. Washington, DC: Smithsonian Institute Press, pp. 95–117.

Stepp, J., S. Cervone, H. Castaneda, A. Lasseter, G. Stocks, Y. Gichon. 2004. Development of a GIS for global biocultural diversity. *Policy Matters* 13: 267–70.

Subbiah, S. 2004. Reaping what they sow: The basmati rice controversy and strategies for protecting traditional knowledge. *Boston College International and Comparative Law Review* 27(2): 529–59.

Thompson, Carol. 2007. Africa: Green revolution or rainbow evolution? Foreign Policy in Focus, July 17, Washington, DC, http://www.fpif.org/articles/africa_green_revolution_or_rainbow_evolution, last accessed November 7, 2010.

Tsegaye, B. and T. Berg. 2007. Utilization of durum wheat landraces in East Shewa, Central Ethiopia: Are home uses an incentive for on-farm conservation. *Agriculture and Human Values* 24(2): 219–30.

Vernooy, Ronnie and Yiching Song. 2004. New approaches to supporting the agricultural biodiversity important for sustainable rural livelihoods. *International Journal of Agricultural Sustainability* 2(1): 55–66.

Veteto, J. and K. Skarbo. 2009. Sowing the seeds: Anthropological contributions to agrobiodiversity studies. *Culture and Agriculture* 31(2): 73–87.

Warner, K. 2008. Agroecology as participatory science: Emerging alternatives to technology transfer extension practice. *Science, Technology and Human Values* 33(6): 754–77.

Wright, J. 2008. *Sustainable Agriculture and Food Security in an Era of Oil Scarcity: Lessons from Cuba*. London: Earthscan.

Wynne, B. 1996. May the sheep safely graze? A reflexive view of the expert–lay knowledge divide. In *Risk, Environment and Modernity: Towards a New Ecology*, edited by S. Lash, B. Szerszynski, and B. Wynne. London: Sage, pp. 44–83.

PART III

Food security and the environment

8

HOW MUCH IS ENOUGH?

KEY TOPICS

- Before we can determine how much food we will need to produce to keep up with growing demands we have to decide exactly what the food will be going towards: food (for humans), feed (for livestock), or fuel (for automobiles)?
- Affluent countries, in an attempt to enhance their own food producing capacities, are acquiring land and water (and cheap labor) from poorer nations to fed growing populations back home – what is known as the global land grab.

Key words

- Conversion ratios (for livestock)
- Ethical vegetarianism
- First generation biofuels
- Food *v.* fuel
- Global land grab
- Health vegetarianism
- Political economy of biofuels
- Second generation biofuels

How much food is enough to feed the world? Given what has been said earlier, this is a bit of a trick question, as there are sufficient calories to feed everyone alive today. The Food and Agriculture Organization (FAO) of the United Nations has run the numbers many times, repeatedly estimating that world agriculture produces roughly 2,700 calories a day for every man, woman, and child in the world – well above their 1,900 calories a day threshold signifying, for those eating less than this, a state of malnutrition.

In Chapter 4, I mentioned that hunger is a distributional problem. This chapter not only gives further weight to this argument, but explores in greater detail why this is. Part of the problem is the fact that we feed a lot more than people with the food we grow. So when talking about how much food is enough we'll need to also ask: *Who and what do we hope to feed?* Will we be using food to feed more than just people? If we plan on using a substantial portion of what we grow to feed, say, cars and livestock, the answer to the question "How much?" is radically different than if we fed our grain and legumes directly (and solely) to humans. This chapter looks at the issue of food security within the context of these trends, whereby agriculture currently feeds a lot more than just people, and some of their underlying driving sociological variables.

Food security: The limitations of technology

If current trends continue, we will need to produce anywhere between 70 percent (FAO 2009) to 100 percent (Schade and Pimentel 2010: 247) more food by 2050. Some 2.3 billion more mouths – for a total of 9 billion – will need to be fed by 2050. Most of this growth will occur in developing countries (like those in Africa), where food security is already tenuous. Annual average global crop production growth is expected to slow from 2.2 percent (as recorded between 1997 and 2007) to 1.3 percent between now and 2030, slowing still further to 0.8 between 2030 and 2050. In developing countries, yield growth looks to slow even faster, from 2.9 percent to 1.5 to 0.9 percent during these respective periods (Bruinsma 2009). Also, for the first time in decades, crop yields for wheat, rice, and soybean are growing slower than population, while yield increases for maize are just barely surpassing population growth rates (Figure 8.1). In sum: if all goes as planned, yield increases

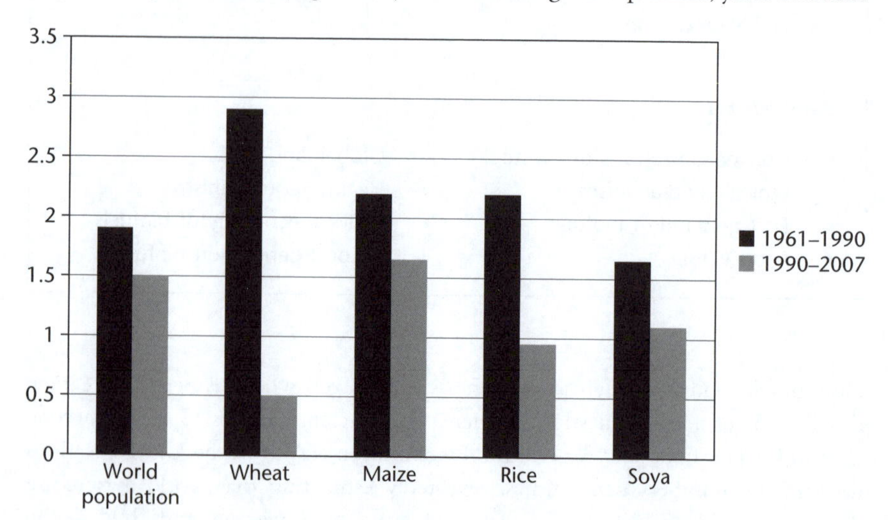

FIGURE 8.1 Annual average growth rates for world population and global yield increases. Based on *The Economist* (2011)

will take care of about *half* of the world's projected needs by 2050 (recognizing that these "needs" are also predicated on feeding more than just humans).

To make up for the yield shortfall we are going to need anywhere between 200 million and 750 million additional hectares of land by 2050 (Schade and Pimentel 2010: 247). A variety of studies have settled on the figure of 1.5 billion as the number of additional hectares available to be brought under cultivation (e.g. Balmford *et al.* 2005; Schade and Pimentel 2010). Much needs to be accomplished, however, before land can be brought into production – land rights have to be settled, credit must be available, and an infrastructure and market must be in place. These constrains explain why arable land worldwide has grown by a net average of 5 million hectares per year over the last two decades (Rabobank 2010). It also means it will be decades until a sizable amount of arable land is prepared for agriculture. More problematic still is the projected slowdown in the annual growth of arable land, as potential arable land becomes increasingly marginal (the land easiest to convert has already been brought into production). The annual growth of arable land will slow from 0.30 percent between 1961 and 2005 to 0.10 percent between 2005 and 2050. This calculates out to an average annual net increase of arable area of 2.75 million hectares per year between 2005 and 2050 (Rabobank 2010); or 120 million additional hectares – well below the most optimistic estimates that claim only 200 million hectares will be needed by mid century to satisfy global food demand. At the same time, arable land in developed and transitional countries will continue to decline, losing an estimated 0.23 percent a year between 2005 and 2050 (Bruinsma 2009).

To put it plainly: increases in yield and arable land will likely not keep up with current rates of demand. *Current rates of demand*: that's a sociologically interesting term. Just what exactly does it mean? We talk as if we need to keep up with these rates. Do we? What I mean is: who and what are driving these rates of demand? And do these "consumers" represent the most optimal use of such limited resources as land, water, and food? Let's begin by looking at the food that's funneled through the livestock sector.

Livestock: Facts and trends

If current trends hold, we are going to have to produce twice as much animal protein by 2050 just to keep up with demand. This means livestock will be consuming basic food to feed the equivalent of 4 billion people. (Four billion, as an aside, was the world's population in the early 1970s.) In other words, if the world continues to eat meat at the rates that we are expecting, the world's *effective* population in 2050 will be 13.5 billion, as opposed to the 9.5 billion as predicted by most models (Tudge 2010). Part of the reason driving current trends is the positive relationship between a country's gross domestic product (GDP) per capita and the average citizen's consumption of animal products, as illustrated in Figure 8.2. Thus, as countries, most notably China, continue to climb the economic ladder, so too should we expect their appetite for meat to grow in equal fashion.

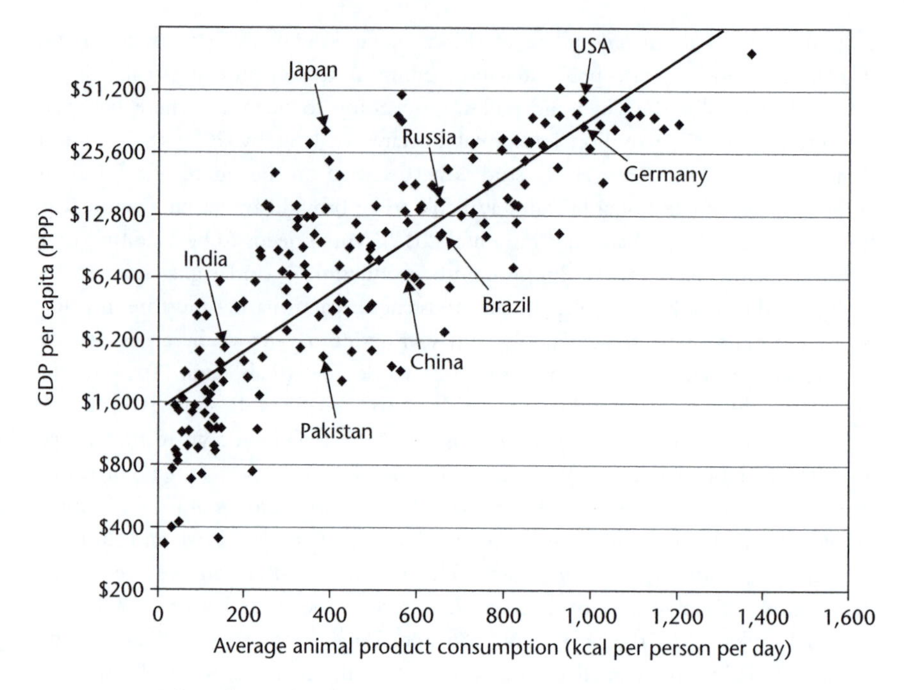

FIGURE 8.2 The relationship between gross domestic product (GDP) per capita and animal product consumption per capita. From Miller (2010) with permission

Let's look at little deeper at these trends. First, while global consumption of all basic foods will continue to increase – which is understandable given population growth alone – the consumption of meat in particular is expected to grow much faster than dairy, cereals, and starchy roots. The global demand for meat, dairy, cereals, and starchy roots since 1960 (and forecasted out to 2050) is illustrated in Figure 8.3.

Looking closer at that category "meat" reveals considerable variation in growth rates when it comes to global consumption of sheep/goats, cattle/buffalo, chickens, and pigs from 1960 to 2010. These different rates are depicted in Figure 8.4. As illustrated, the global consumption of chicken has increased 450 percent (or 4.5 times) relative to 1960 levels, compared with pigs, cattle/buffalo, and sheep/goats, which have increased 250, 60, and 50 percent, respectively.

Why do these trends in the consumption of animal proteins matter? Collectively, cattle, pigs, and poultry consume roughly half the world's wheat, 90 percent of the world's corn, 93 percent of world's soybeans, and close to all the world's barley not used for brewing and distilling (Tudge 2010). And it is not just that livestock consume a lot of basic food that could otherwise be used to feed people directly. We also have to be mindful of how efficiently (or more accurately inefficiently) they convert feed into animal protein. Some animals are clearly better at this than others. Table 8.1 summarizes these conversion ratios, as they are called, for chickens, pigs, and cattle.

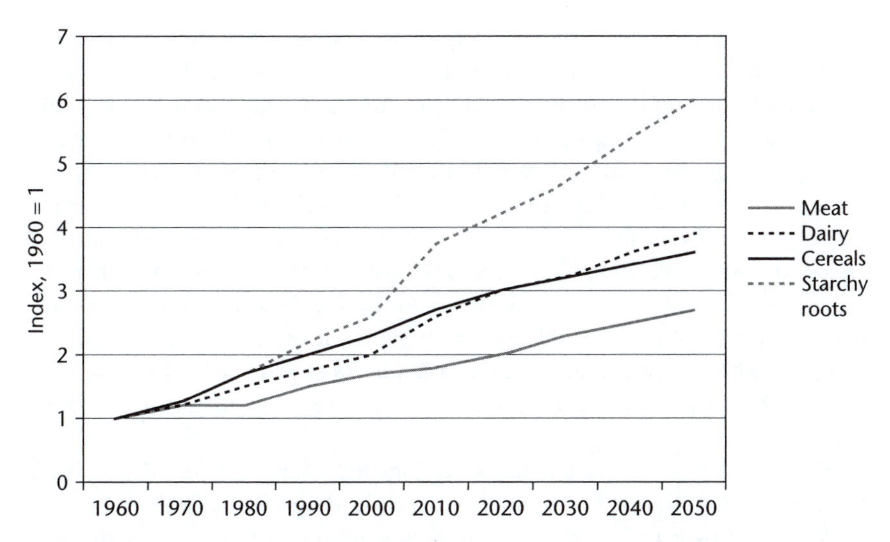

FIGURE 8.3 Global food demand for select commodities since 1960 (forecast 2011–2050). Based on *The Economist* (2011)

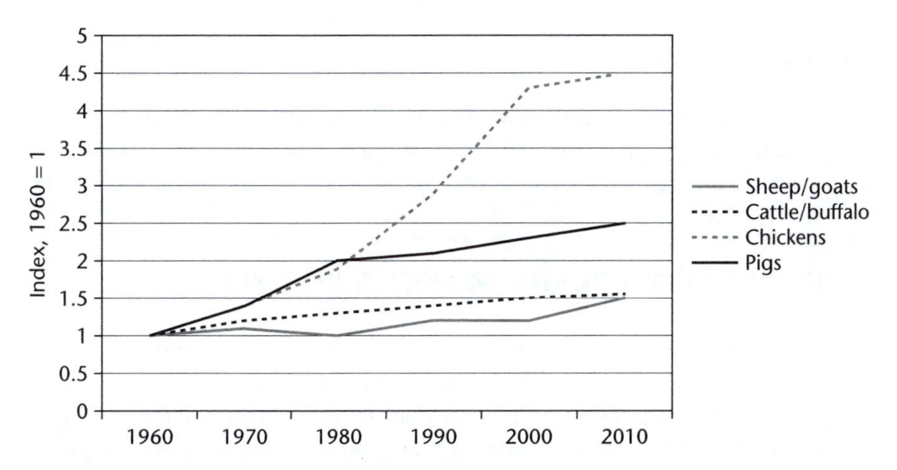

FIGURE 8.4 Changes in global consumption of animals since 1960. Based on Godfray *et al.* (2010)

TABLE 8.1 Feed, calories, and protein needed to produce 1 kilogram (kg) of chicken, pork, and beef

Grain (calories/protein in grams)	⟶	*Animal (calories/protein in grams)*
2 kg feed grain (6,900/200)		1 kg chicken (1,090/259)
4 kg feed grain (13,800/400)		1 kg pork (1,180/187)
8 kg feed grain (27,600/700)		1 kg beef (1,140/226)

Such conversion ratios, however, only apply to animals fed on grain. Professor John Webster at the University of Bristol (UK), and former President of both the Nutrition Society and the British Society for Animal Science, calculates that dairy cows on a 70–80 percent pasture-based diet give back *more* calories than they take in – quite a bit more, in fact (Webster 2010). Webster places the return at 170 percent. Of course, this is due to a wonderful adaptive trait that allows ruminants to digest plant-based foods that we can't. Some regions of the world are well endowed with arable land, like Europe, Russia, and North America. Others are arable land poor but rich in permanent pasture. In Sub-Saharan Africa, for example, only 17 percent of land is arable, while 80.8 percent exists as permanent pasture (Peterson 2009). In places such as this, raising livestock, from a food security perspective, truly makes sense, as livestock may well be the only thing that these lands can sustain.

Monogastrics too have a unique ability. Pigs and poultry are prized worldwide for their ability to thrive on food waste, which, as described in Chapter 10, there is plenty of (Box 8.1). By definition, food *waste* is no longer food, as by this state it is said to have exited the food system. Raising animals – poultry and pigs – at least partially on waste might make good sense from a food security perspective, particularly if there is a lot of organic waste that is going to, well, waste.

Meat on the menu

The previous section was intentionally descriptive. Its purpose was to inform the reader about certain basic facts as they relate to livestock production and consumption; facts that are necessary for anyone to have an informed understanding about issues

BOX 8.1 PIGS, ORGANIC WASTE, AND EGYPT

In May 2009, the Egyptian government ordered the killing of some 300,000 pigs. The rationale behind this declaration was the controlling of swine flu (even though no cases at the time had been reported within the country). Soon thereafter, the organic waste that was previously fed to the pigs began to accumulate in streets in large, smelly piles (Rogers 2009). The trash previously was taken care of by the Zabaleen. The Zabaleen – a primarily Christian community of garbage collectors who reside in slums on the outskirts of Cairo – would go door-to-door and collect organic waste. Their pigs would consume some 60 percent of the waste that was collected. The Zabaleen, in turn, ate some of their animals, while sending others to the market where they brought their keepers a good price. The government originally promised to set aside US$5.4 million to compensate pig owners. Those who have been paid report receiving US$7 for each pig – a fraction of the US$90–180 that a pig sold at market would receive (Stack 2009).

related to food and agriculture in general and food security in particular. Now I will delve sociologically into those trends. *Why* is meat consumption so prized in an increasing number of countries around the world? And *how*, in those countries that currently eat a lot of meat, do people understand and perceive actions related to livestock production and meat consumption? Those questions are the focus of this section.

A number of cultural analyses have identified meat as the food with the highest status in the hierarchy of foods; a finding that appears to hold for developed (Twigg 1984) and developing (Lokuruka 2006) countries alike. Douglas and Nicod (1974) conducted a seminal study of meals among Britons and found meat to be at the center of practically every meal. Its cultural dominance can be seen by the fact that its presence signifies the dish, even when it is just one ingredient of many, whether salads (*chicken* cobb salad), soups (*beef* stew), or casseroles (*tuna* casserole). The dominant position of meat in Western cuisine is even reflected in Western vegetarian culture, as non-animal products are made to appear as much like meat as possible (veggie *hotdogs*, *hamburgers*, etc.) (Gvion-Rosenberg 1990; Box 8.2).

BOX 8.2 SOME OF THE SOCIO-CULTURAL COMPLEXITIES UNDERLYING VEGETARIANISM

Rozin *et al.* (1997) differentiate between what they call "health vegetarianism" and "ethical vegetarianism" (categories that have since been replicated; see, e.g., Lindeman and Sirelius 2001). According to this line of research, health vegetarians avoid meat primarily for health-related reasons. Ethical vegetarians, in contrast, follow a vegetarian lifestyle in order to minimize harm to animals. For health vegetarians, the initial shift to vegetarianism tends to be gradual as they test foods that fit with their new lifestyle. For ethical vegetarians, the shift is typically more abrupt. Environmental concerns are also beginning to turn up as a rationale for vegetarianism. One study finds that while environmentalism is not a primary motivator for vegetarianism among their respondents, ecological justifications for a meat-free diet may become more pronounced as consumers learn more about livestock's impact on the environment (Fox and Ward 2008a). The rise of such ecological rationale may also be the result of vegetarians (and vegans too) seeking still further reasons for their decision (Fox and Ward 2008b).

Social science research also reveals the complexities underlying how the relevant terms are defined. "Vegetarianism," among self-described vegetarians, appears to be a fluid, subjective category (Fox and Ward 2008b). This is evidenced by the fact many people who identify themselves as vegetarian continue to eat foods derived from animals (Hoek *et al.* 2004). It has been found, at least in one study, that vegetarians and vegans define their lifestyles in ways that fit their own practices (Bedford and Barr 2005).

There is some debate, however, over whether meat continues to hold the same privileged position as it did in generations past. One study of meat consumption among residents of Copenhagen, for example, suggests meat's place atop the food hierarchy has reached its end (Holm and Mohl 2000). The researchers note that, if true, some of the factors behind this trend include recent bovine spongiform encephalopathy (BSE) scares, the sizable ecological footprint of industrial livestock production, a growing awareness of animal rights and welfare, and human and public health issues related to the high levels of saturated fat found in certain animal products.

What makes meat – and "red" meat in particular – so symbolically powerful? One important early study on the subject contends that its deep cultural significance lies in the fact that it comes from animals (Twigg 1984). To a certain degree, then, it is blood that makes red meat so prized, as blood has long been associated with virility, strength, and sexuality. In many indigenous cultures, to consume meat is to consume the animal and its vitality. But there are boundaries within the act of meat eating that we cannot transgress, like eating other humans and raw meat.

The cultural significance of cooking cannot be understated. Cooking, and cooking meat in particular, is one of the many things that humans do to symbolically separate themselves from the rest of the animal kingdom (the distinction between "the raw" and "the cooked" has been said to be homologous to that between "nature" and "culture"). The human, like any animal, needs to eat. This need is a constant reminder that we are animals too. Cooking thus acts to separate ourselves from the animal kingdom and appear more "civilized." Cooking, you could say, works to symbolically remove artifacts from the realm of "nature" and places them in the realm of "culture" (Lévi-Strauss 1969).

The "discovery" of the calorie has also given meat (and red meat in particular) cultural significance. Europeans first measured the calorie in 1883. Yet, whereas in Europe the calorie attracted little attention, it quickly established itself in the USA as a major guiding policy principle (Cullather 2007). The calorie allowed food to be talked about without reference, context, or form. Taste, culture, culinary tradition, and variety became secondary qualities when compared to their shared status as carriers of energy. The calorie also allowed for a seemingly "objective" ranking of foods. Grains, meat, and dairy goods ranked very well compared with fruits, leafy vegetables, and fish. Some items in the latter group possessed so few calories that, during the early twentieth century, "they could scarcely be classified as food" (Cullather 2007: 345). The calorie also washed away cultural differences in diet. Scientists argued that, for example, the potatoes and cheese that made up so much of the Irish diet were identical, except perhaps in quantity consumed, to the rice and ghee eaten by those from India and Southeast Asia (Cullather 2007).

With this understanding of "food" in tow, officials began rethinking the provisioning of army, prisons, and schools. The intent of these reforms was to maximize money spent on foods high in calories, while de-emphasizing diets with low caloric values. In terms of improving national diets, it was believed the people of Asia had the most to gain from this new knowledge as their diet lacked the type of energy dense foods – most noticeably red meat – common to North Americans. The Imperial army and navy of Japan began reforming rations in the 1920s, adding "Western" recipes that utilized beef, pork, wheat, and fried batter (tempura) to boost the caloric intake of its soldiers while cutting costs (Cullather 2007).

The later discovery of vitamins and minerals eventually undermined the calorie's identity as the single variable upon which all food is to be measured (and greatly helped rehabilitate the value of fruits and vegetables). Yet the legacy of the calorie remains. The calorie, for a while at least, seemed to objectively solidify meat's place atop the food hierarchy.

Food, including meat, is also gendered. In most Western postindustrial societies, masculine foods have tended to include things like steak, hamburgers, baked potatoes, and beer, while feminine foods have historically included salads, pasta, yogurt, and fruit (Bender 1976; Lupton 1996; Sobal 2005). There is also evidence that men attach special cultural significance to meat by believing a meal is not "real" without meat as the main dish – a finding replicated in such divergent countries as Czechoslovakia (Kiliánová *et al.* 2005), South Korea (Lappé 1971), and the USA (Sobal 2005). The perceived masculinity of red meat might explain why, historically speaking, women have tended to either avoid meat entirely (Lupton 1996) or eat seafood or poultry (Dixon 2002).

While cooking has historically been located within the realm of the feminine, barbecuing has been a thoroughly masculine exercise (which is perhaps why it's never called "cooking"). Smoking and barbecuing contests, where red meat is the star attraction, are dominated by men, as are eating competitions, which too are often centered on the consumption of large quantities of meat. The consumption of meat can be still further masculinized by eating it rare (bloody), with minimal cooking, and with no sauces as they risk feminizing an otherwise masculine dish (Sobal 2005). It has also been argued that men demonstrate their masculinity by dominating other species through hunting and trapping (Lockie and Collie 1999). Men are often characterized by their suppression of emotions, which some scholars claim to be a useful attribute when hunting, killing, and slaughtering animals (Sobal 2005).

There are, of course, exceptions to everything said in the previous paragraphs. When discussing whether or not meat is masculine, we need to be careful not to universalize either gender or the masculinity of meat. As gender scholars point out, and as discussed in Chapter 6, such singular thinking reduces masculinity and femininity to monolithic normative concepts, causing everyone to be judged against fixed gender standards of a society. Yet gender is not fixed. There are multiple ways of expressing masculinity and femininity that go beyond idealized gender types. The same thinking applies to the gendering of food choices. As others have pointed out, it is not universally accurate to equate meat with the male gender just like it is not appropriate to argue that one must eat meat in order to convey masculinity (Probyn 1999).

Biofuels

In addition to being turned into feed, a considerable portion of food is being diverted toward the production of biofuels. World ethanol production has increased 500 percent in the last 10 years, from roughly 17 billion liters in 2000 to just below 87 billion liters in 2010. While a fraction of the world's biofuels, biodiesel production has grown too in the last decade, from less than 1 billion liters in 2000 to approximately 20 billion in 2010 (Figure 8.5).

Yet to meet all the various targets set throughout the world, biofuel production is going to have to increase a lot more. Mandates for blending biofuels into vehicle fuels have been enacted in at least 41 states/provinces and 24 countries and require blending 10–15 percent ethanol with gasoline or 2–5 percent biodiesel with diesel fuel. For example, Brazil, Indonesia, and the European Union (EU) expect to meet 10 percent of their energy demands by 2020 with biofuels. For China, they hope to meet 5 percent of their energy demands by 2020 with biofuels. And for the USA, they have aspirations that biofuels will satisfy 30 percent of their energy needs by 2020 (ethanol currently accounts for 8 percent of the fuel used for vehicles in the USA while consuming almost 40 percent of its maize crop; *The Economist* 2011). Biofuel targets have also been set (or plans are in the works to set such targets) in more than 10 countries in addition to the countries of the EU, which call for significant increases in consumption and production of biofuels in their respective countries (REN21 2010). It has been estimated that if all such targets were met 10 percent of the world's cereal output would be diverted from food to fuels, pushing food prices up anywhere from 15 to 40 percent. Similarly, if all US maize otherwise destined for ethanol plants were instead used

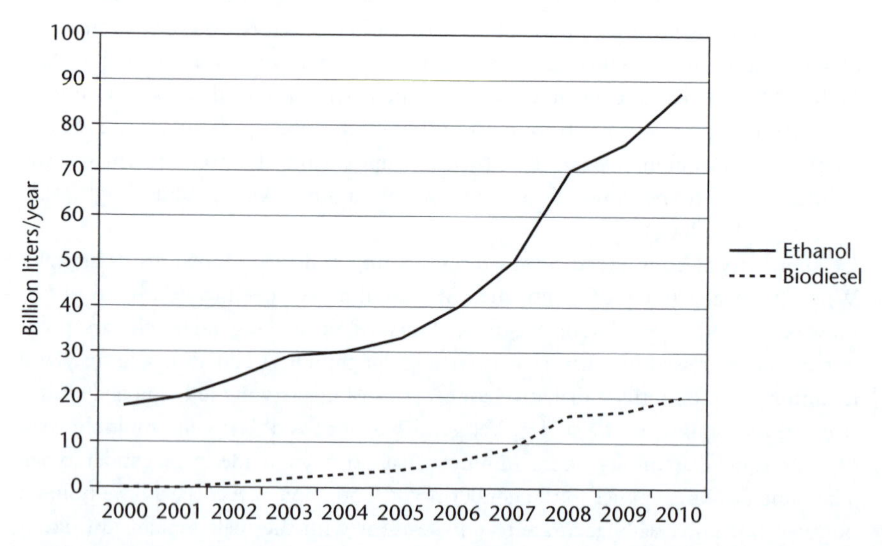

FIGURE 8.5 Ethanol and biodiesel production, 2000–2010. Based on REN21 (2010) and Chazan (2011)

as food, global edible maize supplies would increase approximately 14 percent (*The Economist* 2011).

Corn ethanol, sugar ethanol, and biodiesel constitute the primary biofuels markets, though the markets for biogas for transport and other forms of ethanol (like cellulosic) are growing. Corn represents over 50 percent of the global ethanol market followed by sugarcane ethanol, satisfying another third of world demand. The USA and Brazil make up roughly 90 percent of global ethanol production – the former producing corn ethanol; the latter sugarcane ethanol. The EU leads the world in biodiesel production, representing nearly 50 percent of total output. Biodiesel also accounts for the vast majority of biofuels consumed in Europe (REN21 2010).

Biofuels have a long history as a fuel source. By the turn of the twentieth century, 5–6 percent of all potatoes grown in Germany were going into the production of ethanol. In 1903, Germans consumed 1 million gallons of ethanol to power internal combustion engines, with the government calling for production of alcohol fuel to grow to exceed 3 million gallons. And in Greece at around the same time the use of ethanol had become so great that the government was forced to impose an ethanol fuel tax to compensate for the loss of revenues from petroleum taxes (Carolan 2009a).

The scope of biofuel production and consumption today, however, vastly surpasses anything seen in the past, as evidence by trends illustrated earlier in Figure 8.5. Sociologists have only started to look at the biofuel juggernaut. Nevertheless, the sociological literature on biofuels is growing at a truly remarkable rate.

Rural development

Given the subdiscipline's strong historical links to rural sociology it is not surprising to find food and agriculture scholars interested in biofuels' role in rural development. A fair amount of this research is survey based, revealing not only public and farmer perceptions toward biofuels but also, importantly, expert *mis*conceptions concerning what others think about these alternative fuels.

For instance, it is widely believed biofuels are entering their "second generation." In the case of ethanol, this means a shift from corn-based (grain feedstock) to cellulosic-based (nongrain feedstock) ethanol. One of the items stipulated by the 2007 US Energy Independence and Security Act (EISA) is that 60.56 hm^3 of total US ethanol output be cellulosic ethanol made from biomass (ligno-cellulosic) feedstocks by 2020 – a far cry from current production levels of a little over 1 hm^3. This optimistic level set by the EISA is based heavily on an earlier estimate that placed annual US biomass inventory of approximately 1.42 G dry megatonnes (Mt) of biomass, which presently exists in the form of corn stover (that is, the stalk, leaf, husk, and cob remaining in the field following harvest). In Iowa particularly, as the largest corn producing state in the USA, there is sufficient harvestable corn stover to produce roughly 5.8 hm^3 of cellulosic ethanol, assuming farmers are willing to harvest it. As it turns out, thanks to recent survey research, we now know this is an unrealistic assumption.

Researchers out of Iowa State University found that statewide, only 17 percent of Iowa corn farmers had any interest in harvesting their stover (though 37 percent of those surveyed were undecided; Tyndall *et al.* 2011). Those most interested tended to be younger, planned on farming for at least 10 years, were at least somewhat knowledgeable about corn stover, and farmed large amounts of land. Environmental concerns also appear to be important, as a number of farmers expressed concerns about what the removal of corn stover from their field would mean to environmental quality (e.g. biodiversity, soil erosion). The overall takehome message of the study: "that future supply assessments consider farmer participation more explicitly and forego arbitrary assumptions regarding farmer behavior as previous supply analyses may have overstated the proportion of farmers interested in harvesting stover" (Tyndall *et al.* 2011: 1485).

Another study examines what added benefits (if any) local ownership of ethanol processing plants has upon the communities in which they are located (Bain 2011). A widely popular argument within the rural development literature is that local ownership of firms has an overall net positive effect on local economic and community well-being (a belief underlying the Goldschmidt thesis discussed in Chapter 5). Rather than merely assume local ownership is better than nonlocal ownership, this study explores, by interviewing community and business leaders in a community with a locally own ethanol plant, whether this all-too-common assumption actually holds true in the case of ethanol plants. The study concludes by arguing that "beyond the opportunities presented to individual farmers and investors participants did not consider that local ownership had influenced company practices or its socioeconomic impacts within the community" (Bain 2011: 1405).

In making this argument, the author points out that ethanol plants, even local ones, remain embedded within national and international supply chains and networks. Whereas rural development asserts that firms operate at different scales, with local firms embedded within local supply chains and institutions, locally owned ethanol firms remain connected to national and global networks. As a fuel additive, ethanol cannot be divorced from petroleum's global organizational structure. Even locally owned ethanol plants therefore remain embedded within larger petroleum networks, as its main customers remain just a handful of international petroleum firms. Ethanol plants are but a small player in a supply chain over which they have little control. Consequently, the price offered to local corn growers and returns to investors is largely dictated by major oil firms, the global petroleum market, and government ethanol policies and subsidies.

In a similar study, involving 48 farming and nonfarming individuals involved in switchgrass bioenergy projects in Iowa and Kentucky, respondents held simultaneously positive and negative perceptions toward biofuel's potential as a tool for local and regional development (Rossi and Hinrichs 2011). Some of the strongest skepticism stemmed from observations of the growing concentration of the food system, expecting similar trends and influence to be wielded in the emerging bioeconomy. At the same time, participants were hopeful that biofuels would help revitalize local and regional economies and community structures. This willingness to sacrifice local autonomy in exchange for greater capital investment (by multinational firms) at the local and regional levels reflects some of the tradeoffs that must be negotiated when thinking about biofuel's potential as a strategy for rural development.

When investigating the links between rural development and biofuel processing plants, it is always a good idea to ask "Rural development for *whom*?" One study sought to bring some clarity to this question, concluding that ethanol processing plants, at least in the North Central Region of the USA, are often located in communities and regions least in need of rural revitalization (Goe and Mukherjee 2008). Specifically, the researchers found that the ethanol industry has thus far only reinforced the structural advantage of certain nonmetropolitan localities. To put it another way, the study found that biofuel processing plants tended to locate in communities that *already* possessed thriving economies, low levels of unemployment, less poverty, and lower rates of income inequality. They also learned that these processing plants were more likely to be located in proximity to large corn/grain supplies (in other words, large grain farms) and in regions with a well-developed rail and highway infrastructure.

The political economy of biofuels

There are also significant political and economic factors helping to propel the biofuels juggernaut forward. Arguably, nowhere are these factors greater, in terms of the government supporting this sector of industry (and thus insulating it from market forces), than in the USA. So I'll begin there.

Take, for example, the volumetric ethanol excise tax credit. In 2006, approximately US$2.5 billion was distributed to gasoline blenders through the tax credit, which initially provided a US$0.51 per gallon credit for every gallon of ethanol blended with gasoline. With the signing of the 2008 Farm Bill, Congress reduced the ethanol tax incentive by 6 cents. Yet the US Congress' Joint Committee on Taxation calculates that this US$0.45 per gallon subsidy still costs the US Treasury roughly US$5 billion per year.[1]

To keep ethanol prices artificially high – so as to protect all involved in domestic ethanol production – the US government also has a $0.54 per gallon tariff on imported ethanol (save for a few exemptions [e.g. Ecuador]). The reason for the tariff is simple: countries like Brazil can produce ethanol much more cheaply so a tariff is necessary to make US ethanol cheaper than that coming from other countries. One way around this tariff would be for US firms to import sugarcane and then process it into ethanol. This would make economic sense, if the USDA's estimates are correct that ethanol could be produced from sugarcane at a lower price than if the source material were corn (USDA 2006). US sugar policy, however, has long restricted imports of foreign sugar to protect the domestic sugarcane industry (causing the US price of sugar to be roughly twice the world market price for decades). Ethanol in the USA is also supported by grants and loans, the renewable-fuel standards, and corn subsidies. The General

Accounting Office estimates that over US$19 billion in taxpayer support was given to the ethanol industry between 1980 and 2000. While these subsidies were less than those provided to fossil fuels and nuclear power, they exceeded all government subsidies when calculated in per unit energy terms (General Accounting Office 2000). A more recent (and more comprehensive) calculation estimates that US taxpayers are subsidizing biofuels at a cost of over US$70 billion per year (Koplow 2007).

Though unique as the world leader in ethanol production and ethanol subsidies, the USA is not alone in its aggressive subsidization of biofuels. For example, the Global Subsidies Initiative (2008), a Geneva-based program of the International Institute for Sustainable Development, released a report in 2008 noting that China's actual biofuel subsidies exceed official government estimates. The Chinese government reports total government support for biofuels at RMB 780 million (approximately US$115 million) in 2006, expecting that figure to reach a staggering RMB 8 billion (more than US$1.2 billion) by 2020. In reality, support is much higher, as the official statistics do not include money for such things as feedstocks. The Chinese government, for example, provides RMB 3000 (US$460) per hectare per year, since 2007, for farmers growing feedstock (some of which is converted to biofuels) on marginal land.

As billions (perhaps *hundreds* of billions) of dollars of taxpayer's money worldwide are spent annually to support biofuels, who are the actors benefiting most from this torrent of cash? Chapter 3 reviews the subject of market concentration in the food system. Similar patterns of increasing concentration have been documented in the biofuel industry. By some estimates, there has been an 800 percent increase in venture capital investment in this sector from 2004 to 2007, most of that coming from multinational corporations (Holt-Gimenez 2007).

An inverse relationship has also been reported between market concentration and the state's regulatory capacity – that is, as the former increases, the latter diminishes (Utting and Clapp 2008). This has caused a number of scholars to view corporate social responsibility initiatives among major biofuel firms with skepticism. Structures are in place, for example, for multistakeholder programs for sustainability in soy and palm oil production (like the Roundtable on Sustainable Palm Oil and the Round Table on Responsible Soy), which claim to be interested in the environmental and social impacts of the industries. However, a growing body of research indicates that private sector voluntary programs have not proven particularly effective in regulating activities across transnational corporations as oversight and implementation are weak and sanctions for noncompliance absent. Granted, some legal mechanisms have emerged in recent years in an attempt to make firms more accountable for their actions as they pertain to environmental and social justice. The deep pockets possessed by multinationals, however, allow them to maintain a highly skilled legal department making justice, typically, elusive (Dauvergne and Neville 2010; Utting and Clapp 2008).

Already-dominant players in the agrifood system are taking advantage of their structural position in still other ways. Monsanto, for instance, has developed genetically modified sugarcane that is resistant to its Roundup Ready herbicide which will eventually be marketed heavily in Brazil with its rapidly expanding sugarcane biofuel market. Monsanto also hopes to begin selling an engineered maize variety with abnormally high levels of starch, designed specifically for ethanol production. Monsanto acquired the German company Tinplant Biotechnik, which possess one of the world's largest breeding programs of miscanthus and switchgrass. Another strategy by agribusiness has been to genetically engineer biomass conversion enzymes directly into plants (Torney *et al.* 2007). Engineering plants to carry the microbial cellulose enzymes to facilitate the conversion of fermentable sugars to ethanol will make patented plants appreciably more attractive than nonpatented varieties (Carolan 2009b).

A recent analysis lends support to the conclusion that patents are being used to concentrate control of the biofuels sector into the hands of a few multinational corporations (Glenna and Cahoy 2009; Figure 8.6). As of 2008, the CR3 for cellulosic ethanol patents was approaching 40 percent (in other words, three firms controlled roughly 40 percent of the market). And joint ventures and acquisitions since 2008 – like DuPont's acquisition of Dansico in 2011 and all their patents on enzymes for cellulosic ethanol production – suggest that market concentration in this sector has only increased.

Are biofuels pro-poor?

There is general consensus within the social science literature that raising biofuel crops is a questionable (at best) pro-poor strategy. Biofuel production may provide opportunities for high economic returns for some farmers, especially among those who can afford initial capital investments. Granted, instances have been documented where peasant farmers have profited from biofuel crops, such as when the price

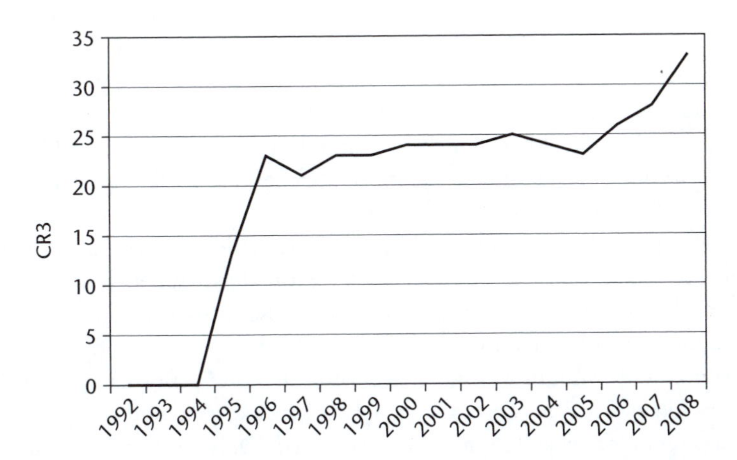

FIGURE 8.6 Patent ownership of cellulosic patents by top three firms, 1992–2008. Based on Glenna and Cahoy (2009)

of palm oil rose from US$570 to over US$1,440 per metric ton from 2007 to 2008 (McCarthy and Zen 2010) or when global sugar prices reached a 30-year high in August 2009 (Richardson 2010). More often, however, the peasant farmers are unable to capitalize on these opportunities.

One crop receiving a lot of attention of late is jatropha, a hardy, drought-tolerant shrub, which explains its wide promotion as a crop ideally suited for marginal lands. In 2008, jatropha was planted on an estimated 900,000 hectares globally, 760,000 hectares of which is located in Asia, 120,000 hectares in Africa, and 20,000 hectares in Latin America (primarily Brazil and Mexico). By 2015, it is estimated that jatropha will be planted on 12.8 million hectares (IRIN 2011).

While jatropha is twice as land intensive as sugarcane, sugarcane is three times more capital intensive as jatropha. Consequently, farmers with greater access to capital and credit have the option of investing in sugarcane and the higher returns per unit of land that this crop brings (Dauvergne and Neville 2010). Sugarcane is a capital intensive crop to establish and as such is arguably most economically efficient at large plantation scales (Richardson 2010). Jatropha farming, conversely, is widely believed to require minimal levels of capital, credit, and water (Arndt *et al.* 2009). This is why biodiesel from jatropha is said to be more pro-poor than bioethanol from sugar, as the former can more easily be produced by small-scale farmers (Arndt *et al.* 2009). Indeed, jatropha is generally viewed as *the* pro-poor biofuels plant (see, e.g., Brittaine and Lutaladio 2010). Let's look closer at this claim.

A popular claim by jatropha proponents is that it grows almost anywhere, including marginal lands where very little else can. This has given it the identity of being a pro-wasteland crop. Let's unpack this identity.

The flexible classification of "wastelands" has allowed government and corporations to easily acquire land under the auspices of this label, including for purposes of jatropha cultivation (Ariza-Montobbio *et al.* 2010). Social scientists have long been suspicious about designations referring to "idle," "waste," or "marginal" land. Such labels inevitably ask the question, "idle" or "marginal" for *who* and in pursuance of *what* end? As one study on the subject notes, "growing evidence raises doubts about the concept of 'idle' land. In many cases, lands perceived to be 'idle', 'underutilized', 'marginal' or 'abandoned' by government and large private operators provide a vital basis for the livelihoods of poorer and vulnerable groups, including through crop farming, herding and gathering of wild products" (Cotula *et al.* 2008: 22–3).

The pro-poor discourse surrounding jatropha is aided by the belief that the crop requires minimal water and labor, making it perfectly suited, it is said, for the small-scale farmer. Framing jatropha as "pro-poor" also helps build consensus around the idea that local rural development is compatible with raising biofuels for industrial economic growth. In reality, analyses of plantations' performance reveals that jatropha irrigation inputs were much higher than expected and the yield of the crop one-tenth of what farmers were promised. Only farmers with access to capital and credit could maximize yields through irrigation and the application of petrochemical inputs. Moreover, the conversion of land to a jatropha monoculture diminished crop diversity. This loss of diversity increased farmers' vulnerability, particularly among small-scale farmers who lacked the resources to mitigate risks with external inputs (Ariza-Montobbio *et al.* 2010).

In a study of jatropha plantations in the Dakatcha Woodlands of Kenya, it was found that the plants require copious amounts of water and nutrients to maximize yields (IRIN 2011). The study also casts jatropha-based biofuels as having a significant carbon footprint. Specifically, the lifecycle of jatropha-based biofuels were said to emit 2.5–6 times *more greenhouse gases* than fossil fuels, due largely to the clearing of the forest (which stores significant quantities of carbon in its vegetation and soil) in order to make room for the plant.

Research looking at the developmental potential of biofuels also notes problems in how "sustainability" is defined. Discussions about the sustainability of biofuels have largely centered on the consequences for small-scale landholders, food security, and lifecycle net carbon emissions. Rarely discussed, however, are issues related to land tenure, nonagricultural forest communities (that are increasingly displaced to make room for biofuel crops), or the indirect deforestation resulting from displaced communities being pushed onto marginal lands (Dauvergne and Neville 2010). When talking about the pro-poor potential of biofuel crops we cannot ignore those small-scale landholders residing in forest communities who are being displaced so we can have fuel for our cars.

Food *v.* fuel

"We are witnessing the beginning of one of the great tragedies of history," writes Lester Brown, founder of the Worldwatch Institute and current Director of the Earth Policy Institute. He continues: "The United States, in a misguided effort to reduce its oil insecurity by converting grain into fuel for cars, is generating global food insecurity on a scale never seen before" (Brown 2008). The Chairman and former CEO of Nestlé, the world's largest food and beverage company, Peter Brabeck-Letmathe, agrees with Brown's assessment, explaining, "If as predicted we look to use biofuels to satisfy twenty percent of the growing demand for oil products, there will be nothing left to eat. To grant enormous subsidies for biofuel production is morally unacceptable and irresponsible" (as quoted in Tenenbaum 2008: A255).

Biofuels received considerable criticism at the height of the 2007–2008 World Food Crisis, when global food prices increased dramatically. To be fair to biofuels, however, that period in time represented a perfect storm (or perfectly terrible storm) of sorts. Not only was biofuels production skyrocketing but severe weather events in some part of the world (wiping out crops), rising global populations, record (or near record) oil prices, and increasing world demand for meat (in Asian countries in particular) all lay claim to some blame for rising world food prices during this period. Proponents of ethanol also point out that distillers grains – a by-product of ethanol production – is recycled back into the food system by being blended into animal feed. Thus, roughly one-third of the original feed value of the corn entering the ethanol process in the USA returns to the food supply as feed for livestock; though, as critics counter, this still means two-thirds of this food is lost that could have otherwise been used to feed people.

Some even went as far as to say that the rising food prices brought on by biofuels is a good thing, at least from a public health perspective – to which I ask, *which* public and *whose* health are they talking about? Their argument goes something like this: increases in food prices could be beneficial to our health as, say, a doubling of corn prices would make corn syrup more expensive, which, in turn, would make things like soda and anything else made from high fructose corn syrup more expensive. And if made more expensive, it would be reasonable to assume that consumers would consume less of these foods (*Businessweek* 2007). That logic may be well and good if you can afford to pay higher food prices and have access to a diversity of food alternatives. For millions of the world's poor (even those residing in affluent nations), however, the claim that rising food prices is good for public health is ignorant and dangerous.

As long as biofuels continue to be derived primarily from food – what are known as first generation biofuels – it is difficult to see their long-term value. I have yet to hear a sound argument in favor of massive increases in global (first generation) biofuel output when so many in the world continue to be malnourished, especially given the massive subsidies required to realize the earlier mentioned production goals. And besides, the more popular these fuels become, the more they will drive up the demand for (and thus the price of) products like corn, which in turn will make their success even more dependent on government subsidies.

The global land grab

The term "global land grab" refers to transcontinental land deals whereby corporations, investment firms, or state-owned enterprises lease or purchase large areas of land in other countries for agricultural purposes (Carolan 2011). Of the nearly 400 projects that have been inventoried (in 80 countries), 37 percent are described as for food production while 35 percent for the production of biofuel crops (Box 8.3) (GRAIN 2010). A recent estimate by the International Food Policy Research Institute places the land "grabbed" at 20 million hectares (49,421,076 acres) – an amount equivalent to one-fifth of all the agricultural land in the EU, with a value of US$30 billion (Bladd 2009).

Typically, land is either practically or literally given away under these deals. Job creation and infrastructure development are viewed as the main economic benefits for developing countries. In Sudan, for example, a feddan (0.42 hectare) can be leased for as little as US$2–3 a year. In Ethiopia, rent was being paid in four out of the six projects examined in an FAO sponsored study, with lease prices ranging from US$3 to US$10 per hectare per year. Madagascar is currently under contract with the biofuel conglomerate Green Energy Madagascar (GEM). Under this contract, GEM pays no rental fees for the leased 450,000 hectares. Instead,

BOX 8.3 BIOFUELS AND THE ACCELERATION OF THE GLOBAL LAND GRAB

With "peak oil" comes "peak soil" as countries rush to find land to increase biofuel output in an attempt to counter rising crude oil prices, slow climate change, and increase domestic energy production. As explained in a recent report on the subject, "the EU is reducing its own emissions by raising emissions in developing countries that produce the feedstock oils (through increased deforestation and land use change, for example) and are not bound by emissions reduction targets, especially Indonesia and countries in Latin America" (Smolker *et al.* 2008: 38). A recent study documents a strong link between deforestation and the international demand for agricultural products (including biofuels) in the 2000–2005 period (DeFries *et al.* 2010). Elsewhere it has been noted that, in an attempt to meet Kyoto Protocol carbon reduction commitments through biofuels, countries may actually be *speeding up* climate change through carbon releases from forest conversion from the millions of hectares "grabbed" in less affluent nations (Danielsen *et al.* 2009).

they promise to build up local infrastructures and hire around 4,500 part-time workers (Cotula *et al.* 2009: 79–80).

How, then, do the countries leasing their lands benefit under contracts so seemingly favorable to the leasee? First, there is little evidence that these deals increase public revenues through taxes, given the "tax holidays" involved. International land deals granted by Ethiopia, for example, typically come with a profit tax holiday for a period of 5 years. It is estimated that this 5-year tax exemption costs the Ethiopian government US$60,276,000 for *each* project (Cotula *et al.* 2009: 80). Instead, benefits are said to come in the form of other commitments. Qatar, for instance, is leasing 40,000 hectares of land in a fertile River Tana Delta on the coast of Kenya. In exchange for the lease Qatar provided the Kenyan government a £2.4 billion loan to construct a port on the Indian Ocean island of Lamu (*Daily Telegraph* 2008). Such benefits, however, depend heavily on whether the lessee actually delivers on its promise. And enforceability of promises depends upon the details of the contract and national legislation. GEM in Madagascar, for example, ended up investing in increased mechanization, despite promises to pursue a labor-intensive business model and create jobs in exchange for 450,000 hectares. Other countries, such as Mali, have legislation allowing for the termination of leases if lessees fail to pay fees or maintain contractual obligations (Cotula *et al.* 2009: 82).

Those who lose the most from these arrangements are the indigenous communities living on the land prior to it being sold or leased to foreign entities. While the land in many of these cases may be technically owned by the state, it is a vital resource to a number of populations, from farmers to hunter-gatherers and herders. In these countries, because land laws do not require the consent of those living on the land in order for it to be leased to another party, most contracts can be enacted without any local participation or notification. Not surprisingly, then, conflicts have erupted between local populations and new landowners (or renters) (*New Scientist* 2011).

Then there is the issue of food security for the (often less affluent) nation leasing their land. Many of the nations handing over land to foreign entities are incredibly food insecure. Reports have surfaced indicating that some investors are pushing for explicit provisions guaranteeing full repatriation of produce, including where this requires amending the national law of the host state. Pakistan's Federal Investment Minister, Waqar Ahmed Khan, is quoted as saying that the government would make sure the investors in corporate farming get the entire crop, even when those crops are located in food insecure countries: "We are negotiating with investors from Gulf states, particularly Saudi Arabia, for investment in corporate farming. Investors will be ensured repatriation of 100 per cent crop yield to their countries, even in the case of food deficit" (as quoted in *The News* 2009). In the end, the Pakistani government had to revise its policy in response to public outrage and now call for "reasonable percentages" of produce to be exported (Cotula *et al.* 2009: 87).

Transition . . .

I have discussed some of the political, economic, social, and dietary realities that threaten food security. In the following chapter, I address how social scientists have dealt with the productive base – what some might call "nature" – that makes agriculture possible. It might surprise some to learn that *social* scientists even concern themselves with a subject that, at its face, seems solely the province of natural scientists. The sociological imagination, it turns out, can tell us a lot about nature in general and agroecology in particular.

Discussion questions

1 Why are GDP per capita and average animal product consumption so strongly correlated to each other (as indicated in Figure 8.2)?
2 Does meat continue to sit atop the food hierarchy or has its prestige been diminished in recent decades? If you think its social position has become diminished, what are some of the factors that have undermined meat's once privilege position?
3 Do you think meat is still gendered?
4 As stated above, biofuels (such as ethanol) have not only been around but have been used for well over a century to power automobiles. Why then have they only recently become so popular?

Suggested reading: Introductory level

Carolan, M. 2009. A sociological look at biofuels: ethanol in the early decades of the twentieth century and lessons for today. *Rural Sociology* 74(1): 86–112.

Lang, T., M. Wu, and M. Caraher. 2010. Meat and policy: Charting a course through complexity. In *Meat Crisis: Developing More Sustainable Production and Consumption*, edited by J. D'Silva and J. Webster. London: Earthscan, pp. 254–74.

Suggested reading: Advanced level

Kubberød, E., Ø. Ueland, Å. Tronstad, and E. Risvik. 2002. Attitudes towards meat and meat-eating among adolescents in Norway: a qualitative study. *Appetite* 38(1): 53–62.

McMichael, P. 2009. The agrofuels project at large. *Critical Sociology* 35(6): 825–39.

Zoomers, A. 2010. Globalisation and the foreignisation of space: Seven processes driving the current global land grab. *Journal of Peasant Studies* 37(2): 429–47.

Note

1 http://www.jct.gov/publications.html?func=startdown&id=3665

References

Ariza-Montobbio, P., L. Sharachchandra, K. Giorgos, and Joan Martinez-Alier. 2010. The political ecology of Jatropha Plantations for biodiesel in Tamil Nadu, India. *Journal of Peasant Studies* 37(4): 875–97.

Arndt, C., R. Benfica, F. Tarp, J. Thurlow, and R. Uaiene. 2009. Biofuels, poverty, and growth: A computable general equilibrium analysis of Mozambique. *Environment and Development Economics* 15(1): 81–105.

Bain, C. 2011. Local ownership of ethanol plants: What are the effects on communities? 35: 1400–7.

Balmford, Andrew, Rhys Green, and Jorn Scharlemann. 2005. Sparing land for nature: Exploring the potential impact of changes in agricultural yield on the area needed for crop production. *Global Change Biology* 11: 1594–605.

Bedford, J. and S. Barr. 2005. Diets and selected lifestyle practices of self defined adult vegetarians from a population-based sample suggest that are more "health conscious". *International Journal of Behavioural Nutrition and Physical Activity* 2(1): 4–14.

Bender, A. 1976. Food preferences of males and females. *Proceedings of the Nutrition Society* 35: 181–9.

Bladd, J. 2009. Call for GCC "land grab" policy to stop. *Arabian Business.com*, September 7, http://www.arabianbusiness.com/566961-call-for-gcc-land-grab-policy-to-stop-experts, last accessed March 22, 2011.

Brittaine, R. and N. Lutaladio. 2010. Jatropha: A Smallholder Bioenergy Crop – The Potential for Pro-Poor Development. Food and Agriculture Organization of the United Nations. Paris: FAO, http://www.fao.org/docrep/012/i1219e/i1219e.pdf, last accessed April 11, 2011.

Brown, L. 2008. Why ethanol production will drive world food prices even higher in 2008. *Environment New Service* January 25, 2008, http://www.ens-newswire.com/ens/jan2008/2008-01-25-insbro.asp, last accessed August 21, 2011.

Bruinsma, Jelle. 2009. The resource outlook to 2050: By how much do land, water and crop yields need to increase by 2050? FAO expert meeting on How to Feed the World in 2050, Rome, ftp://ftp.fao.org/docrep/fao/012/ak971e/ak971e00.pdf, last accessed November 7, 2010.

Businessweek. 2007. Food vs Fuel. *Businessweek* February 5, http://www.businessweek.com/magazine/content/07_06/b4020093.htm, last accessed August 21, 2011.

Carolan, M. 2009a. Ethanol versus gasoline: The contestation and closure of a socio-technical system in the USA. *Social Studies of Science* 39: 421–48.

Carolan, M. 2009b. A sociological look at biofuels: Ethanol in the early decades of the twentieth century and lessons for today. *Rural Sociology* 74(1): 86–112.

Carolan, M. 2011. *The Real Cost of Cheap Food*. London: Earthscan.

Chazan, G. 2011. Biofuels industry battles past bumps in the road. *The Wall Street Journal*, April 7, http://online.wsj.com/article/SB10001424052748703662804576188851797909800.html?mod=googlenews_wsj, last accessed April 8, 2011.

Cotula, L., N. Dyer, and S. Vermeulen. 2008. Fuelling exclusion? The biofuels boom and poor people's access to land. International Institute for Environment and Development (IIED) and FAO, http://www.iied.org/pubs/pdfs/12551IIED.pdf, last accessed April 11, 2011.

Cotula, L., S. Vermeulen, R. Leonard, and J. Keeley. 2009. Land grab of development opportunity: Agricultural investment and international land deals in Africa. Rome, Food and Agriculture Organization of the United Nations (FAO), and Laxenburg, International Institute for Applied Systems Analysis (IIASA), www.ifad.org/pub/land/land_grab.pdf, last accessed March 23, 2011.

Cullather, Nick. 2007. The foreign policy of the calorie. *American Historical Review* 112(2): 337–64.

Daily Telegraph. 2008. Qater to lease 100,000 acres in Kenya in return for port loan. *Daily Telegraph*, December 3, http://www.telegraph.co.uk/news/worldnews/middleeast/qatar/3543887/Qatar-to-lease-100000-acres-in-Kenya-in-return-for-port-loan.html, last accessed March 12, 2011.

Danielsen, F., H. Beukema, N. Burgess, F. Parish, C. Brühl, P. Donald, D. Murdiyarso, B. Phalan, L. Reijnders, M. Struebig, and E. Fitzherbert. 2009. Biofuels plantations on forested lands: Double jeopardy for biodiversity and climate. *Conservation Biology* 23(2): 348–58.

Dauvergne, P. and K. Neville. 2010. Forests, food, and fuel in the tropics: The uneven social and ecological consequences of the emerging political economy of biofuels. *Journal of Peasant Studies* 37(4): 631–60.

DeFries, R., T. Rudel, M. Uriarte, and M. Hansen. 2010. Deforestation driven by urban population growth and agricultural trade in the twenty-first century. *Nature Geoscience* 3(3): 178–81.

Dixon, J. 2002. *The Changing Chicken: Chooks, Cooks, and Culinary Culture*. Sydney, Australia: UNSW Press.

Douglas, M. and M. Nicod. 1974. Taking the biscuit: The structure of British meals. *New Society* 19: 744–7.

Economist The. 2011. The 9-billion people question: A special report on feeding the world. *The Economist*, February 26, http://www.economist.com/surveys/downloadSurveyPDF.cfm?id=18205243&surveyCode=%254e%2541&submit=View+PDF, last accessed April 3, 2011.

FAO. 2009. Food production will have to increase by 70 percent – FAO Convenes High-Level Expert Forum, Food and Agriculture Organization. United Nations, Rome, http://www.fao.org/news/story/0/item/35571/icode/en/, last accessed November 6, 2010.

Fox, N. and K. Ward. 2008a. Health, ethics and environment: A qualitative study of vegetarian motivations. *Appetite* 50(2–3): 422–9.

Fox, N. and K. Ward. 2008b. You are what you eat? Vegetarianism, health, and identity. *Social Science and Medicine* 66(12): 2585–95.

General Accounting Office. 2000. Tax Incentives for Petroleum and Ethanol Fuels. GAO/RCED-00–301R. http://www.gao.gov/new.items/rc00301r.pdf, last accessed April 9, 2011.

Glenna, L. and D. Cahoy. 2009. Agribusiness concentration, intellectual property, and the prospects for rural economic benefits from the emerging biofuel economy. *Southern Rural Sociology* 24(2): 111–29.

Global Subsidies Initiative. 2008. Biofuels, at What Cost? Government Support for Ethanol and Biodiesel in China. The Global Subsidies Initiative of the International Institute for Sustainable Development, Geneva, Switzerland, http://www.globalsubsidies.org/files/assets/China_Biofuels_Subsidies.pdf, last accessed April 10, 2011.

Godfray, H., J. Beddington, I. Crute, L. Hadded, D. Lawrence, J. Muir, J. Pretty, S. Robinson, S. Thomas, and C. Toulmin. 2010. Food security: The challenge of feeding 9 billion people. *Science* 327: 812–18.

Goe, R. and A. Mukherjee. 2008 Correlates of ethanol factory location in the north central region of the US. Paper presented at the annual meeting of the Rural Sociological Society, Radisson Hotel-Manchester, Manchester, New Hampshire, July 28.

GRAIN. 2010. The World Bank in the Hot Seat Over Land Grabbing. GRAIN Policy Briefing, May 7, http://www.commondreams.org/headline/2010/05/07-0, last accessed July 22, 2010.

Gvion-Rosenberg, L. 1990. Why do vegetarian restaurants serve hamburgers?: Toward an understanding of a cuisine. *Semiotica* 80: 61–79.

Hoek, A., A. Pieternel, A. Stafleu, and C. de Graaf. 2004. Food-related lifestyle and health of Dutch vegetarians, non-vegetarian consumers of meat substitutes, and meat consumers. *Appetite* 42: 265–72.

Holm, L. and M. Mohl. 2000. The role of meat in everyday culture: an analysis of an interview study in Copenhagen. *Appetite* 34: 277–83.

Holt-Gimenez, E. 2007. Displaced peasants, higher food prices, and a crutch for the petrol industry. *Global Policy Forum*, 1–9, www.globalpolicy.org/component/content/paper/212/45389.html, last accessed April 10, 2011.

IRIN. 2011. Jatropha – Not Really Green. Report from the UN Office for the Coordination of Humanitarian Affairs, United Nations, New York, http://www.irinnews.org/PrintReport.aspx?ReportID=92267, last accessed April 11, 2011.

Kiliánová, G., O. Danglová, and M. Kanovský. 2005. *Communities in Transformation: Central and Eastern Europe*. London: Transaction Publishers.

Koplow, D. 2007. Biofuels – At What Cost? The Global Subsidies Initiative of the International Institute for Sustainable Development, Geneva, Switzerland, http://www.globalsubsidies.org/files/assets/Brochure_-_US_Update.pdf, last accessed April 9, 2011.

Lappé, F. 1971. *Diet for a Small Planet*. New York: Ballantine.

Lévi-Strauss, C. 1969. *The Raw and the Cooked*. New York: Harper and Row.

Lindeman, M. and M. Sirelius. 2001. Food choice ideologies: The modern manifestations of normative and humanist views of the world. *Appetite* 37: 175–84.

Lockie, S., and L. Collie. 1999. "Feed the man meat": Gendered food and theories of consumption. In *Restructuring Global and Regional Agricultures: Transformations in Austral Asian Agri-Food Economies and Spaces*, edited by D. Burch, J. Goss, and G. Lawrence. Aldershot, UK: Ashgate, pp. 255–73.

Lokuruka, M. 2006. Meat is the meal and status is by meat. *Food and Foodways* 14(3–4): 201–29.

Lupton, D. 1996. *Food, the Body, and the Self.* Thousand Oaks, CA: Sage.

McCarthy, J. and Z. Zen. 2010. Regulating the oil palm boom: Assessing the effectiveness of environmental governance approaches to agro-industrial pollution in Indonesia. *Law and Policy* 32(1): 153–79.

New Scientist. 2011. African land grab could lead to future water conflicts. *New Scientist*, May 26, http://www.newscientist.com/article/mg21028144.100-african-land-grab-could-lead-to-future-water-conflicts.html, last accessed August 21, 2011.

News The. 2009. Corporate farming raises concerns among local growers. *The News*, January 28, http://www.thenews.com.pk/print1.asp?id=159380, last accessed April 23, 2011.

Peterson, E. Wesley. 2009. *A Billion Dollars a Day: The Economics and Politics of Agricultural Subsides.* Malden, MA: Wiley-Blackwell.

Probyn, E. 1999. Beyond food/sex: Eating and an ethics of existence. *Theory, Culture and Society* 16(2): 215–28.

Rabobank 2010. Sustainability and Security of the Global Food Supply Chain. Rabobank Group, The Netherlands, http://www.rabobank.nl/images/rabobanksustainability_29286998.pdf?ra_resize=yes&ra_width=800&ra_height=600&ra_toolbar=yes&ra_locationbar=yes, last accessed November 7, 2010.

REN21. 2010. *Renewables 2010: Status Report, Renewable Action Policy Network for the 21st Century (REN21).* Paris: REN21 Secretariat.

Richardson, B. 2010. Big sugar in Southern Africa: Rural development and the perverted potential of sugar/ethanol exports. *Journal of Peasant Studies* 37(4): 917–38.

Rogers, Stephanie. 2009. Egypt regrets killing trash-eating pigs. *The Mother Nature Network*, September 22, http://www.mnn.com/lifestyle/health-well-being/stories/egypt-regrets-killing-trash-eating-pigs#, last accessed March 5, 2010.

Rossi, A. and C. Hinrichs. 2011. Hope and skepticism: Farmer and local community views of the socio-economic benefits of agricultural bioenergy. *Biomass and Bioenergy* 35: 1418–28.

Rozin, P., M. Markwith and C. Stoess. 1997. Moralization and becoming a vegetarian: The transformation of preferences into values and the recruitment of disgust. *Psychological Science* 8: 67–73.

Schade, Carleton and David Pimentel. 2010. Population crash: Prospects for famine in the twenty-first century. *Environment, Development and Sustainability* 12(2): 245–62.

Smolker, R., B. Tokar, and A. Petermann 2008. The Real Cost of Agrofuels: Impacts on Food, Forests, Peoples and the Climate. Global Forest Coalition and Global Justice Ecology Project, http://www.globalbioenergy.org/uploads/media/0805_Global_Forest_Coalition_-_The_real_cost_of_agrofuels.pdf, last accessed April 11, 2011.

Sobal, J. 2005. Men, meat, and marriage: Models of masculinity. *Food and Foodways* 13: 135–58.

Stack, Liam. 2009. For Egypt's Christians, pig cull has lasting effects. *Christian Science Monitor* September 3, http://www.csmonitor.com/World/Middle-East/2009/0903/p17s01-wome.html.

Tenenbaum, D. 2008. Food vs. fuel: Diversion of crops could cause more hunger. *Environmental Health Perspectives* 116(6): A254–7.

Torney, F., L. Moeller, A. Scarpa, and K. Wang. 2007. Genetic engineering approaches to improved bioethanol production from maize. *Current Opinion in Biotechnology* 18: 193–9.

Tudge, Colin. 2010. How to raise livestock – and how not to. In *The Meat Crisis: Developing More Sustainable Production and Consumption*, edited by Joyce D'Silva and John Webster. London: Earthscan, pp. 9–21.

Twigg, J. 1984. Vegetarianism and the meanings of meat. In *The Sociology of Food and Eating*, edited by A. Murcott. Aldershot: Gower Publishing, pp. 18–30.

Tyndall, J., E. Berg, and J. Colletti. 2011. Corn stover as a biofuel feedstock in Iowa's bio-economy: An Iowa farmer survey. *Biomass and Bioenergy* 35: 1485–95.

USDA. 2006. The Economic Feasibility of Ethanol Production from Sugar in the United States, http://www.usda.gov/oce/EthanolSugarFeasibilityReport3.pdf, last accessed April 8, 2011.

Utting, P. and J. Clapp (eds). 2008. *Corporate Accountability and Sustainable Development*. Delhi, India: Oxford University Press.

Webster, J. (2010) The Meat and Dairy Crisis. Earthcast, 11 October, sponsored by Earthscan, www.earthscan.co.uk/Earthcasts/tabid/101760/Default.aspx, last accessed November 1, 2011.

9

AGROECOSYSTEMS AND THE NATURE OF "NATURES"

KEY TOPICS

- The concept of agroecology has broadened over the twentieth century, from a concept that initially referred to a marriage between agronomy and ecology to one that today brings together disciplines from all around the university (including sociology).
- An example of a meso-scale agroecological approach is the Community Capitals Framework, with its emphasis on the inter-relationship between natural, built, financial, political, social, human, and cultural capitals.
- A new strand of literature emerged in the 1990s, interested in the subjective experience of farmers, particularly in terms of how they understand the natural world, known as the social construction nature.
- Actor-network theory (ANT) is a methodological and conceptual approach that attempts to blur the lines between the so-called natural and social worlds.

Key words

- Actor-network theory
- Agroecology
- Built capital
- Community capitals
- Financial capital
- Human capital
- Molecularization
- Multifunctionality
- Natural capital
- Political capital
- Privatization
- Productivism
- Social capital
- Social constructionism
- Triple bottom line

A couple of times a year I'll have a student ask whether or not sociologists pay attention to ecology. My response usually goes something like, "Yes, sociologists of food and agriculture are interested in the natural world; but we're also careful not to draw too thick a line between 'society' and 'nature'." The sociology of food and agriculture literature is actually quite diverse in terms of how scholars have dealt with the material realm, as this chapter makes clear.

Agroecology: A brief introduction

The term "agroecology" dates back to at least the late 1920s (see, e.g., Klages 1928), although it is not difficult to find antecedents that spoke of similar agro-ecological processes without actually using the term (King 1911). Initially, the term referenced the coming together of two disciplines: agronomy and ecology. Later, other disciplines were brought into the fold as the scope and scale of agroecology evolved (Figure 9.1), such as zoology, botany and plant physiology, animal science, pest ecology, and sociology. As the website Agroecology explains:

> Agroecology is:
>
> - The application of ecology to the design and management of sustainable agroecosystems.
> - A whole-systems approach to agriculture and food systems development based on traditional knowledge, alternative agriculture, and local food system experiences.
> - Linking ecology, culture, economics, and society to sustain agricultural production, healthy environments, and viable food and farming communities.
>
> *www.agroecology.com*

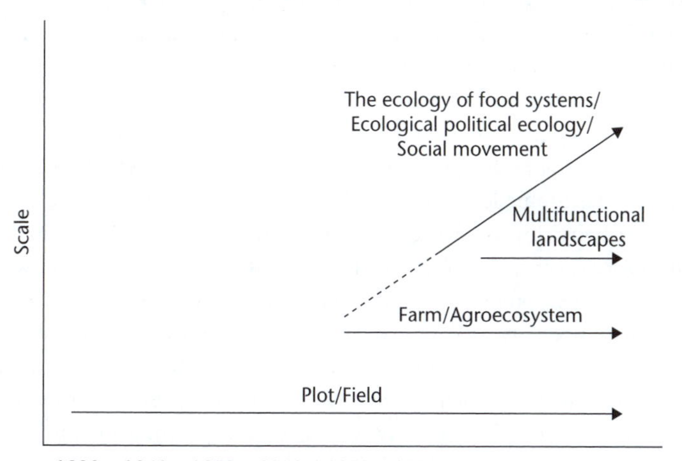

FIGURE 9.1 Changes (and increasing diversity) in scale and scope of agroecology research. Developed in consultation with Buttel (2002) and Wezel *et al.* (2009)

To understand the broadening of the term over the twentieth century, it is helpful to understand the ultimate purpose agroecology seeks to serve. Buttel (2002: 4–5) concisely sums this up by noting that "agroecology is both a *critique* of productivism, molecularization, and privatization as well as a set of *responses* to their shortcomings" (emphasis in original). Productivism refers to the strong consensus that emerged during the twentieth century among agricultural scientists, government officials, and agribusiness making increased productivity and/or output *the* principal goal of agricultural research. Molecularization speaks to modern agriculture's reductionist and thus highly abstract approach to the world. Monocultures, for example, are cast in a considerably more favorable light when cropping systems are abstracted from their ecological productive base. This abstraction helps mask the destruction monocultures bring to that base in the form of, among other things, biodiversity loss and the various nefarious effects associated with pesticide use (as monocropping typically necessitates the use of pesticides). Finally, privatization makes it harder to think ecologically, as it encourages people to only "see" up to their property line. At the farm level, for example, this makes it harder to implement, say, Integrated Pest Management (IPM) strategies (discussed in Chapter 7).

To take the most narrow definition, as a method of agricultural production agroecology is highly efficient and, yes, even highly productive (see, e.g., Altieri 1995; Gliessman 1998). It is well documented that small diverse farms that raise grains, fruits, vegetables, and livestock can easily out-produce, in terms of harvestable products per unit area, large, specialized (monoculture) operations by 20–60 percent (Altieri and Nicholls 2008). Gliessman (1998) notes that a 1.73-hectare plot of land in Mexico planted in corn produces as much food as a smaller 1-hectare plot planted with a mixture of corn, squash, and beans. In Brazil, fields planted in both corn (at 12,500 plants per hectare) and soybeans (at 150,000 plants per hectare) exhibited a 28 percent yield advantage over a comparable soybean monoculture (Altieri 1999). This inverse relationship between farm size and productivity has been attributed to a more efficient use of land, water, biodiversity (e.g. keeping intact internal ecological pest controls thus eliminating the need for pesticides), and other agricultural resources by small farmers. Small, diverse farms are also less vulnerable to catastrophic loss due to environment events – an attribute of growing importance in light climate change related weather risks (Alexander 2008).

Agroecological principles also lie at the center of many organic farms, which explains, when managed properly, their favorable comparison to conventional methods of food production. Examining approximately 300 studies from all around the world, Badgley *et al.* (2007) conclude that organic agriculture has the potential to feed the world. They further conclude that, after comparing yields between organic and conventional systems, organic systems have the potential to produce yields comparable to or *better than* conventional operations.

As noted in Figure 9.1, the term "agroecology" today can refer to any number of scales and involve any number of disciplines. In their review of the last 80 years of published work on the concept, Wezel *et al.* (2009) note that agroecology initially focused on the plot or field scale (1930s–1960s). The scope and scale of research notably expanded in the 1970s to landscape agroecosystems, expanding still further in recent years to food systems scales (though plot and field scale approaches persist up to the present). They further note in their review that the term can refer to, at least in some circles, a social movement, often subsumed within the larger umbrella term "sustainable rural development." The Brazilian Association of Agroecology (ABA), for example, was created in 2004. For the ABA, agroecology is the ecology of food systems *and* a social movement (Wezel *et al.* 2009). "Higher" scale conceptions of agroecology clearly allow for a greater role for the sociologist. Nevertheless, even traditional approaches leave space for the social scientist. While the sociologist would not be directly involved in the analysis of, say, a test plot they could provide invaluable insight into understanding the challenges and barriers farmers experience, thus keeping them from adopting agroecological principles.

Moreover, agroecological principles work because they are context-based. To put it another way, they work *with* natural, ecological principles rather than by trying to *bend* those principles to fit the needs of the mass-produced commodities associated with the green revolution. Agroecology thus puts the farmer first, to refer to a subject discussed previously in Chapter 7. Finding ways to communicate this knowledge "up" from farmer to university scientist or "horizontally" from farmer to farmer offers space for the social scientist to work.

Figure 9.1 also references, following Buttel (2002: 5), the "third most long-standing form of agroecology," what "can be referred to as *ecological political economy*." The principal focus of this tradition of agroecology is a political–economic critique of conventional agriculture, while also paying close attention to its ecological shortcomings. The dashed lines in the figure indicate early research in sociology of agriculture providing such a focus, without directly referencing to the term "agroecology," like the Mann–Dickinson thesis (see Chapter 2) and, later, looking beyond the farm-gate, the food regime approach (see Chapter 3). This tradition remains strong within the sociology of food and agriculture literature.

Finally, the figure also references multifunctional landscapes, which Buttel (2002: 6) calls "the most recent variety of agroecology." The central point of this approach is its denial of the primacy of the farm and the agricultural enterprise as the basic unit of production and thus of analysis. It views agriculture and the food systems more generally as embedded in landscapes and a broader socio-institutional complex. Multifunctionality explains that, in addition to the production of food and fiber, agriculture produces or maintains a number of other, mostly non-commodity, "outputs," like flood control, the maintenance of ecosystem services, rural development, and agricultural heritage and culture (Box 9.1).

Macro-level agroecological approaches are discussed elsewhere in this book, such as the food regime (see Chapter 3) and social movements (see Chapter 11). Let's turn our attention with the next section to a more meso-level framework: the Community Capitals Framework.

BOX 9.1 AGROECOLOGY: "GROWING" MORE THAN FOOD AND FIBER

When we think about agriculture we tend to view it as exclusively a site of food and fiber production. This is a terribly (and perhaps even dangerously) narrow view of all that is provided by this sector. Agriculture is important to us in such additional ways as:

Diet

- Enhancement of nutritional quality and the cycling of trace elements
- Food security

Public health

- Protection of the health of farm workers and consumers
- Suppression of vectors of human disease (e.g. mosquitoes, snails, ticks)

Environmental sustainability

- Preservation of biodiversity
- Protection of wildlife
- Preservation of our productive capacity against erosion, salinization, acidification, compaction
- Maintenance of an ecological community of natural enemies of pests and diseases of crops
- Protection of the general environment against runoff, eutrophication, volatization of nitrites, and dust in atmosphere
- Protection of water resources and quality

Rural development

- Supporting employment, farm income, and rural life
- Support for the economic independence of women

National development and sovereignty

- Contribute to the international balance of payments
- Defense of national sovereignty against possible dumping or political blackmail backed up by economic blockade
- Food sovereignty

Based on Carolan (2011) and Levins (2006)

Community Capitals Framework

Cornelia Flora (2001a) makes the case that the agroecosystem concept cannot be divorced from human communities. As she explains:

> Agriculture and forestry represent attempts of human communities to use the perceived potential of a local landscape to extract value and maintain human communities: rural, suburban, and urban. The systems by which agriculture and forestry transform the ecosystem into an agroecosystem are varied. Some systems deplete the natural capital of place, whereas others replenish it. The agroecosystems that emerge are not simply natural outgrowths of humans and landscapes with productive potential, but the product of human communities mediated by culture and technology.
>
> *C. Flora (2001a: 5)*

Ecosystem health is thus the product of a triple bottom line, involving economic, social, and environmental accountability (C. Flora 2008). These three sustainability pillars come together when a variety of capitals are present in sufficient degree. Those include natural, built, financial, political, social, human, and cultural capitals (Figure 9.2).

Though equally important in the achievement of the aforementioned triple bottom line, research looking at the interactions of these capitals shows some to be better starting points toward this end than others. This "community capitals"

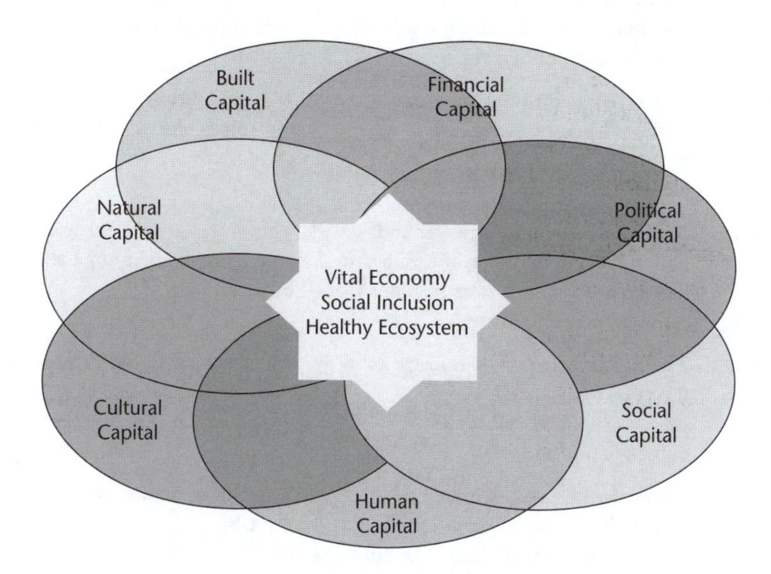

FIGURE 9.2 The seven capitals that constitute ecological, social, and economic sustainability. Developed by Cornelia Flora and Jan Flora (reprinted with their permission)

research, as it's called, points to social and human capitals as the ideal entry points, leading eventually to a "spiraling up" effect whereby all capitals are enhanced over time (Emery and Flora 2006). Beyond analytically teasing apart the various capitals at play in rural development, this approach, as discussed in Table 9.1, focuses on the interaction among the seven capitals and how they build upon each other. Using this framework, scholars have shown how investments in

TABLE 9.1 Seven capitals and their role in creating sustainable agroecosystems and communities

Capital	*Definition*	*Role sustainable agroecosystems and communities*
Natural	The natural biophysical assets of any given locale – can include natural resources (e.g. water, soil, air, minerals), amenities (e.g. trout streams and sandy beaches), and natural beauty	These assets, when utilized in a manner cognizant of ecological limits, represent the ecological productive base for the long-term growth of other capitals
Cultural	Institutionalized (widely shared) cultural symbols – attitudes, preferences, beliefs – that shape how we see the world, what we take for granted, and possible alternatives for social change	By investing in cultural diversity (and including those who are traditionally excluded) biodiversity and different ways of approaching change can be utilized to the enhancement of all capitals
Human	Includes the skills, knowledge, and abilities of the people within a community to enhance local as well as access outside resources	Increasing the knowledge base of a community will help that community bolster its other capitals
Social	The social glue of a community – includes levels of mutual trust, reciprocity, and a sense of share identity and future	A social lubricant in that it makes the enhancement of the other capitals considerably easier
Political	Access to structures of power and power brokers as well as the ability to influence the rules and regulations that shape access to resources	This access greatly enhances a community's ability to bolster other capitals
Financial	The financial resources available to invest in things like community capacity building and social entrepreneurship	These resources can help pay for the maintenance and accumulation of other capitals
Built	Infrastructure (also includes built "natural" areas, like reconstituted wetlands, ski runs, and artificial coral reefs)	This infrastructure supports other capitals

Informed by C. Flora (2008) and C. Flora *et al.* (2009).

human capital at the community level through, say, leadership training can impact financial capital as those leaders employed their newly acquired skills to acquire new funds and better manage existing funds, all of which, in turn, bolstered political capital (improved access), social capital (as social networks expanded), and so forth (Flint 2010; C. Flora *et al.* 2007). A community capitals approach expands our understanding of "return on investment," noting that it should be measured in terms of an increase in *all* capitals – not just financial. Furthermore, community capitals research firmly believes in involving people so they can direct change from within a community (Box 9.2).

There is considerable evidence linking social capital to equitable and sustainable solutions to local development problems (Pretty and Ward 2001). Social capital can lower the cost of working together, as it facilitates cooperation and reduces the likelihood of selfish action that comes at the expensive of social and environmental well-being (Coleman 1988; J. Flora 1998). Investments in natural capital alone rarely seem to produce long-term attitudinal and behavioral change. Thus, while regulations, laws, and economic incentives are commonly used to encourage a change in behavior (e.g. adoption of practices to reduce soil erosion), there is little evidence suggesting that they also elicit a change in attitudes, which is why farmers, for example, tend to revert to less-than-best management practices when incentives disappear or regulations go unenforced (Pretty and Smith 2004). These "deeper" attitudinal changes, while not occurring overnight, come from repeat participation in a social network with sufficient levels of mutual trust, bridging ("thin" relationships) and bonding ("thick" relationships) social capital, and opportunities for social learning.

At the same time, we must be aware that not all forms of social capital enhance individual, community, or environmental well-being. Shared rules and norms can entrap people within nonoptimal (and maybe within oppressive) arraignments, depleting human capital and all other capitals in the process. Too much homogeneity within a community has also been associated with xenophobic tendencies, groupthink, and the general exclusion of anyone not thought of as "one of us" (Daly and Silver 2008). The Mafia, for example, is held together by, among other things, very high levels of social capital. Yet I would not hold this organization up as the ideal model for rural economic and agroecological development.

Social capital is also gendered, "as men and women often have different networks and different ways of building and generating trust, due to their different material and cultural position" (C. Flora 2001b: 43). Indeed, it is quite fair to say that all capitals are gendered, as distribution and access to each varies, on average, considerably between men and women. The case has been repeatedly made that this gendered asymmetry in capitals must be redressed in the context of sustainable rural development. For example, research found that women in the Siquijor and Leyte provinces of the Philippines were significantly more inclined to protect common resources than men, as those resources represented the base for their domestic and market activities (e.g. women used local rivers when washing their family's clothes). Yet if those resources were not equally important to men they were repeatedly

BOX 9.2 COMMUNITY CAPITALS AND CHILDHOOD OBESITY

Obesity, as previously detailed (see Chapter 4) is an ecological phenomenon. Elaborating on this, C. Flora and Gillespie (2009) set out to explain how excess food intake (often in the form of calorically dense foods) and low levels of physical activity are products of one's community environment and the capitals therein implied. Flora and Gillespie argue that the present "choice architecture" helps perpetuate (and exacerbate) childhood obesity, referring to those structures that enhance or inhibit healthy choices (in terms of both caloric intake and energy expenditure). This architecture varies from community to community, noting that among inner city families this "choice" is much more constrained when compared to that available to suburban families.

According to Flora and Gillespie, market, state, and civil society actors need to invest in the seven capitals to reconfigure that architecture, thereby making it easier for institutions, families, and individuals to make healthier choices. At the community level, people need to organize to open those "choices" up. Flora and Gillespie contend that the community capitals framework helps focus mobilization efforts by providing a tool for identifying local capitals that can be invested in to reduce the rates of childhood obesity. Some of their conclusions include the following.

- Built capital that provides opportunities for recreation (e.g. community soccer fields, bike and hiking trails) cannot happen without local social capital influencing political capital, which together would enhance financial capital.
- A number of community initiatives have successfully utilized social capital to impact built capital (e.g. the construction of a playground) or to change a local school district's policies related to the availability and sale of unhealthy foods. These organizations may be temporary but because they are built on existing social capital they leave a solid base to take effective action to change political capital for a healthier environment for children.
- Getting healthier food into schools requires the mobilization of social capital to get decision makers to redefine "good value" from being about cost minimization to about healthy food and healthy students.
- Community efforts to reduce childhood obesity must also involve making healthy food, notably fruits and vegetables, affordable. This includes building cultural and social capital to convert natural capital to community gardens and increasing human capital in terms of increased knowledge of gardening and preparation of fresh fruits and vegetables.

destroyed in the name of economic growth – the developmental trajectory typically favored by those in positions of power (who continue to be overwhelmingly men) (Shields *et al.* 1996).

The subjective turn: From nature to "nature"

Before New Rural Sociology, in the words of Buttel *et al.* (1990: 44), the "social psychological-behaviorist era" ruled rural sociology. Beginning in the 1940s, this research tradition "would remain unchallenged in rural sociological studies of agriculture until the early to mid-1970s" (Buttel *et al.* 1990: 46). The 1980s brought with it a turn away from individual experience as approaches rooted in agrarian political economy dominated the literature. As Wilson notes (2001: 85), the prevalence of the agrarian political economy approach in the 1980s and 1990s produced a wealth of research with "a heavy emphasis on the importance of the state and policies, a strong focus on the importance of macro-economic factors in actor decision-making." By the early 1990s, despite close to two decades of theorizing social change in agriculture, we still knew very little about how and what farmers thought. It is therefore no surprise that a growing literature began to emerge during this period to fill that gap – a "turn," you might say, to subjective experience.

The social construction of nature

A significant literature has been assembled looking into how people (most notably farmers) understand nature – a process known as the social construction of nature. As early as the 1970s, Newby *et al.* (1978) made clear that the roots of these different understandings were derived, at least in part, from the experience of farmers in their daily interaction with the land. Given farmers' diminishing presence in the countryside, it might not seem relevant to focus on how they socially construct the term "nature." Yet, remember, while a small minority of the population, farmers have significant influence over the "look" (and touch, smell, etc.) of the countryside. And given the symbolic significance of the countryside for populations at large (such as its role in shaping national identity in some countries [Lowenthal 1991]), we must not be too quick to dismiss the perceptions of farmers. What they think matters.

The term "social construction" is a bit of a misnomer, or at least it can be easily misunderstood. To speak of the social construction of nature (or the social construction of anything) is not to deny the existence of a material world. The material phenomena we call "wetlands" or "corn" (to give just two examples) would still exist, for the social constructivist, if humanity suddenly vanished from the face of the earth. The point, rather, is that the *meaning* we give to those biophysical artifacts would cease to exist in the event of our extinction. To speak of the social construction of nature is to remind ourselves that the meaning we give to artifacts we deem "natural" is a sociological expression, which is why different people, groups, and cultures often perceive the same physical thing in different ways. As meaning-making animals, we all confer meaning onto – which is to say we all socially construct – the material world around us.

We also know that perception is a reflection of material circumstances; though just how far we take this statement is a topic of debate among social theorists. Greider and Garkovich (1994) point out that often there are power struggles between groups about *whose* construction of nature will be held as "true." In other words, knowledge does not have to be objectively true to have influence (McHenry 1997; Figure 9.3; Box 9.3). What matters is power, as getting people and organizations to perceive your knowledge claims as representing the truth is often what really matters.

Conservation and the "good farmer"

Research reveals that farmers may resist change on the basis of an anticipated change to their identity and/or social or cultural rewards that were traditionally conferred through conventional farming practices (Burton 2004); further evidence still of why it matters what farmers think. There are normative pressures to be a "good farmer," as defined by the social networks producers locate themselves within (Carolan 2005; Stock 2007). For example, woodlands approaches that encourage farmers to also become foresters and leisure providers have been met with considerable resistance as they ask farmers to essential become – identity-wise – something they are not: foresters and leisure providers (Lloyd *et al.* 1995). These findings were corroborated by Allison (1996), with his survey of farmers in the National Forest in the North Midlands of England. For them, it was *farming* that gave them their identity as well as a sense of accomplishment. Burgess *et al.* (2000: 125) nicely sum up this point, noting that "first and foremost, farmers saw

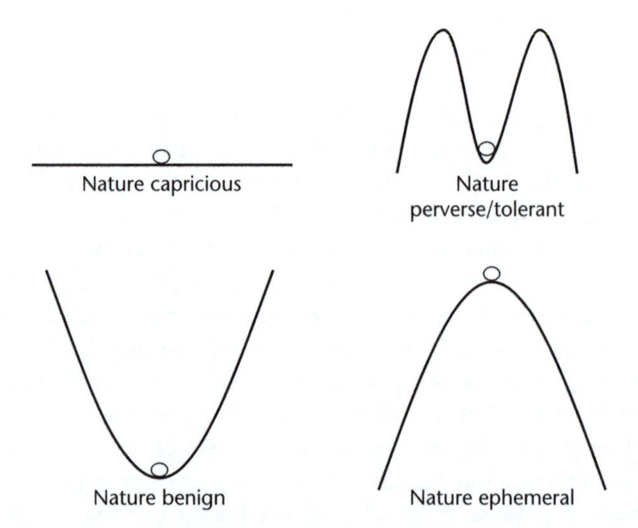

FIGURE 9.3 The four "myths of nature." Based on Schwarz and Thompson (1990)

BOX 9.3 FOUR "MYTHS" OF NATURE

Understanding assumptions about nature also help us better understand why individuals do what they do in terms of perceived links between those actions and ecosystems. Some scholars have categorized our understandings of "nature" into one of four conceptual "boxes" – what are called the four "myths of nature" (Schwarz and Thompson 1990). Using the movement of a ball in a landscape as an analogy for relations between human activity and the broader environment, it quickly becomes clear that what "myth" one ascribes to will drastically shape the perceived ecological consequences of their actions (Figure 9.3).

Nature benign

This is when nature is viewed as being capable of recovering quickly from significant disturbances (e.g. a large pollution event). It is a robust view of nature. Adherents to this myth often come to rely on technocratic and managerial approaches for speeding up or assisting nature's recovery. This view also encourages economic growth with little care for environmental consequences, as nature's self-renewal capabilities are thought to be strong enough to withstand most of what humanity can throw at it.

Nature ephemeral

This emphasizes the vulnerability of natural systems, where practically any human ecological footprint is regarded as potentially damaging to the integrity of natural systems. This view has close links with the "deep green" environmentalists. It is also consistent with advocacy of the precautionary principle – a better-safe-than-sorry regulatory approach that takes a guilty until proven innocent view of technology (in contrast to risk analysis, which takes an innocent until proven guilty stance).

Nature perverse/tolerant

Recognition of both nature's vulnerability and its resilience is incorporated in this myth. This view of nature places a premium on expert knowledge to ensure that natural limits are not exceeded. With this myth, it is acceptable to place pressure on natural ecosystems, as long as those systems are not overly abused.

Nature capricious

Gives credence to views of nature that emphasize its unpredictability. According to this myth, we have no way to predict the consequences of our actions on natural ecosystems. Nature is not law abiding; instead, randomness rules. As with nature benign, nature capricious discourages government regulations, yet for different reasons. The former favors no regulation because nature is viewed as immensely robust. Nature capricious, conversely, favors no regulation for the simple reason that we cannot regulate what we cannot understand.

themselves as food producers." Hence, to be a "good farmer" one need not be a "good forester." Indeed, in some networks, as long as food production remains a requisite of a farmer's identity, any actions that do not result in this end could ultimately diminish their status among other producers.

Similarly, research indicates that farmers are more open to conservation practices if those practices do not conflict with their farmer (as food-producer) identity. "Biodiversity," therefore, that does not threaten their bottom line – like barn owls, pheasants, and song birds – are therefore welcome. Conversely, foxes and out-of-place plants ("weeds") are not entitled to conservation in the eyes of many farmers. However, if one's social relationships (and thus identity) include individuals and organizations more sympathetic to organic and/or sustainable agricultural practices then those 'weeds' may be perceived as 'wildflowers' and thus just as much a component of a field's biodiversity as any other living organism (Carolan 2006; McHenry 1997).

Cary (1993) speaks of "symbolic conservation" to describe that while many surveyed farmers attributed environmental benefits to their behaviors, those actions were primarily the result of more instrumental motives. In other words, conservation was only done when it was profitable and when it "fit" with already held conceptions of what it meant to do "good farming." The concept "symbolic conservation" is useful as it reminds us that farmers negotiate the concept of "conservation." As farmers have become increasingly aware of needing to minimize their ecological footprint, at least when such activities don't cut into profits, they have begun to impute conservation motives to many activities that they would have previously done under the guise of "good farming" (McHenry 1997). This finding is important as it reminds us that farmers are motivated by more than just a need to be good stewards. Not surprisingly, farmers are also motivated by profit, which means conservation practices must be properly incentivized in order to achieve their widescale adoption. Also, as conceptions of being a "good farmer" shape what practices are considered acceptable within social networks, attempts could also be made to redraw the parameters of this concept, thereby making room for previously unacceptable practices, like forestry.

Actor-network theory

David Goodman famously argued back in 1999 that the "theoretical purview and contemporary political relevance of agro-food studies are significantly weakened by their methodological foundations, still largely unexamined, in modernist ontology" (Goodman 1999: 17). The problem with this ontology, which is held widely throughout social science, is that the distinction between nature and society is simply *assumed*. This assumption places "nature" outside and wholly separate from the "social" domain yet still somehow shaped by it. "The modernist ontology," Goodman continues, "thus supports the objectification of nature and its de-politicization, undermining coherent engagement with the bio-politics and ethical principles of environmental organizations and Green movements" (Goodman 1999: 17).

Social constructionism has a similar failing, in that it does not provide us with the conceptual tools to talk about nature and society as inter-related. To talk about socially constructing something presupposes there is something out there – *separate* from us – to impart meaning onto. Moreover, social constructionism tends to paint what we construct as passive. We give meaning to *it*. And as an "it" it is thoroughly incapable of having a role in that meaning-making process.

One way to bridge this ontological divide is offered by Actor Network Theory (ANT). ANT rejects such categorical divisions like "nature" and "society," seeing the world instead as made up as heterogeneous collective associations (Latour 1993). The one "sided" notions of agency (whereby only humans have it) is rejected by ANT, which "insists that social agency is never located in bodies and bodies alone, but rather that an actor is a patterned network of heterogeneous relations. [. . .] Hence the term actor-network – an actor is also, always, a network" (Law 1992: 384). ANT therefore represents, say its proponents, a more "symmetrical" approach, by equally considering the role of nonhumans in social order. The rationale for this is simple: when human actors attempt to create networks they are often very disappointed when nonhumans do not perform as anticipated (Callon 1986). So why not say nonhuman actors act too.

In "Where are the missing masses? Sociology of a few mundane artifacts," Latour (1992), a key ANT theorist, writes about, among other things, doors. Latour sets out to explore the potent effect of doors – and other technological artifacts – upon social order. His reference to the "missing masses" (a word play upon the black matter – the missing mass – in physics) allows him to point to something missed in conventional analyses of order: the power of things. Material artifacts, by bringing forth new relations, produce an *effect* that is often indistinguishable from normative, moral, and legal control. Some examples: a door shapes the speed and direction we can walk through walls; a speedgun fastened to a stop light enforces certain traffic codes similar to a human policeperson; and a seat belt that automatically slides over a person after ignition implements seat belt laws. In technology, then, lies the ability to delegate actor-like status to these material artifacts, which, in turn, means we delegate power to them – making then, in a word, powerful.

Take the debate over whether guns or people kill people. Given the assumptions of Western political thought, with its focus on states rather than relations, we seemingly have only these two choices – that either guns or people kill. But clearly a gun cannot shoot a bullet without the involvement of a human and a human cannot shoot a bullet without the involvement of a gun. The conventional framing of the debate misses a third interpretation, involving a "citizenweapon" or "weapon-citizen" (Latour 1999: 179). Thinking in terms of relations allows us to see that a citizen-with-a-gun-in-hand is no longer the same as an unarmed citizen, just as a loaded-weapon-in-hand is no longer the same object as an unloaded-gun-in-a-safe. ANT therefore not only offers a novel way to bridge the so-called divide between society and nature, it also opens up new ways to talk about what is and what ought to be.

Criticized a decade ago for failing to move agrifood studies beyond the aforementioned modernist ontology (Goodman 1999, 2001; Lockie and Kitto 2000), recent applications of ANT have spawned novel challenges to how we conventionally think about the food system. Take, for one example, a recent ANT-informed investigation into how we understand and respond to foodborne illness (Stuart 2010). While bacteria are typically blamed for these outbreaks, this article contends we are being too hasty in assigning guilt. Rather, we need to explore the larger context in which these bacteria emerge. What is being ignored is how these organisms are given a space act, thanks to the design of industrial agriculture and food processing systems. Upon closer inspection, human and nonhuman actors appear equally involved in industrial food outbreaks.

Eighty percent of outbreaks and 98.5 percent of all foodborne illnesses have been linked to bagged produce rather than unprocessed heads or bunches of greens. The source of the *E. coli* O157:H7 is believed to be animal manure. But the associations are a bit more complex than that. As ANT reminds us, causality is anything but black and white.

Antimicrobials, a necessity in large-scale animal confinement feeding operations, have been shown to enhance *E. coli* O157:H7 production. In addition, evidence strongly suggests that recent changes in cattle diets – more starch (corn) – further increases the presence and prevalence of *E. coli* O157:H7 in the gut of these animals. Thus, as feeding grain to cattle has become more widespread, manure today has higher concentrations of pathogenic bacteria than manure from cattle under a more traditional feeding regime (e.g. pasture-based diet). And the recent move to biofuels may further exacerbate the problem, as feedlot cattle fed distillers grain (a by-product from ethanol production) can shed up to six times more *E. coli* O157:H7 than cattle not eating this by-product. So the industrial cow – and a society that thinks it needs "cheap" meat – is part of the problem.

Another accusatory finger needs to be pointed at the industrialization of the salad industry. Manure used to fertilize fields (or fecal matter from birds, rats, rabbits, or anything else that has visited the field) can easily be sucked up with the mowed greens. In order to improve "efficiency" and reduce costs, labor is substituted for capital, which, in turn, allows for increases in the size of lettuce farms. But the mechanization of harvesting also comes at a cost, as machines cannot be as discriminatory as a human, as evidenced by the fact that machines also "harvest" things that humans never would, like fecal matter. The contaminated harvested greens are then sent to centralized plants where they are mixed together during processing, thereby spreading the fecal matter out over otherwise clean plants. And while the harvested greens are washed in a water–chlorine mixture to kill any bacteria, it is readily acknowledged that the legal levels of chlorine cannot kill all the bacteria that may be present – especially if a sizable amount of manure has contaminated the batch.

The author of the study also interviewed members of the produce industry. Most shared in the belief that the bag holding processed salad greens also has an important role in encouraging the growth of pathogenic *E. coli*. Specifically, they point to how the bag creates a highly hospitable environment in which bacteria can grow; growth that could then be further enhanced by any increases in temperature as it is trucked across the country. Furthermore, after being purchased, bags left in warm automobiles or kitchens for extended periods further increase the risk of bacteria growth.

In sum, the article argues that if we are serious about minimizing the risk of foodborne illness we need to look at the *system* that makes these outbreaks possible. Blaming bacteria or the owners of the farm said to be the source of the bacteria does nothing to inform our understanding of the root cause of these outbreaks. As the author of the study explains:

> Other industrial networks also need to be addressed. As stated earlier, confined animal operations, the use of antibiotics and cattle feeding regimes are likely to be manufacturing the *E. coli* O157:H7 that are now so problematic throughout industrial food systems. US agricultural subsidies for corn play an important role by providing cheap cattle feed. [. . .] It remains unlikely that *E. coli* O157:H7 contamination in food processing can be fully addressed without first re-evaluating the industrial livestock sector. However, in this case industrial actors are again looking for quick fixes such as cattle vaccines for *E. coli* O157:H7 rather than altering the design of production systems.
>
> *(Stuart 2010: 11)*

Transition . . .

This chapter hints strongly at an ecological critique of conventional agriculture and the related global system of food provisioning more generally. We hear repeatedly about how the current way of doing things – food and agriculture-wise – is unsustainable. Even proponents of the conventional system recognize that there are glaring environmental inefficiencies in how the world is presently fed (see, e.g., DeGregori 2001). The next chapter will help illuminate the subject of our ecological *food*print.

Discussion questions

1 Can you think of a scenario where social capital actually hinders rural development, at least for a certain segment of a community's population? How do you differentiate between "good" and "bad" social capital?
2 Recall the various myths of nature. How might they shape the way a farmer manages her/his land?
3 How does the ANT approach differ from social constructionism?

Suggested reading: Introductory level

Agroecology website at www.agroecology.com

Emery, M. and C. Flora. 2006. Spiraling-up: Mapping community transformation with community capitals framework. *Community Development* 37(1): 19–35.

Wezel, A., S. Bellon, T. Dore, C. Francis, D. Vallod, and C. David. 2009. Agroecology as a science, a movement, and a practice: A review. *Agronomy for Sustainable Development* 29(4): 503–15.

Suggested reading: Advanced level

Emery, M. and C. Flora. 2006. Spiraling-up: Mapping community transformation with community capitals framework. *Community Development* 37(1): 19–35.

Greider T. and L. Garkovich. 1994. Landscapes: The social construction of nature and the environment. *Rural Sociology* 59: 1–24.

Pretty, J. and D. Smith. 2004. Social capital in biodiversity conservation and management. *Conservation Biology* 18(3): 631–8.

Wezel, A., S. Bellon, T. Dore, C. Francis, D. Vallod, and C. David. 2009. Agroecology as a science, a movement, and a practice: A review. *Agronomy for Sustainable Development* 29(4): 503–15.

References

Alexander, William. 2008. *Resiliency in Hostile Environments: A Comunidad Agricola in Chile's Norte Chico*. Cranbury, NY: Rosemount Publishing.

Allison, L. 1996. On planning a forest: Theoretical issues and practical problems. *Town Planning Review* 67(2): 131–43.

Altieri, M. 1995 (1987). *Agroecology: The Science of Sustainable Agriculture*. Boulder, CO: Westview Press.

Altieri, Miguel. 1999. Applying agroecology to enhance the productivity of peasant farming systems in Latin America. *Environment, Development and Sustainability* 1: 197–217.

Altieri, Miguel and Clara Nicholls. 2008. Scaling up agroecological approaches for food sovereignty in Latin America. *Development* 51(4): 472–80.

Badgley, C., J. Moghtader, E. Quintero, E. Zakem, M. Chappell, K. Aviles-Vazquez, A. Samulon, and I. Perfecto. 2007. Organic agriculture and the global food supply. *Renewable Agriculture and Food Systems* 22: 86–108.

Burgess, J., J. Clark, and C. Harrison. 2000. Knowledges in action: An actor network analysis of a wetland agri-environment scheme. *Ecological Economics* 35: 119–32.

Burton, R. 2004. See through the "good farmer's" eyes. *Sociologia Ruralis* 44(2): 195–215.

Buttel F. 2002. Envisioning the Future Development of Farming in the USA: Agroecology between Extinction and Multifunctionality? Paper developed for workshop titled, The Many Meanings and Potential of Agroecology Research and Teaching, May 29–30, Madison, WI, http://www.agroecology.wisc.edu/downloads/buttel.pdf, last accessed April 20, 2011.

Buttel, Fredrick, Olaf Larson, and Gilbert Gillespie. 1990. *The Sociology of Agriculture*. New York: Greenwood Press.

Callon, M. 1986. Some elements in a sociology of translation: domestication of the scallops and fishermen of St. Brieuc Bay. In *Power, Action, Belief*, edited by J. Law. London: Routledge, pp. 19–34.

Carolan, M. 2005. Barriers to the adoption of sustainable agriculture on rented land: An examination of contesting social fields. *Rural Sociology* 70(3): 387–413.

Carolan, M. 2006. Do you see what I see? Examining the epistemic barriers to sustainable agriculture. *Rural Sociology* 71(2): 232–60.

Carolan, M. 2011. *The Real Cost of Cheap Food.* London: Earthscan.

Cary, John. 1993. The nature of symbolic beliefs and environmental behavior in a rural setting. *Environment and Behavior* 25(4): 555–76.

Coleman, J. 1988. Social capital and the creation of human capital. *American Journal of Sociology* 94: S95–120.

Daly, M. and H. Silver. 2008. Social exclusion and social capital: A comparison and critique. *Theory and Society* 37(6): 537–66.

DeGregori, T. 2001. *Bountiful Harvest: Technology, Food Safety, and the Environment.* Washington, DC: CATO Institute.

Emery, M. and C. Flora. 2006. Spiraling-up: Mapping community transformation with community capitals framework. *Community Development* 37(1): 19–35.

Flint, R.W. 2010. Seeking resiliency in the development of sustainable communities. *Human Ecology* 17(1): 44–57.

Flora, C. 2001a. *Interactions Between Agroecosystems and Rural Communities.* Boca Raton, FL: CRC Press.

Flora, C. 2001b. Access and control of resources: Lessons from SANREM CRSP. *Agriculture and Human Values* 18: 41–8.

Flora, C. 2008. Social capital and community problem solving combining local and scientific knowledge to fight invasive species. *Learning Communities: International Journal of Learning in Social Contexts* 2: 30–9.

Flora, C., C. Bregendahl, and S. Fey. 2007. Mobilizing internal and external resources for rural community development. In *Perspectives on 21st Century Agriculture: A Tribute to Walter J. Armbruster*, edited by In R. Knutson, S. Knutson, and D. Ernstes. Chicago, IL: The Farm Foundation, pp. 210–20.

Flora, C. and A. Gillespie. 2009. Making healthy choices to reduce childhood obesity: Community capitals and food and fitness. *Community Development* 40(2): 114–22.

Flora, C., M. Livingston, I. Honyestewa, and H. Koiyaquaptewa. 2009. Understanding access to and use of traditional foods by Hopi women. *Journal of Hunger and Environmental Nutrition* 4: 158–71.

Flora, J. 1998. Social capital and communities of place. *Rural Sociology* 63(4): 481–506.

Gliessman, Stephen. 1998. *Agroecology: Ecological Process in Sustainable Agriculture.* AnnArbor, MI: AnnArbor Press.

Goodman, D. 1999. Agro-food studies in the "age of ecology": Nature, corporality, bio-politics. *Sociologia Ruralis* 39(1): 17–38.

Goodman, D. 2001. Ontology matters: The relational materiality of nature and agro-food studies. *Sociologia Ruralis* 41(2): 182–200.

Goodman, M. 2011. Towards visceral entanglements: knowing and growing the economic geographies of food. In *The SAGE Handbook of Economic Geography*, edited by R. Lee, A. Leyshon, L. McDowell and P. Sunley. London: Sage, pp. 242–57.

Greider T. and L. Garkovich. 1994. Landscapes: The social construction of nature and the environment. *Rural Sociology* 59: 1–24.

King, F.H. 1911. *Farmers of Forty Centuries.* Madison, WI: F.H. King.

Klages, K.H.W. 1928. Crop ecology and ecological crop geography in the agronomic curriculum. *Journal of the American Society of Agronomy* 20: 336–53.

Latour, B. 1992. Where are the missing masses? Sociology of a few mundane artifacts. In *Shaping Technology, Building Society: Studies in Sociotechnical Change*, edited by W. Bijker and J. Law. Cambridge, MA: MIT Press, pp. 225–58.

Latour, B. 1993. *We Have Never Been Modern.* Brighton: Harvester Wheatsheaf.

Latour, B. 1999. *Pandora's Hope: Essays on the Reality of Science Studies.* Cambridge, MA: Harvard University Press.

Law, J. 1992. Notes on the theory of actor network: Ordering, strategy, and heterogeneity. *Systems Practice* 5(4): 379–93.

Levins, H. 2006. A whole-system view of agriculture, people, and the rest of nature. In *Agroecology and the Struggle for Food Sovereignty in the Americas*, edited by Avery Cohn, Jonathan Cook, Margarita Fernández, Rebecca Reider, and Corrina Steward. London: International Institute for Environment and Development (IIED), pp. 34–49. http://pubs.iied.org/pdfs/14506IIED.pdf, last accessed August 22, 2011.

Lloyd, T., C. Watkins, and D. Williams. 1995. Turning farmers into foresters via market liberalization. *Journal of Agricultural Economics* 46(3): 361–70.

Lockie, S. and S. Kitto. 2000. Beyond the farm gate: production-consumption networks and agri-food research. *Sociologia Ruralis* 40(1): 3–19.

Lowenthal, D. 1991. British national identity and the English landscape. *Rural History* 2: 205–30.

McHenry, H. 1997. Wild flowers in the wrong field are weeds: Examining farmers constructions of conservation. *Environment and Planning A* 29: 1039–53.

Newby, H., C. Bell, D. Rose, and P. Saunders. 1978. *Property, Paternalism and Power: Class und Control in Rural England.* London: Hutchinson.

Pretty, J. and D. Smith. 2004. Social capital in biodiversity conservation and management. *Conservation Biology* 18(3): 631–8.

Pretty, J. and H. Ward. 2001. Social capital and the environment. *World Development* 29(2): 209–27.

Schwarz, M. and M. Thompson. 1990. *Divided We Stand: Redefining Politics, Technology and Social Choice.* London: Harvester Wheatsheaf.

Shields, M., C. Flora, B. Thomas-Slayer, and G. Buenavista. 1996. Developing and dismantling social capital: Greener and resource management in the Philippines. In *Feminist Political Ecology: Global Perspectives and Local Experience*, edited by D. Rocheleau, B. Thomas-Slayter, and E. Wangari. London: Routledge, pp. 155–79.

Stock, P. 2007. Good farmers as reflexive producers: An examination of family organic farmers in the Midwest. *Sociologia Ruralis* 47(2): 83–102.

Stuart, D. 2010. Nature is not guilty: Foodborne illness and the industrial bagged salad. *Sociologia Ruralis* 51(2): 158–74.

Wezel, A., S. Bellon, T. Dore, C. Francis, D. Vallod, and C. David. 2009. Agroecology as a science, a movement, and a practice: A review. *Agronomy for Sustainable Development* 29(4): 503–15.

Wilson, G. 2001. From productivism to post-productivism ... and back again? Exploring the (un)changed natural and mental landscapes of European agriculture. *Transactions of the Institute of British Geographers* 26(1): 77–102.

10

FOOD, AGRICULTURE, AND THE ENVIRONMENT

KEY TOPICS

- The costs of the current food system are not adequately reflected in food's retail price.
- The concept of food miles has become a popular, albeit gravely misunderstood, measure of food's ecological footprint.
- The lifecycle analysis arguably gives a more complete (compared with, say, food miles) picture of food's overall environmental impact.
- Waste, abuse, and overuse is a problem that not only plagues our food system, but also threatens the food security of nations (water use/abuse, food waste, and fertilizer overuse are discussed).

Key words

- Energy/emissions intensities
- Food miles
- Food waste
- Footprint shifting
- Industrial freshness
- Lifecycle analysis
- Lifecycle hotspot
- Metabolic rift
- Peak phosphorus
- Tons per miles traveled
- Vehicle miles
- Virtual water
- Water privatization

The retail price of our food comes nowhere close to approximating its real cost (Carolan 2011b). One calculation places the *annual* external costs of agriculture (to say nothing of the broader food system's ecological footprint) in Germany at US$2 billion, in the UK that figure is US$3.8 billion, and in the USA the total is US$34.7 billion. These average out to an additional cost of approximately US$81–343 per hectare of arable land and pasture (Pretty *et al.* 2001). Other analyses concentrate on specific components of production agriculture by calculating, for instance, the real cost of pesticides. Two highly cited estimates, summarized in Table 10.1, come from Leach and Mumford (2008) and Pimentel (2005). Note the considerable costs these estimates attribute to pesticide use in agriculture, which Pimentel places (for the US case alone) at a staggering US$9.6 billion annually.

Yet, to play devil's advocate, it is simply not possible to feed the world's 6.8 billion people, 45 billion chickens, 11.7 billion sheep and goats, 3 billion cattle, and 0.94 billion pigs (Carolan 2011b) without impacting social and ecological

TABLE 10.1 Estimates on the real costs of pesticides from Leach and Mumford (2008) and Pimentel (2005), in US$

Costs	*UK*	*USA*	*Germany*
Adapted from Leach and Mumford (2008)			
Pesticides in sources of drinking water	287,444,066	1,126,337,798	180,522,199
Pollution incidents, fish deaths & monitoring costs	20,360,620	161,720,798	51,355,453
Biodiversity/wildlife losses	29,942,089	207,383,142	6,224,903
Cultural, landscape, tourism, etc.	118,570,678	insufficient data	insufficient data
Bee colony losses	2,395,367	144,597,420	1,556,225
Acute effects of pesticides to human health	2,395,367	167,428,591	28,012,065
Total	*461,108,190*	*1,807,467,750*	*267,670,848*
Adapted from Pimentel (2005)			
Public health impacts	not calculated	1,140,000,000	not calculated
Domestic animals deaths and contaminations	not calculated	30,000,000	not calculated
Cost of pesticide resistance and loss of natural enemies	not calculated	2,020,000,000	not calculated
Honeybee and pollination losses	not calculated	334,000,000	not calculated
Crop losses	not calculated	1,391,000,000	not calculated
Bird and fishery losses	not calculated	2,260,000,000	not calculated
Groundwater contamination	not calculated	2,000,000,000	not calculated
Government regulations to prevent damage	not calculated	470,000,000	not calculated
Total	*not calculated*	*9,645,000,000*	*not calculated*

systems. The goal isn't (is it?) to produce food without *any* costs to ecological and social systems. Rather, it is a question of tradeoffs: what are we willing to give up so we can have the food we (think we) need? The aforementioned pricing analyses remind us of some of what we are currently sacrificing for our food; quite a bit as it turns out. This ought to lead us to ask: do the benefits of the current food system outweigh its costs?

Complicating matters further is the fact that this question cannot be answered objectively, by tallying up both sides (costs and benefits) and choosing whichever figure is the largest. Value judgments inevitably have to be made. Let's go back to the subject of pesticides to tease this point out. It is generally held that each US dollar invested in pesticide control returns about US$4 in protected crops, as long as, and this is an important qualifying statement, those costs described in Table 10.1 remain externalized. This calculates out to a net gain of roughly US$40 billion per year in favor of pesticides. Throw any of the currently externalized costs into the bottom line, however, and pesticides fare less well (throw them all in and pesticides fare really poorly). Next, consider further that some things just don't translate nicely into a dollar figure, like a human life. Pimentel (2005: 246) captures this tension nicely when he explains, "assuming that pesticide-induced cancers numbered more than 10,000 cases year and that pesticides returned a net agricultural benefit of $32 billion year, each case of cancer is 'worth' $3.2 million in pest control." Is US$3.2 million in pesticide benefits worth one person dying from cancer? But even this question is not entirely accurate, as it fails to ask if similar ends can be achieved without the use of these chemicals. If we can successfully substitute, say, agroecological controls for pesticides (see Chapter 9), then one cancer death for each US$3.2 million in pesticide benefits seems an unnecessary price to pay.

My point is that food production, distribution, and consumption is all about tradeoffs. Any food system, real or imagined, has an ecological footprint. What, then, does the dominant food system's ecological footprint look like and what are some of the sociological consequences that can be attributed to it?

Food miles (kilometers)

The concept of food miles has a simple elegance to it: food that travels great distances obviously consumes more energy during transportation than food grown locally. It has been calculated that every calorie of food produced ends up consuming over its entire lifecycle about 10 calories of fossil fuel (Frey and Barrett 2007). According to the Japan Ministry of the Environment (2009), Japan not only imports more food by volume than the USA, UK, South Korea, Germany, and France, but that imported food also travels on average further than food coming into those other nations (Figure 10.1).

When assessing the ecological benefits of local food we need to first have a discussion about tradeoffs. For as long as we choose to eat foods out of season and/or that are non-native to local agroecological conditions then we will be

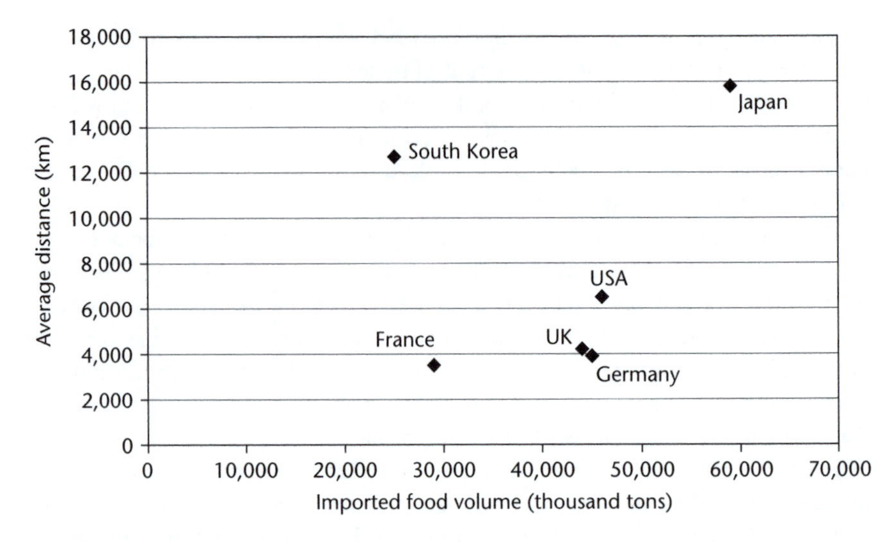

FIGURE 10.1 Food imports by volume and average distance traveled by food for select countries. Based on Japan Ministry of the Environment (2009)

continually presented with uncomfortable choices. Do we want, for example, to incur the ecological cost associated with food miles or would we rather produce something locally, even if doing so consumes copious levels of energy and water (levels far higher than what would be required to produce the commodity in a non-local region with conditions better suited for its production)? It is also important to bear in mind that fruits and vegetables account for roughly one-quarter of exports in some developing countries. And undoubtedly some of those producers are peasants and other smallholder farmers. How, then, would a shift over to an entirely local diet in affluent nations impact the developmental trajectories of less developed countries? It's a question local food advocates rarely bother to ask.

Another point to keep in mind is that the "conclusions" of any food miles analyses hinge on how the concept is operationalized. For example, there are "vehicle miles": the distance traveled by vehicles carrying food regardless of the amount transported. There are also "tons per miles traveled": the distance multiplied by load. Let us say there is a large container ship transporting 1 million mandarin tangerines from Japan (where one-quarter of the world's crop is grown) to San Francisco. The trip is roughly 8,200 kilometers (5,100 miles). How one chooses to operationalize food miles widely changes the outcome of the analysis. Using vehicle miles, then each tangerine is attributed 5,100 miles upon arriving in San Francisco. Alternatively, utilizing tons per miles traveled, you might attribute to each tangerine a paltry 0.0000051 food miles.

Complicating matters further is that not all forms of transportation are equal. Table 10.2 lists the energy and emissions intensities for different modes of freight transportation. As illustrated, air transport requires the most energy to move 1 ton of food 1 kilometer – 15,839 kilojoules (kJ) to be exact. It also

TABLE 10.2 Energy and emission intensities for different modes of freight transportation

Transportation type	*Energy consumption (kJ/ton-km)*	*Total emissions (g/ton-km)*				
		CO_2	*Hydrocarbons*	*VOCs*	*NOx*	*CO*
Water	423	30.00	0.04	0.10	0.40	0.12
Rail	677	41.00	0.06	0.08	0.20	0.05
Road	2,890	207.00	0.30	1.10	3.60	2.40
Air	15,839	1,260.00	2.00	3.00	5.50	1.40

CO, carbon dioxide; NOx, nitrogen oxide; VOC, volatile organic compound.
Based on Hill (2008).

emits the most greenhouse gases (among other noxious gases) per ton-kilometer traveled, which explains why air transport accounts for 11 percent of Britain's CO_2 emission even though only 1 percent of the country's food arrives via airplane (Smith *et al.* 2005).

Yet even these transportation categories are too general. As illustrated in Table 10.3, the utilization of highly efficient trucks does not necessarily signify an ecologically efficient distribution operation because fuel efficiency can be offset by either a poor utilization of the vehicle's capacity or an overall low hauling capacity. This is why some suggest that a better measure of energy efficiency in food distribution is "energy intensity," which is an expression of fuel consumption on a pallet-kilometer or ton-kilometer basis. There are energy economies of scale that come from being able to transport large amounts of food in a single haul. This no doubt helps explain why cargo ships fare so well in Table 10.2, as each ship has the capacity to carry thousands of tons of food. These economies of scale can be so great that they can even offset a vehicle's

TABLE 10.3 Average energy efficiency and intensity by vehicle type

Vehicle type	*Average fuel efficiency (km/liter)*	*Average load (pallets)*	*Average payload (tones)*	*Average energy intensity by volume (ml/pallet-km)*	*Average energy intensity by weight (ml/ton-km)*
7.5–18 ton truck	3.87	5.78	2.25	33.0	83.8
Truck greater than 18 ton	2.91	8.69	7.41	31.8	37.1
32 ton semi-trailer	3.35	14.38	10.37	19.1	26.4
38–44 ton semi-trailer	2.79	17.11	11.83	18.0	26.0

Based on Defra (2008b).

poor performance when assessed on a kilometer per liter (or miles per gallon) basis. That is in fact what we see in Table 10.3 – that the carrying capacity of large semi-trailers can, if fully loaded when on the road, offset their engines' poor fuel efficiency.

Looking at energy demands of transportation in aggregate, compared with other sectors of the food system, shows that the real ecological problem lies someplace other than in food miles. The research is pretty clear: greenhouse gas emissions associated with food can be attributed significantly to the *production phase*. In other words, *what* is produced matters more than *how far* food travels after leaving the farm gate. One analysis found that transportation represents only 11 percent of lifecycle greenhouse emissions for food (Weber and Matthews 2008: 3508). The authors also note that food groups vary considerably in terms of the greenhouse gas intensity, where, for example, red meat is about 150 percent more intensive than either chicken or fish. A change in diet, according to this research, would thus be a considerably more effective means of lowering one's greenhouse gas footprint than buying local. They explain that "[s]hifting less than one day per week's worth of calories from red meat and dairy products to chicken, fish, eggs, or a vegetable-based diet achieves more GHG [greenhouse gas] reduction than buying all locally sourced food" (Weber and Matthews 2008: 3508).

The retail sector also possesses a considerable carbon footprint. The overall level of greenhouse gas emissions on a per kilogram basis is quite high for things like refrigerator displays in the grocery store compared to transportation. For example, according to a 2008 study from the UK, while frozen peas have a gCO_2e per kg of 16.2 in transport (using a 38+ ton semi-trailer) that figure is a staggering 903.4 during refrigeration in the grocery store. The in-store figure includes not only energy used to operate the refrigeration cabinets but also energy leakage whenever those cabinets are opened by customers and employees of the store (Defra 2008b).[1] According to this lifecycle analysis, *90 percent* of the total greenhouse gas emissions from the transportation stage onward for frozen peas occurring during the grocery store refrigeration phase. Thus, to bring the subject back to local food, the localness of food matters little if that food then spends a great deal of time in a refrigerated or frozen state. Clearly, food miles do not give us a complete picture of a food's ecological footprint.

Finally, while concerns for the environment are employed at the discursive surface of these discussions, the real value of the food miles concept in some instances seem to be to protect domestic sectors of a country's food system. The British-based think tank the Africa Research Institute released a report on Kenya's horticulture industry, where it is explained how a recent push by UK supermarkets for local food labels is the direct result of intense lobbying from European farm lobby groups (Gikunju 2009). The hope among these groups was that by presenting air-freighted produce from countries like Kenya as environmentally unsustainable they would be able to increase the consumption of domestically raised food. The Africa Research Institute's report show just how counterintuitive food miles, as an indicator of sustainability, can be. According to this report, certain foods flown from Kenya into Britain can actually be produced with a *smaller* energy footprint than if they were raised domestically. Average annual temperatures in Kenya, which hover between 20 and 25 Celsius year round, eliminate the need for energy intensive greenhouses, while the small-scale nature of Kenya farms (on average 2 acres) eliminate the need for tractors. African horticultural imports represent 0.1 percent of all of Britain's carbon emissions. Total flights from Nairobi represent 0.5 percent of all inbound flights into Heathrow (a major London airport). As the report goes on to note, Britain's government buildings emit more carbon than all of Kenya.

The lifecycle analysis

Traditional approaches to environmental impacts tend to focus on a single plant or link in the commodity chain and on one or two pollutants. Doing this, however, can lead to "footprint shifting," where ecological efficiencies might be achieved at one stage of the lifecycle only to increase environmental impacts in another. Examining the entire lifecycle of a product also allows "hotspots" – links where impacts are high and where reductions can be achieved at minimal cost and effort – to be identified. The lifecycle analysis (LCA), which crept into the discussion in the previous section, is a tool to assess overall ecological footprints. It has been utilized to inform our understanding of many different aspects of the food system.

A recent LCA, for example, examines a British food staple: the cottage pie (this is a meat-filled pie with a crust constructed out of mashed potatoes) (Defra 2008a). The ready-to-eat (store bought) pie modeled contained more than 20 different ingredients, although 70 percent of the commodity was mashed potatoes and cooked beef. The production of the commodities represented a major emission hotspot, contributing 65 percent to the lifecycle greenhouse gas emissions, due primarily to greenhouse gas emissions associated with beef production. The manufacturing, retail, consumer-use, and disposal stages contributed 12 percent, 12 percent, 9 percent, and 2 percent, respectively. The study also found that method of household preparation significantly influenced the overall ecological footprint of the meal. Using a microwave instead of an electric fan oven to heat the cottage pie cuts drastically the greenhouse gas footprint of the consumer-use stage, from 9 percent to 2 percent of the total emissions.

Alternatively, LCA also provide "big picture" views of the food system. A report by the US Department of Agriculture (USDA) offers some insight into the amount of energy that goes into making food in the USA – over 17,000 calories on a per capita daily basis – and the allocation of that energy (Canning *et al.* 2010). Over half of those 17,000+ calories are consumed to make highly processed ("junk") food (Figure 10.2). A third goes into the making of animal products like meat, eggs, and milk. One-sixth goes to grains, fruits, and vegetables. In short, our energy investments are the inverse of what we should be eating. Eating well, it turns out, appears to be less energy intensive than eating poorly (Bomford 2011).

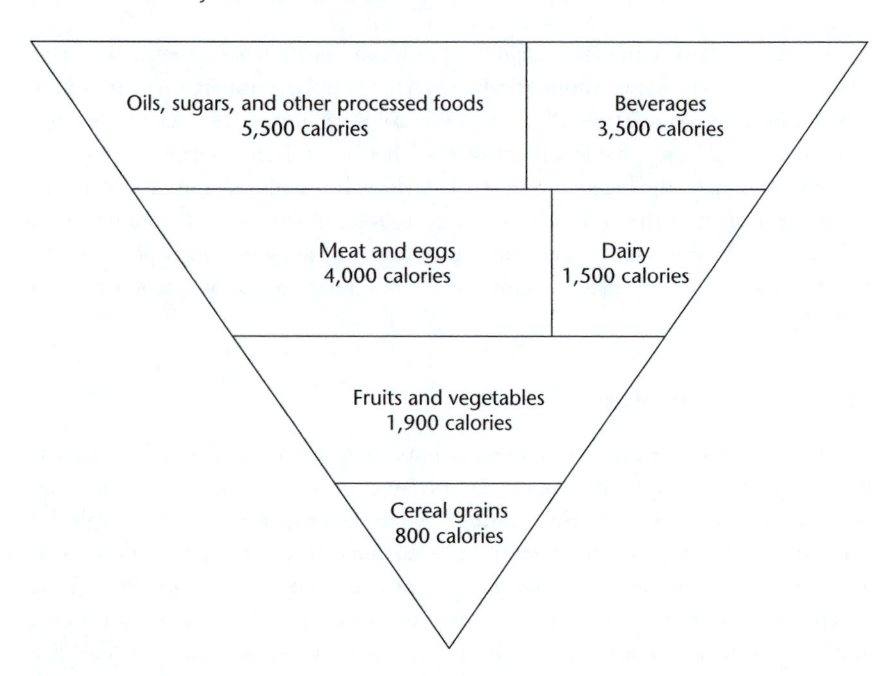

FIGURE 10.2 Breakdown of the amount of energy (17,000+ calories per day per capita) consumed by US food system. Based on Canning *et al.* (2010) and Bomford (2011)

Virtual water

The United Nations predicts water scarcity (rather than arable land) will be the number one constraint on food security for many years to come (UNDP 2007). Global demand for water has increased 300 percent since the 1950s (Hanjra and Qureshi 2010). The per capita water requirement for food in China has increased fourfold since 1961, due largely, as indicated in Figure 10.3, to increases in the

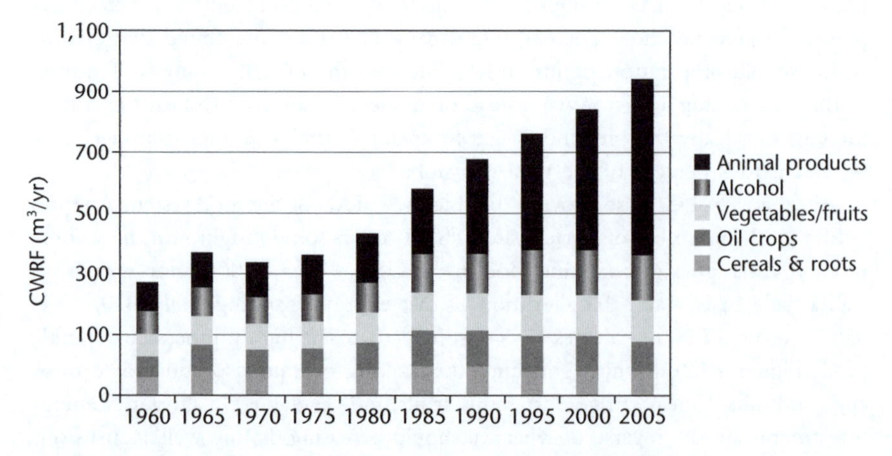

FIGURE 10.3 Annual per capita water requirements for food (CWRF) in China. Adapted from Liu and Savenije (2008: 891)

consumption of animal products (Liu and Savenije 2008). Further increases in water demands due to food seem inevitable as the average Chinese diet continues to eat more animal products per capita. However, China is not usual in its water demands. In fact, its water needs remain considerably below that of other countries. For example, while China's current annual per capita water requirements for food are over 900 cubic meters (m^3) per year, for the USA that figure is roughly 1,800 m^3 (two-thirds of which come from the production of animal products) (Liu and Savenije 2008).

Regardless of future projected water needs, China is water-stressed *today*. Eighty percent of the country's major rivers are so polluted they no longer support life (Barlow 2007). It has been estimated that 300 million Chinese drink contaminated water and 70 million drink water that is sufficiently polluted to put their health at risk (Kill 2010). According to the country's Ministry of Water Resources, over 70 billion tons of wastewater are released annually, 45 billion of which are pumped "raw" (untreated) into lakes and rivers (*Asia News* 2006). The country has 40 times more people than Canada, yet not as much fresh water. Since 1965, the water table beneath the city of Beijing has fallen about 60 meters (or more than 200 feet) (Ranjan 2010).

Discussions about local food must be informed by an understanding of local water availability. It would be very hard for a highly populated country to embrace local food policies if they are water poor. A per capita freshwater supply of below 1,000 cubic meters places countries in the category of "severe water stress" (Qadir *et al.* 2007). As documented in Table 10.4, a number of countries fall well

TABLE 10.4 Renewable water resources (RWR) per capita for select countries, 2005 and 2030

Country	*2005 (m^3/year)*	*2030 (m^3/year)*
Kuwait	7	5
United Arab Emirates	48	37
Saudi Arabia	94	56
Libya	94	56
Singapore	137	122
Jordan	157	104
Yemen	191	81
Israel	254	190
Oman	331	191
Algeria	435	324
Tunisia	458	372
Rwanda	604	387
Egypt	779	534
Morocco	919	682
Kenya	919	734
Lebanon	1170	938
Somalia	1257	553
Pakistan	1382	820
Syria	1410	915
Ethiopia	1483	865

Adapted from Qadir *et al.* (2007).

within, or dangerously close to, that category. And their futures look to bring only greater thirst.

The term "virtual water" is widely utilized by food scholars. The logic behind the concept is fairly simple: it is far more efficient to import food than the water to produce those foods domestically. Virtual water refers to the embedded water within food, to either the water actually used or to the water that would have been used had the product been produced domestically. The world's top virtual water exporter, as detailed in Table 10.5, is the USA (exporting almost three times more water than the second place nation), while Sri Lanka is the leading virtual water importer (Hoekstra and Chapagain 2007).

The amount of virtual water a country imports is to a degree a factor of the types of foods they import. Not all foods, as revealed in Figure 10.4, have the same (virtual) water footprint. The relative thirstiness of ground beef reflects the beef cow's consumption of grains, at a 9 : 1 kg (feed to meat) conversion ratio when "finished" in an industrial feeding operation. The modern cow takes on average 3 years before it is slaughtered, producing roughly 200 kg of meat. During its life it consumes approximately 1,300 kg of grains, 7,200 kg of roughage, 24 m^3 of water for drinking, and an additional 7 m^3 of water for servicing. This translates into roughly 155 liters of water for drinking and servicing, plus an additional

TABLE 10.5 Top virtual water exporting and importing countries (annual net volume)

Top exporting (billions of cubic meters)	*Top importing (billions of cubic meters)*
USA (758.3)	Sri Lanka (428.5)
Canada (272.5)	Japan (297.4)
Thailand (233.3)	Netherlands (147.7)
Argentina (226.3)	Korea (112.6)
India (161.1)	China (101.9)
Australia (145.6)	Indonesia (101.7)

Data from Hoekstra and Chapagain (2007).

FIGURE 10.4 Virtual water (in liters) embedded in 1 kilogram of various commodities. Based on Hoekstra (2008) and Pearce (2006)

15,000–20,000 liters to grow feed and roughage (this figure still excludes things like the water polluted due to the leaching of fertilizers to grow the feed) (Hoekstra 2008).

How crops are watered also matters. The efficiency of traditional gravity irrigation, a method used widely in the developing world, is about 40 percent. Sprinkler systems have an efficiency range of 60–70 percent and drip irrigation systems are 80–90 percent efficient (Seckler 1996). There are very good reasons why farmers in less affluent countries rely upon the less expensive, albeit highly inefficient, gravity method. Most lack access to credit to make the necessary capital investments (for reasons tied to, as discussed in earlier chapters, macroeconomic policies imposed upon these countries by entities like the World Bank). And often the governments of these countries are just as strapped for cash, making massive public sector investments to infrastructure impossible.

Water as a right *v.* water rights

Roughly one in five child deaths – about 1.5 million annually – is the result of diarrhea. Diarrhea kills more children annually than AIDS, malaria, and measles combined (WHO 2009). An American taking a 5-minute shower uses more water than a typical person in a developing country slum uses over an entire day (United Nations 2006). People living in the slums often pay 5–10 times more per liter of water than wealthy people living in the same city (United Nations 2006). As more countries dump untreated household and industrial water into their rivers and lakes (as in the case of China), that sewage inevitably makes its way into the food system. A global survey conducted in 2004 on the practice of wastewater irrigation finds that 10 percent of the world's crops are watered with largely untreated sewage (Barlow 2007).

These statistics become all the more remarkable when one realizes the impressive returns on investments that can be realized when a country puts money towards its water infrastructure. According to a report by the United Nations (2009), safe drinking water, proper sanitation, and efficient irrigation systems contribute to economic growth while also saving governments money. Each US$1 invested returns somewhere between US$3 and US$34 – that's a return of between 300 and 3,400 percent – depending on the region and technology being invested in (United Nations 2009).

The reasons that have gotten us to this state are many. In the case of agriculture, the green revolution certainly has not helped things along. Since its beginning, in the aftermath of World War II, the amount of land under irrigation has tripled. Global agriculture today grows twice as much food as it did a generation ago but it uses three times more water to grow it. Two-thirds of all the fresh water in the world withdrawn from the ground goes to irrigate crops (Pearce 2006). At a minimum, 10 percent of world's grain is grown with groundwater not being replenished (Barlow 2007). This helps contextualize why many of the world's major rivers can no longer be counted on to reach the sea, like the Colorado in the USA and Australia's Murray River. The lower reaches of the Nile used to carry 32 billion m^3 of water annually; that figure is now down to 2 billion m^3. Same holds for the Indus in Pakistan ("Asia's Nile"), which has lost 90 percent of its volume over the last 60 years (Lean 2006).

Besides the green revolution, there is a broader political economic context in which water use must be situated within. For example, many less affluent countries are exporting vast sums of water in the form of thirsty commodities like grain, fruit, vegetables, coffee, and cut flowers. Their indebtedness to such entities as the World Bank and International Monetary Fund (IMF), both of whom were discussed in Chapter 4, requires the production of these high value commodities to pay down the country's debt. For example, Lake Naivasha in Kenya is being destroyed to grow cut flowers for export. To quote Maude Barlow, National Chairperson of the Council of Canadians and President of the Food and Water Watch Board of Directors, after a visit to the lake: "We saw pipes pumping water from the lake to the flower greenhouses and a ditch where waste water drained back into the lake. Pesticides and fungicides were plainly visible in a storage facility on the property. If action isn't taken immediately, the lake will not only be polluted, it will be drained" (quoted in Food and Water Watch 2009: 1).

There has also been a tremendous push by international monetary organizations like the World Bank in recent decades to privatize water. The practice of privatizing water, however, has its critics. As some argue, it is absurd at its face to take a free liquid that falls from the sky and sell it for as much as four times what we pay for gas (Lazarus 2007). There must be sound logic driving this push to privatize. So what is it?

The thinking goes something like this: incorrectly priced water causes people to use that resource irresponsibly. The choice, however, is not between free or privatized water but between government subsidized or privatized water. The latter scenario, however, cares nothing about affordability, as the goal under privatization scenarios is cost recovery of the investment *plus* a little more (namely, profit). Never mind that most affluent countries have public water services and would never dream of handing control of them over to a profit seeking corporation. Also never mind that water is no less "incorrectly priced" in affluent nations. The cost recovery for US Bureau of Reclamation irrigation projects averages at around 10–20 percent total costs. Irrigation subsidies for the semi-arid states west of the Missouri River exceed US$500 million a year (Peterson 2009). And let's not forget, much of this subsidized water in the USA is used to raise subsidized crops like corn, rice, wheat, oats, and soybeans – a double subsidy.

In 1990, fewer than 51 million people globally received their water from Northern-based water firms. By 2015 that number is predicted to hit 1.16 billion. To ensure this shift toward water privatization within developing countries these nations are now unable to borrow from the World Bank or IMF without having in place a domestic water privatization policy (Goldman 2007). Consequently, most recent public utility privatization deals in the Global South are the result of the direct participation of these international monetary institutions. As Michael Goldman explains:

> That participation comes in the form of a threat, since every government official knows that the Bank/IMF capital spigots can always be shut off for those governments refusing to conform to their loan conditions. As overwhelming debt burdens have put tremendous pressure on borrowing-country governments and created dire social conditions in their countries [...] the Bank and IMF are using the carrot of debt relief to foist water policy reform on borrowing country governments.
>
> *Goldman (2007: 794)*

Unfortunately, the promises of privatization have proven largely hollow. No profit seeking firm can expect to expand their services to a population that cannot pay. So the water situation for the poor remains deplorable in many countries. Moreover, this aggressive push to privatize water utilities has caused previous capital streams (which used to support government-funded utilities) to dry up. Thinking private firms will provide all the necessary capital has produced a significant drop in donor spending on infrastructure. In 2002, for example, World Bank lending for water and sanitation for sub-Saharan African countries was one-quarter of what it was just 5 years earlier (Barlow 2007).

Recognizing that they are failing in their contractual commitments to provide water for all, water firms have started redefining the language in their legal contracts. In the city of La Paz, Bolivia, for example, to connect a large slum to the water system, the corporate-owned utility argued that "connection" would no longer mean a "piped connection" but "access to a standpipe or tanker." Curiously, this "new" definition perpetuates the very conditions that Northern elites earlier used to make the case for water privatization, as developing countries were previously criticized for erecting an inadequate water infrastructure (Goldman 2007). In other instances, private firms are requiring that residents of poor communities supply free labor to help build the connecting infrastructure. Yet how can any situation where a profit seeking firm seeks out and uses free labor not be exploitative? Moreover, it is plainly evidence of the failings of water privatization schemes, where only with free labor can these firms live up to their contractual obligations while ensuring a return for their shareholders (Goldman 2007).

Food waste

Between 30 and 40 percent of all food in affluent and less affluent countries is lost to waste (Godfray *et al.* 2010). Think for a moment about what that means, especially in light of the earlier discussion of future food demand trends. All this talk about needing to increase food production between now and 2050 *also* means food waste will double during that time period (though as food waste is increasing faster than demand we will likely see more than a doubling of food waste between now and 2050). The food waste generated in the USA alone constitutes sufficient nourishment to pull approximately 200 million people out of hunger. If you were to include all the feed that went into producing the meat and dairy thrown away annually by consumers, retailers, and food services in the USA and UK, the number increases to 1.5 billion people (Stuart 2009). Up to 50 percent of all food in the USA is wasted, costing the American economy at least US$100 billion annually (Jones 2005). US per capita food waste has increased by 50 percent since 1974, to more than 1400 kcal per person per day, or 150 trillion kcal per year (Hall *et al.* 2009: 1). Among affluent nations, the USA does not stand alone in the amount of food waste it generates. British consumers discard approximately 7 million tons of food, which represents one-third of all the food they purchase. At a retail cost of around £10.2 billion (US$19.5 billion), this waste has a CO_2 equivalent of 18 million tons, which equals the annual emissions of one-fifth of Britain's total car fleet (Desrochers and Shimizu 2008). A study by the George Morris Center estimates that Canada wastes a total of CAN$27 billion in food annually (Gooch *et al.* 2010).

As a percentage of their total food, developing countries do not fair much better in terms of wasting food – what varies is *where* in the food system those loses are incurred. As illustrated in Figure 10.5, whereas the majority of food in developed nations is lost in retail, food service, and while at the home, in the developing world losses are attributable largely to things like an inadequate food system infrastructure and insufficient storage facilities on the farm. Thus, in India, for example, 35–40 percent of fresh produce is lost annually as neither wholesalers nor retail stores have adequate cold storage facilities (Godfray *et al.* 2010).

There are many factors underlying the wastefulness of our current food system. Some of these might have the initial appearance of being purely technological, as in the case of developing nations lacking a proper infrastructure to store and transport their food. But the reasons *why* these infrastructural deficiencies exist in the first place are thoroughly sociological (and by extension

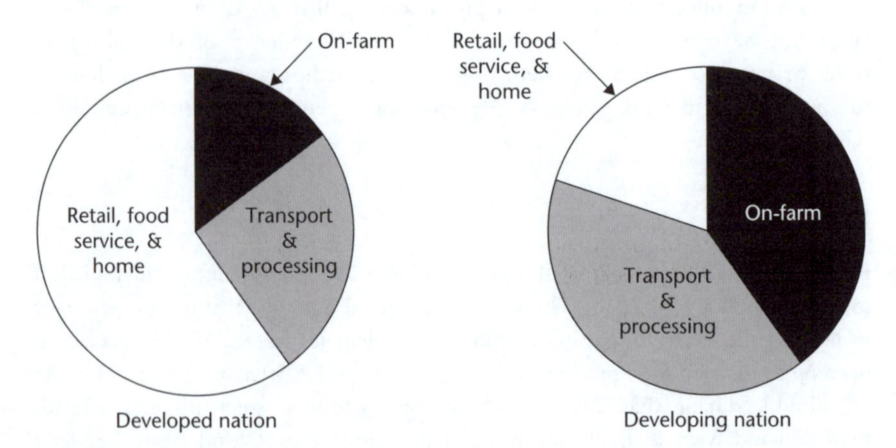

FIGURE 10.5 Makeup of total food waste in development and developing countries. Based on Godfray *et al.* (2010) and Stuart (2009)

economic and political), for a variety of reasons that I have already touched on. In other cases, the sociological factors are more apparent. The Chinese, for example, have a long-held cultural practice of providing more than their guests can eat as hosts seek to, as it is called, "gain face" (Smil 2004: 107). Rising costs for waste disposal caused one Shanghiai branch of the roast duck restaurant Quanjude to offer its customers a 10-percent discount if they would finish their food, as many of its customers were eating less than half of what they ordered (Smil 2004).

Another factor explaining waste, especially in affluent nations, is aesthetics. The state of Queensland, Australia, disposes over 100,000 tons of bananas annually due purely to the fact that the fruits fail to meet national cosmetic retail standards – in other words, they just don't look right (Hurst 2010). Globally, the amount of bananas wasted due to failing to meet such aesthetic standards represent 20–40 percent of the annual harvest (Stuart 2009: 118). Perhaps the most absurd story comes to us from Bristol, UK, involving a fruit seller who was told he was breaking the law by selling kiwi that were 2 grams below the 62 gram minimum prescribed by EU law (*Daily Mail* 2008). Evidence of the law's absurdity was apparently not lost on EU officials who recently wiped it from the books.

Susanne Freidberg (2009), in her captivating book *Fresh: A Perishable History*, describes the importance of aesthetics for food companies when it comes to getting people to buy their products, noting that this focus goes back generations. Fruit growers' associations have a history of sponsoring contests among store owners to see who could come up with the most pleasing fruit display (Freidberg 2009: 142–3). Visually rich advertising campaigns were used at the turn of the twentieth century to convince consumers of the quality of the industrial food products (see Carolan 2011a: 30). Some pear growers in the Montreuil region of France in the late 1800s went so far as to wrap each immature fruit in a paper bag while it still hung from the tree. Though immensely labor intensive this method protected the fruit from weather damage and the snout moth. Yet it also, just as importantly, created a pear with brilliant hues by increasing the fruit's photosensitivity once the bag was removed 2–3 weeks before harvest (Freidberg 2009: 148).

The industrialization of the food system has also altered our aesthetic sensibilities toward food (Carolan 2011a; Freidberg 2009). Freidberg (2009: 2) calls it "industrial freshness": a freshness that travels well – freshness *because of* transportation and processing rather than *in spite* of it. This "freshness" requires fruit that can withstand the brutalities of mechanical harvesting and transportation and can ripen during transport (as certain tomatoes, among other things, are now bred to do) rather than while on the vine. This change in our collective understanding of "freshness" is evident, according to Freidberg (2009: 156), in the "fresh cut" apples that populate fast-food restaurants, which are made possible thanks to a calcium citrate formula that keeps the fruit looking "fresh" for up to 28 days.

Another factor contributing to food waste is market concentration in the supermarket sector. As mentioned in Chapter 3, while supermarkets can select from a range of processors and farmers to work with, the inverse is less the case, giving retailers tremendous leverage when it comes to the writing up of contracts. This allows firms like Birds Eye, for example, to contractually forbid growers from selling their peas to anyone else, even those rejected by the company for not meeting certain standards and leaving growers with no choice but to either feed their surplus to livestock or to till it back into the soil (Stuart 2009).

Retailers are also very good at redistributing risk in ways that encourage overproduction and waste among consumers. Take, for example, the buy-one-get-one (BOGO) promotional offers, which represent nothing more than attempts to push the cost of waste onto households (though wisely framed by retailers as a "deal" to the consumer). However, most consumers do not consume the "free" one, as researchers in Canada recently concluded (Gooch *et al.* 2010).

Metabolic rift

To offer a more explicitly theoretical approach to thinking about the ecological footprint of food systems I conclude the chapter by reviewing the metabolic rift thesis. The writings of Karl Marx inform this framework. Within this growing body of literature, scholars elaborate on Marx's argument that the development of capitalism alienates humans from the natural environment while disrupting traditional practices (see, e.g., Clark and York 2008; Foster 1999).

Mid-nineteenth century environmental crises – from declining agricultural soil fertility to rising levels of sewage in cities – and the equally deplorable living conditions of urban workers were linked, according to Marx, to a disruption (a "rift") in a previously sustainable socio-ecological metabolism. This rift, for Marx, is tied to the expansion of capitalist modes of production and urbanization (the latter being made possible because of the rise of the former and in particular the displacement of small-scale agriculture). This process created a rift in ecological systems leading to environmental degradation at points of production and consumption.

Let's make this argument more concrete with the problem of what was called "soil exhaustion." Concerns over soil exhaustion began in the early 1800s in Britain and later in North America and Continental Europe in parallel with their emerging capitalist economies. John Bellamy Foster (1999) provides a thorough account of early market responses to this phenomenon. Loss of natural soil fertility resulted in exponential increases in bone and Peruvian guano imports. Over time, however, inputs such as bone and the dung of sea birds gave way to artificial fertilizers. Initially, these fertilizer inputs remained in short supply. Yet, once a process to produce synthetic nitrogen was developed in 1913 artificial inputs became abundant.

Early disruptions to the soil nutrient cycle were of great concern to Marx. Marx (1963: 1962–3) notes that "every moment the modern application of chemistry is changing the nature of the soil [...]. Fertility is not so natural a quality as might be thought; it is closely bound up with the social relations of the time." Drawing inspiration from German chemist Justus von Liebig and his (later in life) ecological critique of "modern" agricultural methods, Marx (1981: 949) notes how the expansion of capital "produces conditions that provoke an irreparable rift in the interdependent process of the social metabolism, a metabolism prescribed by the natural laws of life itself."

Soil exhaustion during this period has been linked to the expansion of capitalism, which drew people into the cities, forced specialization, and increasingly made individuals dependent upon the (low) wages garnered by the selling of their labor (Foster 1999). Disconnecting people from the land caused major disruptions in the soil nutrient cycle in the form of too few nutrients in the countryside and far too many in the cities, often in the form of sewage – the summer of 1858 in London was famously known as the summer of the Big Stink as the Thames' smell was so foul (from excess "nutrients") that lawmakers were forced to flee Parliament for the countryside. Such disruptions reeked ecological and social havoc; a point Marx witnessed firsthand in London where "they can do nothing better with the excrement produced by 4½ million people than pollute the Thames with it, at monstrous expense" (Marx 1981: 195). And the "solution" to this problem – was to repair the rift by bringing agricultural practices in line with ecological limits? No; the solution was to exacerbate the rift through artificial fertilizers. This solution may have relieved certain tensions in the short term but it failed to deal with the root of the rift – namely, producing food in ways that ignore ecological limits. In Marx's words:

> [A]ll processes in increasing the fertility of the soil for a given time is a progress toward ruining the more long-lasting sources of that fertility. [...] Capitalist production, therefore, only develops the techniques and the degree of combination of the social process of production by simultaneously undermining the original sources of all wealth – the soil and the worker.
>
> *Marx (1976: 638)*

We are living with the consequences of this rift. The first European Nitrogen Assessment (ENA) was released in 2011. It documents that nitrogen pollution is costing each person in Europe around £130–650 (€150–740) annually. The ENA represents the first time that the multiple threats of nitrogen pollution, including its impact on climate change and biodiversity, have been valued in economic terms at a continental scale. The study, carried out by 200 experts from 21 countries and 89 organizations, calculates that the annual cost of damage caused by nitrogen throughout Europe is £60–280 billion (€70–320 billion); a figure that is more than double the income gained from using nitrogen fertilizers in European agriculture (Sutton *et al.* 2011).

Likewise, global agriculture has applied phosphorus over the twentieth century with near abandon. This has resulted in, among other things, the eutrophication (overfertilization) of many watersheds. This occurs when excess phosphorus leaches into freshwater stores and feeds algal blooms that in turn starve fish of oxygen and produce "dead zones" (the Gulf of Mexico is a famous dead zone). The production process itself comes with a sizeable ecological footprint, as each ton of phosphate processed from phosphate rock generates 5 tons of phosphogypsum (which is radioactive) (Cordell *et al.* 2009).

Just as troubling is the fact that our phosphorus reserves are running out (Box 10.1). The International Geological Correlation Program estimated in 1987 that there could be as much as 163,000 million metric tons of phosphate rock remaining in the planet. This represents more than 13,000 million metric tons of phosphorus, which at current rates would be enough to last hundreds of years. The problem with this estimate, however, is that it includes types of rocks that are

BOX 10.1 PHOSPHORUS – A STRATEGIC GEOPOLITICAL RESOURCE

As the carrying capacity of any ecosystem is conditioned upon available phosphorus, the phosphorus "haves" will be able to produce their own food, while the phosphorus "have-nots" will have to rely upon other countries for their food. So who falls within these categories? The "haves" include Morocco, China, the USA, and South Africa, possessing 45, 21, 7, and 5 percent of the global resource base, respectively (US Geological Survey 2009). In anticipation of future food and thus phosphorus demands, China (the world's largest producer of phosphorus) has imposed, in mid-2008, a 135 percent tariff on phosphorus exports. Roughly 45 percent of global reserves are controlled by a single country, Morocco – the Saudi Arabia of phosphorus (Vaccari 2009). But much of this reserve is located on the Western Sahara, which Morocco has occupied since 1975 in violation of a decision by the International Court of Justice. In 2004, the USA signed a bilateral free trade agreement with Morocco giving them long-term access to its phosphate. It is therefore not surprising that, as a permanent member of the UN Security Council, the USA has consistently vetoed any resolution requiring Morocco to leave Western Sahara. At the other extreme – the phosphorus "have nots" – are France, Spain, and India (in that order), who are the world's three largest importers of phosphorus. India possesses no domestic supply of phosphate rock.

Complicating matters further is the availability of sulfuric acid. Sulfuric acid is produced primarily in developed countries (the USA alone accounts for approximately 30 percent of world production), in addition to China. It takes, on average, roughly 3 tons of sulfuric acid and 3.5 tons of phosphate rock to produce 1 ton of phosphoric acid, the basic ingredient of phosphate fertilizers. Consequently, the price of sulfuric acid directly determines the price of phosphate fertilizer. Thus, by controlling the supply and price of sulfuric acid developed nations can in effect control the price of the phosphate they import (Rosemarin *et al.* 2009).

either impractical to mine or that lie in environmentally sensitive areas. When looking at phosphate rock that is actually recoverable, the general consensus is that we have 75–100 years of phosphorus left, *at current rates of consumption* (Christen 2007; Vaccari 2009). Yet global demand for phosphorus is expected to increase somewhere between 50 and 100 percent by 2050 as a result of changing diets (more meat), increases in biofuel productivity, and increased food demand (Cordell *et al.* 2009). In short, it looks as though we will be reaching the state of "peak phosphorus" well before this century comes to an end.

Phosphorus and oil have been historically linked. In the past, cheap oil helped to make phosphorus cheap and abundant, as it allowed the nutrient to be cheaply shipped all around globe. It is therefore interesting that rising oil prices have actually *increased* global demand for phosphorus. The rise of biofuel crop production has dramatically increased demand for phosphate fertilizers (White and Cordell 2008: 1).

Some of the proposed solutions to the ticking phosphorus clock seek to at least partially repair the metabolic rift created through modern agricultural techniques. Suggestions include leaving the inedible biomass in the field, to be returned to the soil (while second generation biofuels seek to strip fields of this precious biomass). Another option is to make better use of animal waste, as both manure and animal bones are phosphorus sources. Or human waste: half the phosphorus humans excrete is in their urine (it is also a rich source of nitrogen). Recovering that phosphorus, however, is more difficult than it sounds. In cities around the world, urban planners decided to develop infrastructures that mix human excreta with industrial wastewater streams. Unfortunately, because the latter carry contaminants not fit for food systems, this essentially nullifies the value of the entire wastewater stream. Those systems would therefore need to be reconfigured. Moreover, if urine is not mixed with fecal matter in the toilet, the urine (as an essentially sterile compound) can be used safely with simple storage. An illustrative case of this involves two towns in Sweden. They have mandated that all new toilets be urine-diverting, sending the phosphorus rich liquid to either a communal urine storage tank or separate tanks under each house. Local farmers then collect the urine and apply it on their fields (Cordell *et al.* 2009).

Transition. . .

Marking the beginning of a new section – titled, simply, "Alternatives" – the following chapter speaks of cases that challenge, amend, and outright refute the conventional model of industrial food production. Its limitations at this point in the book ought to be clear. The question remains, however: are the proposed "solutions" any better?

Discussion questions

1 What are some differences between lifecycle analyses (discussed in this chapter) and the commodity systems approach (discussed in Chapter 3).

2 I mentioned that infrastructural deficiencies in less affluent countries contribute substantially to food waste. How have development policies rooted in neoliberalism (e.g. the Washington Consensus) contributed to these "deficiencies?"

3 Earlier I mentioned that the metabolic rift perpetuates, and is perpetuated by, a false dichotomy between "humans" and "nature." How is capitalism's survival predicated on the continuation of these dichotomies?

Suggested reading: Introductory level

Barlow, Maude. 2007. *Blue Covenant: The Global Water Crisis and the Coming Battle for the Right to Water*. New York: The New Press.

Clark, B. and R. York. 2008. Rifts and shifts: Getting to the root of environmental catastrophe. *Monthly Review* 60(6): 13–24.

Stuart, Tristram. 2009. Waste: *Uncovering the Global Food Scandal*. New York: Norton.

Suggested reading: Advanced level

Foster, J.B. 1999. Marx's theory of metabolic rift: Classical foundations for environmental sociology. *American Journal of Sociology* 105(2): 366–405.

Griffin, M., J. Sobal, and T. Lyson. 2009. An analysis of a community food waste stream. *Agriculture and Human Values* 26(1–2): 67–81.

Mariola, M. 2008. The local industrial complex? Questioning the link between local foods and energy use. *Agriculture and Human Values* 25(2): 193–6.

Mollinga, P. 2008. Water, politics and development: Framing a political sociology of water resources management. *Water Alternatives* 1(1): 7–23.

Note

1 gCO_2e is an expression of all greenhouse gasses converted to CO_2 equivalent units.

References

Asia News. 2006. Some 60 percent of the Yellow River, the cradle of Chinese civilization, is dead. *AsiaNews.it* December 14, http://www.asianews.it/news-en/Some-60-per-cent-of-the-Yellow-River,-the-cradle-of-Chinese-civilisation,-is-dead-7995.html, last accessed April 16, 2011.

Barlow, Maude. 2007. *Blue Covenant: The Global Water Crisis and the Coming Battle for the Right to Water*. New York: The New Press.

Bomford, M. 2011. Beyond Food Miles. Post Carbon Institute, March 9, Santa Rosa, CA, http://www.postcarbon.org/article/273686-beyond-food-miles#_edn9, last accessed April 12, 2011.

Canning, P., A. Charles, S. Huang, K. Polenske, and A. Waters. 2010. Energy Use in the US Food System. United States Department of Agriculture, Economic Research Service, Report number 94, March, Washington, DC, http://ddr.nal.usda.gov/bitstream/10113/41413/1/CAT31049057.pdf, last accessed April 12, 2011.

Carolan, M. 2011a. *Embodied Food Politics*. Burlington, VT: Ashgate.

Carolan, M. 2011b. *The Real Cost of Cheap Food*. London: Earthscan.

Christen, Kris. 2007. Closing the phosphorus loop. *Environmental Science and Technology* April 1: 2078.

Clark, B. and R. York. 2008. Rifts and shifts: Getting to the root of environmental catastrophe. *Monthly Review* 60(6): 13–24.

Cordell, Dana, Jan-Olof Drangert, and Stuart White. 2009. The story of phosphorus: Global food security and food for thought. *Global Environmental Change* 19: 292–305.

Daily Mail. 2008. EU forces market trader to pulp thousands of kiwi fruit because they're ONE MILLIMETERE too small. *Daily Mail* June 27, http://www.dailymail.co.uk/news/article-1029715/EU-forces-market-trader-pulp-thousands-kiwi-fruit-theyre-ONE-MILLIMETRE-small.html, last accessed September 3, 2010.

Defra (Department for Environment, Food and Rural Affairs, UK). 2008a. PAS2050 Case Study. Department for Environment, Food and Rural Affairs, UK, http://randd.defra.gov.uk/Document.aspx?Document=FO0404_8191_OTH.PDF, last accessed January 19, 2011.

Defra (Department for Environment, Food and Rural Affairs, UK). 2008b. SID5 Research Project Final Report. DEFRA, http://randd.defra.gov.uk/Document.aspx?Document=FO0405_8189_FRP.pdf, last accessed November 19, 2010.

Desrochers, Pierre and Hiroko Shimizu. 2008. *Yes, We Have No Bananas: A Critique of the 'Food Miles' Perspective*. Policy Primer No. 8, Mercatus Center, George Mason University, Washington DC.

Food and Water Watch. 2009. Lake Naivasha withering under the assault of international flower venders. The Council of Canadians, Ottawa, Canada, http://www.canadians.org/water/documents/NaivashaReport08.pdf, last accessed April 16, 2011.

Foster, J.B. 1999. Marx's theory of metabolic rift: Classical foundations for environmental sociology. *American Journal of Sociology* 105(2): 366–405.

Freidberg, S. 2009. *Fresh: A Perishable History*. Cambridge, MA: Harvard University Press.

Frey, Sibylle and John Barrett. 2007. Our health, our environment: The ecological footprint of what we eat. Paper prepared for the International Ecological Footprint Conference, Cardiff, May 8–10, http://www.brass.cf.ac.uk/uploads/Frey_A33.pdf, last accessed September 1, 2010.

Gikunju, J. 2009. Kenya's flying vegetables. Africa Research Institute, London, UK, http://www.africaresearchinstitute.org/files/policy-voices/docs/Kenyas-Flying-Vegetables-Small-farmers-and-the-food-miles-debate-0V6S400WZM.pdf, last accessed April 15, 2011.

Godfray, H., J. Beddington, I. Crute, L. Hadded, D. Lawrence, J. Muir, J. Pretty, S. Robinson, S. Thomas, and C. Toulmin. 2010. Food security: The challenge of feeding 9 billion people. *Science* 327: 812–18.

Goldman, Michael. 2007. How "what for all" policy become hegemonic: The power of the World Bank and its transnational policy networks. *Geoforum* 38: 786–800.

Gooch, M., A. Felfel, and N. Marenick. 2010. Food waste in Canada. George Morris Center, University of Guelph, Guelph, Canada, http://www.vcmtools.ca/pdf/Food%20Waste%20in%20Canada%20120910.pdf, last accessed April 16, 2011.

Hall, K., J. Guo, M. Dore, and C. Chow. 2009. The progressive increase of food waste in America and its environmental impact. *PLoS One* 4(1): 1–6.

Hanjra, Munir and Ejaz Qureshi. 2010. Global water crisis and future food security in an era of climate change. *Food Policy* 35: 365–77.

Hill, Holly. 2008. Food miles: Background and marketing. National Sustainable Agriculture Information Service, Butte, MT, http://www.attra.org/attra-pub/PDF/foodmiles.pdf, last accessed April 13, 2011.

Hoekstra, A. 2008. The water footprint of food. Water for Food Report, http://www.waterfootprint.org/Reports/Hoekstra-2008-WaterfootprintFood.pdf, last accessed April 15, 2011.

Hoekstra, A. and A. Chapagain. 2007. Water footprints of nations: Water use by people as a function of their consumption pattern. *Water Resources Management* 21(1): 35–48.

Hurst, Daniel. 2010. Growers go bananas over waste. *Brisbane Times* January 7, http://www.brisbanetimes.com.au/business/growers-go-bananas-over-waste-20100106-lu7q.html, last accessed July 30, 2010.

Japan Ministry of the Environment. 2009. Annual Report on the Environment, FY2008. Japan Ministry of the Environment, Tokyo, http://www.env.go.jp/en/wpaper/2008/fulltext.pdf, last accessed April 15, 2011.

Jones, Timothy. 2005. How much goes where? The corner on food loss. *BioCycle* July: 2–3.

Kill, M. 2010. *Understanding Environmental Pollution*. Cambridge, UK: Cambridge University Press.

Lazarus, David. 2007. Spin the (water) bottle. *San Francisco Chronicle* January 17, http://www.sfgate.com/cgi-bin/article.cgi?f=/c/a/2007/01/17/LAZ.TMP, last accessed April 16, 2011.

Leach, A. and J. Mumford. 2008. Pesticide environmental accounting: A method for assessing the external costs of individual pesticide applications. *Environmental Pollution* 151: 139–47.

Lean, G. 2006. Rivers: A drying shame. *The Independent* March 12, http://www.independent.co.uk/environment/rivers-a-drying-shame-469598.html, last accessed April 16, 2011.

Liu, J. and H. Savenije 2008. Food consumption patterns and their effects on water requirements in China. *Hydrology and Earth System Science* 12: 887–98.

Marx, K. 1963 (1847). *The Poverty of Philosophy*. New York: International.

Marx, K. 1976 (1867). *Capital*, Vol. 1. New York: Vintage.

Marx, K. 1981 (1863–65). *Capital*, Vol. 3. New York: Vintage.

Pearce, F. 2006. The parched planet. *New Scientist* 2540(February 25): 32.

Peterson, E. Wesley. 2009. *A Billion Dollars a Day: The Economics and Politics of Agricultural Subsides*. Malden, MA: Wiley-Blackwell.

Pimentel, David. 2005. Environmental and economic costs of the application of pesticides primarily in the United States. *Environment, Development and Sustainability* 7: 229–52.

Pretty, Jules, Craig Brett, David Gee, Rachel Hine, Chris Mason, James Morison, Matthew Rayment, Gert Van Der Bijl, and Thomas Dobbs. 2001. Policy challenges and priorities for internalizing the externalities of modern agriculture. *Journal of Environmental Planning and Management* 44(2): 263–83.

Qadir, M., B. Sharma, A. Bruggeman, R. Choukr-allah, and F. Karajeh. 2007. Non-conventional water resources and opportunities for water augmentation to achieve food security in water scarce countries. *Agricultural Water Management* 87: 2–22.

Ranjan, A. 2010. Beijing's threat to India's water security. Center for the Study of Globalization, Yale University, New Haven, CT, http://yaleglobal.yale.edu/content/beijings-threat-indias-water-security, last accessed April 16, 2011.

Rosemarin, A., G. De Bruijne, and I. Cladwell. 2009. The next inconvenient truth: Peak phosphorus. *The Broker* 56(2) http://www.thebrokeronline.eu/en/Magazine/articles/Peak-phosphorus#f6, last accessed April 18, 2011.

Seckler, D. 1996. The New Era of Water Resources Management: From 'Dry' to 'Wet' Water Savings. Research Report, International Irrigation Management Institute, Colombo, Sri Lanka, http://www.iwmi.cgiar.org/Publications/IWMI_Research_Reports/PDF/pub001/REPORT01.PDF, last accessed September 2, 2010.

Smil, V. 2004. *China's Past, China's Future: Energy, Food, Environment*. New York: Rutledge.

Smith, A., P. Watkiss, G. Tweddle, A. McKinnon, M. Browne, A. Hunt, C. Treleven, C. Nash, and S. Cross. 2005. The Validity of Food Miles as an Indicator of Sustainable Development. Final report produced for Defra, Report number ED50254, http://www.igd.com/download.asp?id=3&dtid=2&did=1633, last accessed April 13, 2011.

Stuart, Tristram. 2009. *Waste: Uncovering the Global Food Scandal*. New York: Norton.

Sutton, M., C. Howard, J. Erisman, G. Billen, A. Bleeker, P. Grennfelt, H. van Grinsven, and B. Grizzetti. 2011. *The European Nitrogen Assessment: Sources, Effects and Policy Perspectives*. Cambridge: Cambridge University Press.

UNDP. 2007. Human Development Report 2006 – Beyond Scarcity: Power, Poverty and the Global Water Crisis. United Nations Development Programme, New York, http://hdr.undp.org/en/media/HDR06-complete.pdf, last accessed April 1, 2011.

United Nations. 2006. Human Development Report 2006. United National Development Program, New York, http://hdr.undp.org/en/media/HDR06-complete.pdf, last accessed April 16, 2011.

United Nations. 2009. Water in Changing World. 3rd UN World Water Development Report, http://www.unesco.org/water/wwap/wwdr/wwdr3/pdf/WWDR3_Water_in_a_Changing_World.pdf, last accessed April 19, 2011.

US Geological Survey. 2009. Phosphate Rock. Mineral Commodity Summaries, http://minerals.usgs.gov/minerals/pubs/commodity/phosphate_rock/mcs-2009-phosp.pdf, April 18, 2011.

Vaccari, David. 2009. Phosphorus famine: The threat to our food supply. *Scientific American* June 3.

Weber, Christopher and H. Scott Matthews. 2008. Food miles and relative climate impacts of food choices in the United States. *Environment, Science and Technology* 42: 3508–13.

White, Stuart and Dana Cordell. 2008. Peak Phosphorus: The Sequel to Peak Oil. Information Sheet 2, Global Phosphorus Research Initiative, http://phosphorusfutures.net/peak-phosphorus, last accessed November 18, 2010.

WHO 2009. Diarrhea: Why Children are Still Dying and What Can be Done. United Nation's Children Fund, World Health Organization, Geneva, Switzerland, http://whqlibdoc.who.int/publications/2009/9789241598415_eng.pdf, last accessed April 19, 2011.

PART IV

Alternatives

11

ALTERNATIVE AGROFOOD NETWORKS

KEY TOPICS

- The sociological literature has extensively studied the rise of organics, from the "ends" of production as well as consumption, for well over a decade, providing us with a number of important insights.
- The Slow Food movement has slowly gained momentum in recent years, yet its overall impact on the dominant food system remains unclear.
- As perhaps the world's most important transnational social movement, *La Via Campesina* has made important inroads toward building food sovereignty in developing nations.

Key words

- Biodynamic agriculture
- Committed conventional farmer
- Committed organic farmer
- Decision map
- Environmental but not organic farmer
- Food sovereignty
- *La Via Campesina*
- Organic agriculture
- Peasant *v.* farmer
- Pragmatic conventional farmer
- Pragmatic organic farmer
- Slow Food
- Sustainability (social, environmental, economic)
- Virtuous globalization

"Bread, it is bread that the Revolution needs!" (Kropotkin 2011). These words come from Peter Kropotkin (1842–1921), written at the turn of the twentieth century. Our daily interaction with food, at such an intimate level (we literally put it into us), makes food political. Kropotkin's point: little stirs one's heart and mind like an empty stomach. Yet I would read still further into Kropotkin's comments. Not just empty stomachs but food *itself* drives people, especially recently as millions strive to reconfigure food systems to make them more sustainable and just. This chapter reviews three recent challenges to the dominant food system: the rise of organic food, the Slow Food movement, and *La Via Campesina*. Their success can be measured in widely different ways, as each is driven toward different ends by different interests, goals, and segments of the population.

Organic agriculture

Organic agriculture is often traced to the beginning of the twentieth century. Early luminaries to promote organic farming methods include Austrian philosopher Rudolf Steiner (who, as discussed in Box 11.2, is also linked closely with biodynamic agriculture), Sir Albert Howard and Lady Eve Balfour (who together led the founding of the British Soil Association), and the Swiss couple Maria and Dr Hans Müller, and the German doctor H.P. Rusch. Despite some differences between the approaches, the main aim of organic farming – its *spirit*, you might say – is to create a sustainable agricultural production system. "Sustainable" here defined in a broader sense than just economic sustainability, including also environmental and social sustainability (Box 11.1). It is also worth mentioning that while "organic" to biochemists refers simply to anything containing carbon, to early twentieth century thinkers it held a different meaning. Organic, in this other sense, compelled farmers to think of their farm as an organism, that is, as a self-regulating whole that must be managed accordingly (rather than as a bunch of substitutable parts).

Statistics on organic agriculture have been compiled for 141 countries (Willer and Kilcher 2009). The results of this global survey on certified organic farming show:

- A total of 32.2 million hectares of agricultural land are managed organically by more than 1.2 million producers. In addition to the agricultural land, there are 0.4 million hectares of certified organic aquaculture.
- About one-third of the world's organically managed land – almost 11 million hectares – is located in developing countries.
- Almost two-thirds of the land under organic management is grassland (20 million hectares). The cropped area is 7.8 million hectares – a figure that is rapidly increasing (in some instances at the expense of organic grassland).
- International sales of organic food are approaching US$50 billion annually. The vast majority of this demand is concentrated in North America and Europe, constituting 97 percent of global revenues.

As illustrated in Figure 11.1, a region's designation as a hotspot for organic land does not necessarily mean it is equally an organic farmer's and rancher's hotspot.

BOX 11.1 THE SOCIOLOGY OF "SUSTAINABILITY"

Sustainability (like in the context of "sustainable" agriculture) is often likened to a stool, held up by the three legs of social, environmental, and economic sustainability. I have found this an effective metaphor as it emphasizes the equal value that must be placed on each of the three "legs" – ignore, say, social variables and the chair of sustainability invariably will tip over. True sustainability can thus only be attained with social, environmental, and economic sustainability. But what, precisely, does that mean? The problem with "sustainability," even the three-legged type, is that it's a terribly vague term.

Social, environmental, and economic sustainability are hotly contested terms. *What* they mean depends on *who* you ask. When talking about social sustainability, for example, are we talking about improving social capital or social justice? Or environmental sustainability: do we mean striving for zero ecological impact or minimal impact – and if the latter *who* decides what "minimal impact" means? Or economic sustainability: does this mean simply that firms and farms need to be profitable or does it also suggest a desire to reduce, say, economic equalities. And what *scale* are we talking about: the farm/firm, the community, or the entire food system – and *who* decides this?

In the end, talk of the *science* of sustainability puts the cart before the horse. We can use science to help us figure out how best to reach certain ends we define as "sustainable." How we determine those ends, however, is a process best left for a little thing called democracy.

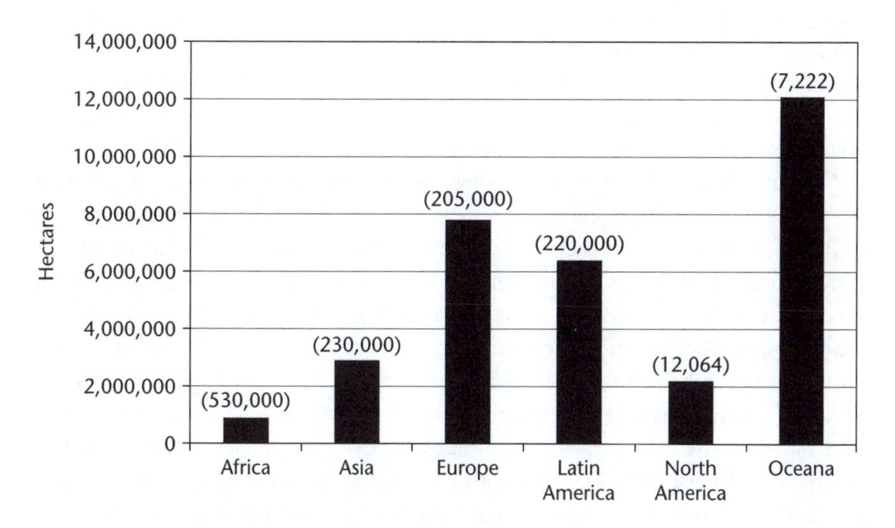

FIGURE 11.1 Global distribution of organic land (number of producers for each region in parenthesis). Based on Willer and Kilcher (2009)

Indeed, Africa, with roughly 900,000 hectares of organic land – just 3 percent of the world's organic land – is home to almost half of the world's organic producers. At the other extreme: 7,222 producers in Oceania (including Australia and New Zealand as well as the islands of Fiji, Papua New Guinea, Tonga, and Vanuatu) manage over 12 million hectares of organic land.

Who produces organic food

What attitudes and background information correlate positively (and strongly) with organic farmers? Organic farmers are likely to have an urban background, have higher levels of academic education, and be younger (and thus have less farming experience than their conventional counterparts) (see, e.g., Burton *et al.* 1999; Duram 1999; Lockeretz 1997; Tovey 1997), though more recent research looking at Norwegian organic dairies shows that newcomers (those who converted in 2000 or later) were less educated than the early organic entrants who converted in 1995 or earlier (Flaten *et al.* 2006). Gender also appears to be correlated with the decision to convert to organic farming, as female farmers seem more open to the practice (see, e.g., Ashmole 1993; Fisher 1989; Padel 2001). Research from the UK, for example, surveyed organic and conventional producers finding a higher proportion of women growers to be employed by organic operations (Burton *et al.* 1999). There is also evidence that organic growers tend to express higher levels of care for the environment and animal welfare than their conventional counterparts (Lockie *et al.* 2006; Padel 2001).

An exemplary (albeit somewhat dated) review of the literature is offered by Padel (2001). Padel looks not only at how European organic farmers differ from their conventional counterparts, but also at how motives differ *among* organic producers. In particular, she compares newer organic farmers with well-established organic producers. The former, Padel discovers, are more likely to come from the ranks of conventional agriculture, operate larger than average farms, and are slow to convert their entire operation over to organic. The implications of this mean that as organic agriculture becomes more widespread, the profile of organic growers will become less distinct from conventional growers.

Knowing that certain attitudes and demographic variables correlate strongly with the adoption of organic farming methods still tells us little about the decision process itself that causes a farmer to choose to convert over to organic. A study of farmers in northeastern Austria gives some insight into this process (Darnhofer *et al.* 2005). Specifically, the researchers sought to understand why farmers chose to either keep their operations conventional or to convert them over to organic. What's novel about this research is its focus on *both* organic and conventional farmers. Focusing solely on organic growers can lead one (falsely) to believe that those farmers who have not converted do not share the motives, attitudes, and sentiments expressed by organic farmers. Including conventional farmers illuminates both rationale and constraints associated with conventional farmers who have considered organic farming but decided to not make the switch. This research culminated in a "decision map" showing the various decision criteria respondents went through to decide whether or not organic farming was for them (Figure 11.2). The research team was also able to categorize farmers into different "conventional" and "organic" agriculture types by the motivations that lie behind their farming practices.

Figure 11.2 differentiates between five farming types: committed conventional, committed organic, environmental-conscious but not organic, pragmatic organic, and pragmatic conventional. "Committed conventional" farmers are characterized by giving no thought to organic farming. They do not view it as more environmentally friendly than conventional production methods, nor do they believe the health claims made about organic foods or perceive that organic production is technically and/or economically feasible. Their attitudes and values also reflect minimal levels of concern for the environment. "Pragmatic conventional" farmers do not oppose organic farming *per se*. Their concern lies in the amount of risk and/or uncertainty associated with the conversion to organic, which they are unwilling to confront unless they can expect certain financial rewards. They are therefore open to conversion but only after these uncertainties have been resolved and a market for organic products better established. "Environment-conscious but not organic" farmers are committed to environmentally friendly farming practices but are not currently operating organically certified farms – some were, however, operating self-declared organic farms. Some of these farmers are choosing to retain the flexibility of not being certified (e.g. to use some synthetic inputs on crops based on need), while others might be following the organic standards but choose to not involve themselves in the bureaucracy and paperwork that is involved in certification. The costs of running a formally certified operation – including certification and the associated record-keeping – were also a factor for some of these farmers. And since they have an established clientele willing to pay a premium without a formal organic label (perhaps trust between grower and consumer makes the label unnecessary), they do not see the need to go this extra step.

Among "pragmatic organic" farmers, the ability to obtain a premium price in the market outweighed health, ethical, or sustainability concerns. Of the sample included in the study, most of farmers of this type have been involved in organic agriculture less than 10 years. This lends support, coincidently, to Padel's (2001) earlier mentioned argument that the profile of organic growers is becoming less distinct from conventional growers. Lastly: the "committed organic" farmers. These individuals are deeply committed to the founding philosophy of organic farming and are therefore willing to forgo some of their income to live by these commitments. Seventy-six percent of the organic farmers who have been farming longer than 10 years are "committed organic." For them, organic farming cannot be reduced to a set of processes but requires that one embrace a distinct way of life.

Who consumes organic food

In the last two decades there has been a flurry of studies looking for variables that correlate with organic food consumption. This research has done little to spell out

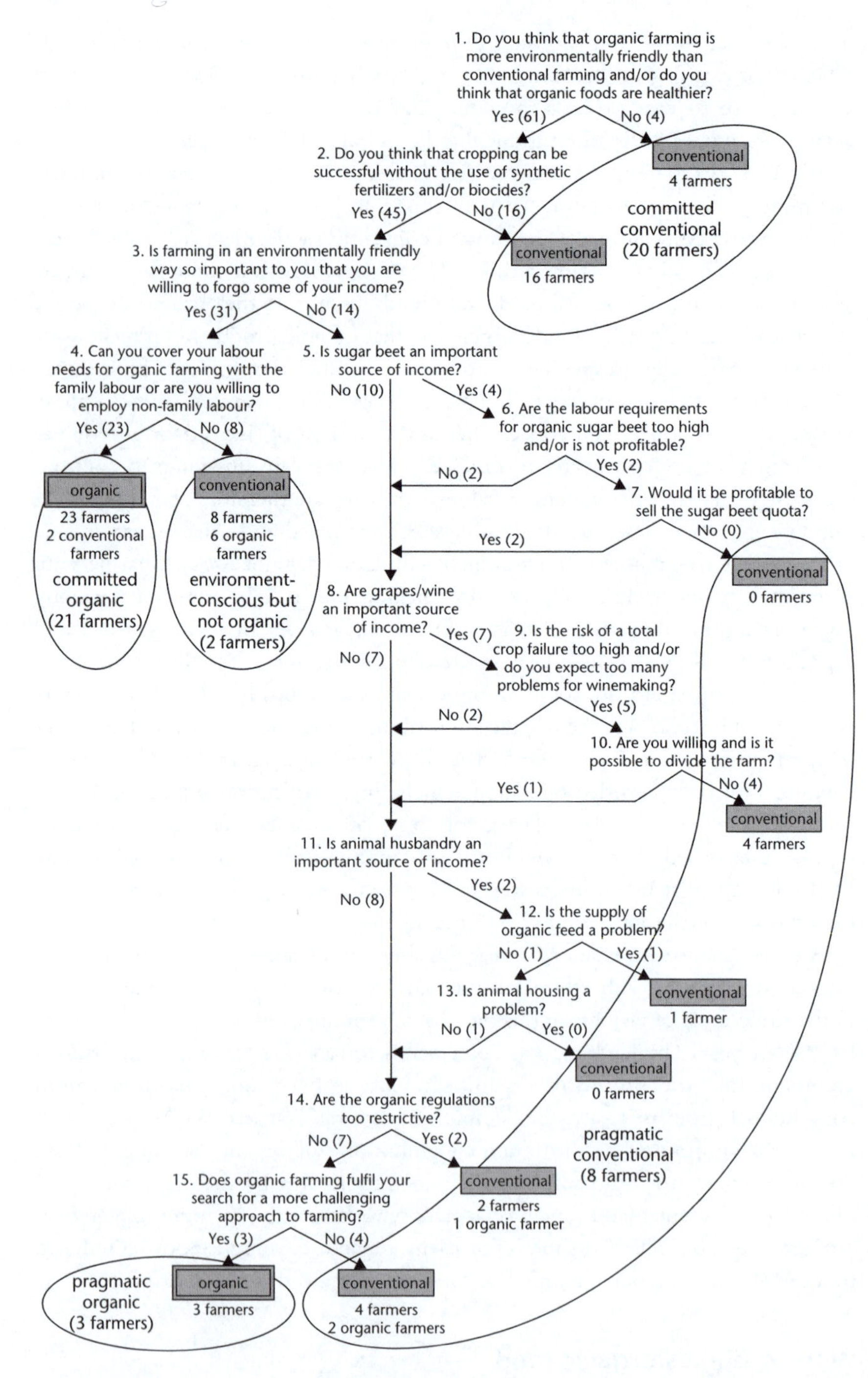

FIGURE 11.2 Decision-tree depicting the criteria considered in the choice between organic and conventional farm management in northeast Austria. Based on Darnhofer *et al.* (2005) with help from Ika Darnhofer

clearly who the people are that eat organics (and what their underlying motivations are), though it has helped dispel some myths. To review this literature efficiently, the following discussion is organized around emergent themes that are common in studies with this scope. These themes are health, taste/sensory experience, green/universal values, income, gender, family/children, age, education, and socio-structural.

- *Health*: Care for one's health and the health of one's family have repeatedly been shown to be a strong motive for purchasing organic food (see, e.g., Aertsens *et al.* 2009; Botonaki *et al.* 2006; Chen 2009; Chryssohoidis and Krystallis 2005; Shepherd *et al.* 2005).
- *Taste/sensory experience*: A study of Swedish consumers finds taste to be the primary reason for the consumption of organics (Magnusson *et al.* 2001). Similar results were obtain by researchers from Ireland (Roddy *et al.* 1996), the Netherlands (Schifferstein and Oude Ophuis 1998), Scotland (McEachern and McClean 2002), and Greece (Fotopoulos *et al.* 2003). In an analysis of factors influencing food choice among Australian consumers who had consumed at least some organic food in the preceding 12 months, the researchers found that "concern with the naturalness of food and the sensory and emotional experience of eating were the major determinants of increasing levels of organic consumption" (Lockie *et al.* 2004: 135).
- *Green/universal values*: In one study, respondents who possess high levels of understanding, appreciation, tolerance, and protection for the welfare of all living things also expressed positive sentiments towards organic food (Dreezens *et al.* 2005). Strong "green" attitudes have repeatedly been shown to be a strong predictor of pro-organic food beliefs (see, e.g., Aertsens *et al.* 2009; Lockie *et al.* 2004; Loureiro *et al.* 2001; Padel and Foster 2005). The data are not conclusive on this relationship, however. In a study from Greece, expressions of care for the environment were not statistically linked to organic food consumption (Chryssohoidis and Krystallis 2005). Elsewhere, among sampled UK consumers, there was no connection between pro-organic food attitudes and pro-environmental attitudes, while among Germans sampled such a link existed (Baker *et al.* 2004).
- *Income*: The link between income and consumption of organic food is mixed, dispelling a common myth about organic food – that it is for only the wealthy. Thus, on the one hand, some studies conclude that income does have a significant role in explain organic food purchased (see, e.g., Arbindra *et al.* 2005; Gracia and de Magistris 2007), while on the other there seems to be an equal amount of research that finds no such link (see, e.g., Zepeda and Li 2007). Elsewhere, among Australian consumers, income and organic food consumption were slightly correlated, but only up until a point (Lockie *et al.* 2002). The highest income earners (those earning US$50,000 and over annually) actually purchased less organic food than those in the second highest income category (US$35,000–49,999 annually). Even among the lowest income earners (those making less than US$20,000 annually), organic food consumption levels were surprisingly close to those in the highest income category.

- *Gender*: Women are consistently more likely to hold greater levels of pro-organic attitudes – and report buying more organic food – than men (Aertsens *et al.* 2009; Arbindra *et al.* 2005; Hughner *et al.* 2007; Lockie *et al.* 2002, 2004; Storstad and Bjorkhaug 2003).
- *Family/children*: Families with children are more likely to buy organics (Hughner *et al.* 2007; Lockie *et al.* 2004; McEachern and Willock 2004). A number of studies also report that following a childbirth experience mothers changed their purchasing patterns, buying more organic products, not only for their child but often for the whole family (Aertsens *et al.* 2009).
- *Age*: Here too findings are mixed (Aertsens *et al.* 2009). Some research finds UK organic consumers to be older than the average population (Geen and Firth 2006). Other research, conversely, notes that younger respondents are more likely to purchase organic foods and have higher levels of pro-organic sentiments (Arbindra *et al.* 2005; Magnusson *et al.* 2001). Other studies find no relationship between age and levels of organic consumption and pro-organic sentiments (O'Donovan and McCarthy 2002).
- *Education*: More mixed results, though more research finds a link between education and organic consumption than does not. Some studies have found a clear positive relationship between education and organic food consumption and pro-organic attitudes (see, e.g., Lockie *et al.* 2002; Wandel and Bugge 1997). Other studies conclude that the link is minimal (Lea and Worsley 2005) or nonexistent (Arbindra *et al.* 2005). Lockie *et al.* (2002) also find organic food consumption and science education to be positively correlated. In their own words:

  > Contrary to yet another stereotype of organic consumers – as possessed by an irrational fear, or inability to understand, technological development – it seems that the more training people have in the critical evaluation of knowledge claims the more likely they are to consume organic food.
  >
  > *Lockie* et al. *(2002: 36)*

- *Socio-structural*: Other studies remind us that pro-organic attitudes do little good if the food is not available. In one study, for instance, "convenience" was highlighted as an important factor shaping organic purchasing decisions (Hjelmar 2011). Admittedly, it is difficult to know, in hindsight, which came first – sentiments or market share – though undoubtedly the two grew together. As one researcher explains in a study of organic food consumption in Europe:

  > [T]he three countries with the highest organic market shares in Europe, and in the World, Switzerland, Austria, and Denmark, all benefited from large retailers making organic food available and affordable to broad segments of consumers early in the development of the organic market.
  >
  > *Thøgersen (2010: 182)*

Local versus organic

Australia has had national standards for organic and biodynamic products in place since 1992 (Box 11.2). The European Union (EU) developed comprehensive organic legislation in 1992. In the USA, organic standards took effect in 2002. While in New Zealand, a national organic standard was launched in 2003. Organics are truly making a worldwide impact.

Some argue that organic's tremendous market success in recent years says more about its conventionalization (see Chapter 2) than indicating any real structural changes to the food system. To support this claim, they'll point to, among other things, the internationalization of process-based organic standards – standards that focus on the *processes* of growing, harvesting, and preparing the foods. Process-based standards still allow for input substitution practices, which have allowed capital to shape the organic sector in much the same image as the convention food sector.

There is some evidence that the significant increase in popularity of so-called local food in recent years reflects a growing discontent toward "organic" food as currently defined by national, process-based, standards (Lockie *et al.* 2006). To quote

BOX 11.2 BIODYNAMIC AGRICULTURE

The roots of biodynamic farming can be traced back to a series of lectures given in 1924 by Rudolf Steiner, an Austrian writer and philosopher. Biodynamic agriculture can be likened to, one might say, "organic plus" (whereas conventionalized organic systems have been called by critics "organic lite" [Guthman 2004]). As one sourcebook explains: "Biodynamic farming is a combination of biological and dynamic practices; it also involves animal manures, crop rotations, care for animal welfare, looking at the farm entity and local distribution systems" (Padmavathy and Poyyamoli 2011: 387). The term "dynamic practices" in that definition is a bit controversial, at least in the scientific community, which perhaps is why it was not defined in any detail. The Biodynamic Farming and Gardening Association write of this dynamism as follows:

> [Biodynamics is a] type of organic farming that incorporates an understanding of "dynamic" forces in nature not yet fully understood by science. By working creatively with these subtle energies, farmers are able to significantly enhance the health of their farms and the quality and flavor of food.
>
> *http://www.biodynamics.com/biodynamics*

The "biodynamic" label is trademarked by Demeter International, the largest certification organization for biodynamic agriculture in the world.

one food activist following the release of the US Department of Agriculture's (USDA) final rules on organic certification in the USA:

> When we said organic we meant local. We meant healthful. We meant being true to the ecologies of the region. We meant mutually respectful growers and eaters. We meant social justice and community. In other words, industrial organic farming isn't really organic.
>
> *quoted in Adams and Salois (2010: 332)*

Local food also seems to evoke a greater willingness to pay among consumers than an organic label (Adams and Salois 2010; Toler *et al.* 2009). One study reports that California residents view locally grown to be a "moderately important" factor for shopping at farmers' markets, while organically grown is only a "slightly desirable" factor (Wolf *et al.* 2005). The study also found that shoppers believed local food to be fresher, better quality, and more affordable than organic food. Elsewhere, researchers found that having food locally grown is said to be either a "very important" (49 percent) or "somewhat important" (31.5 percent) reason by the vast majority of those interviewed who shop at farmers' markets, while an organic designate was said to be "very important" or "somewhat important" far less frequently (15.8 percent and 19.9 percent of the time, respectively) (Gallons *et al.* 1997).

This is not to suggest that consumers no longer desire organic food. Clearly they do. In 2010, the US organic industry alone was valued at just under US$30 *billion*. But consumers also seem to want something more than what the organic label currently offers, as evidenced by the growth of farmers' markets and community-supported agriculture, which grew in the USA from 1,755 in 1994 to 4,685 in 2008 and from 50 in 1985 to 2,500 in 2008, respectively.

This move toward post-organic sentiments among consumers can be better understood against the structural backdrop of the organic food sector, which has been undergoing radical changes since the 1990s. The average size of certified organic farms in the USA, for example, has more than doubled from 189 acres in 1995 to 477 acres in 2005 (Adams and Salois 2010). In Finland, the average size of organic farms has more than tripled between 1990 and 2006. The average Finnish organic farm is now 5 percent larger than the average size of all farms in the country.[1]

Large firms have entered the organic market either by way of acquiring old organic brands or introducing new ones. As detailed in one recent study, the acquisition strategy happened early on, when there were still small organic brands to acquire (Howard 2009). From here, top food processors began developing their own organic lines, such as M&M Mars' Dove Organic, which was created specifically for retail giant Walmart as they worked to increase their organic offerings in 2006. Another more infamous example comes from Anheuser-Busch and their Wild Hop and Stone Mill organic beers, which were stealthily introduced as if coming from independent micro-brews until the *San Francisco Chronicle* revealed otherwise (Howard 2009). In the USA, roughly 80 percent of the organic food market is handled by just two large distributors (Adams and Salois 2010).

Local food, however, does not appear immune to the forces of conventionalization. In 2009, Frito-Lay launched their "Lay's Local" campaign. The chips are made utilizing potatoes grown near their factories. Consumers can even go to the company's website (to their "Chip Tracker"), enter the first three digits of the product code on the bag and their ZIP code and, presto, out pops the location of the plant that made the snack. Or take Walmart's recent move to become a major local food player. In 2008, the company announced their commitment to source more local produce. Yet *who* the retail giant gets its local food from is a handful of very large farms. That is because only large farms can afford the "minimum" requirements required by Walmart, such as UPC bar code technology, US$2 million in commercial liability insurance, and a financial stability rating (Adams and Salois 2010). Then there is the question what is "local" food anyway? For Whole Foods, a product can be defined as local if the total transport time from farm to store is 7 hours or less by truck, whereas Walmart defines local as food coming from within a state's border (Adams and Salois 2010).

Clearly, the concept of "local food" needs further scrutiny. Yet it is also an incredibly thorny issue. Knowing this, I will wait until the next chapter to tackle it, where I devote significant space to this topic.

Slow Food

The Slow Food movement was launched in 1987 by a group of Italian writers and journalists with the aim of preserving traditional and regional cuisine at the points of production (e.g. farming), preparation (e.g. cooking), and consumption (e.g. taste). The term "slow" still evokes quizzical looks, though there is a lot of talk of late around the concept by social and political theorists, journalists, and activists – a backlash, arguably, against the ever-quickening pace of everyday life (Wolin 1997; MacKenzie 2002; Connolly 2003). Slow food, slow living, slow knowledge, slow politics: just a few examples of where "slowness" is being evoked for purposes of social change.

Aspects of Slow Food were introduced in Chapter 6, most notably the Ark of Taste. Building on that, it is worth mentioning the international reach of the movement, which today spans over 150 nations in Europe, North, South and Central America, Asia, Oceana, and Africa. This translates into more than 100,000 members, joined in 1,300 *convivia* (local chapters).[2] Slow Food also has an influential lobbying presence in the EU on matters of trade and agricultural policy. A publishing collaboration between Slow Food and the popular UK-based *Ecologist* magazine in 2004 marks a strategic move to broaden the movement's rhetoric to include environmental issues. Other accomplishments of Slow Food include the promotion of more than 500 initiatives annually, giving support to 10,000 small producers, the promotion and conservation (e.g. through the Ark of Taste catalog) of 903 products at risk of extinction, and 1,300 food education activities and 350 school gardens in 100 countries.

Take the example of Tuscarora White Corn, also known as Iroquois White Corn. Tuscarora is an heirloom variety long prized as *the* staple food crop of the Iroquois Nation (Native Americans who originally resided largely in upstate New York). The corn is famous for its large white kernels that are of the perfect consistency for grinding into cornmeal for breads and soups. Tuscarora White Corn is an important material medium through which the Iroquois are able to practice their culture. And it was disappearing at an alarming rate. Prior to its inclusion in the Ark of Taste, fewer than 100 acres of this corn were in cultivation in the USA.[3] The cry "Give me Tuscarora or give me death!" might sound a bit over-the-top, until you understand the truly existential nature of Tuscarora White Corn. Without it, aspects of the Iroquois's culture will die, literally.

Critics of the food system status quo have a lot to like about Slow Food, as its theoretical underpinnings seem to complement what critics of capitalism have been saying since Karl Marx (Meneley 2004). The movement's symbol is of all things – a snail. This rejection of speed has a long lineage in social thought, as critical theorists have for over a century lambasted the wage labor structures of capitalism that regulate the bodies of the workers at neck-breaking speeds; a pace becoming matched gastronomically, as we turn increasingly to fast, convenient, portable food. Sidney Mintz (1986) has famously written about the centrality of sugar in working-class diets due to its "fit" with the ever-quickening pace of factory life. A food ideally suited for the working class, energy-dense sugary foods replaced those that were more nutritionally sound but which required too much time to prepare and therefore did not fit with capitalist work schedules. The last couple of centuries have made life a blur, Slow Food advocates argue. It's time we slowed down.

The Slow Food project has also been dubbed by some of its founders as representing "virtuous globalization," in that it "promotes itself as providing a model for imagining alternative modes of global connectedness, in which members of minority cultures – including niche-food producers – are encouraged to network and thrive" (Leitch 2009: 45). This term could easily be passed over, mistaken for hollow marketing jargon. Beside, as a student once asked me, "Isn't 'virtuous globalization' a contradiction in terms?" If we spend just a little time unpacking the term our effort will yield something interesting.

This talk of virtuous *globalization* tells us that Slow Food is less interested in food *revolution* as it is in food *reform*. What this suggests, then, is that Slow Food is seeking, at least in part, to utilize some of the same pathways employed by conventional agriculture to gain support and grow (Lotti 2010). Michael Pollan (2003) attempts to champion virtuous globalization when he writes: "Paradoxically, sometimes the best way to rescue the most idiosyncratic local products and practices is to find a global market for them. This is what Slow Food means by 'virtuous globalization,' a simple but powerful idea that throws a wrench of complexity into the usual black-or-white arguments over free trade." Globalization scholar and anti-corporate activist Starr (2010) even recently wrote about the potential virtues of globalization (though she never used the term "virtuous") when talking about alternative food movements:

> Anticapitalists generally spurn market projects as ideologically inadequate pragmatism. [...] But anticapitalists have innovated precious few strategies engaging to the alienated, individualistic, fearful consumer culture we find ourselves organizing in. If the market is where society increasingly spends its time and attention, then we need to learn to organize in the market, keen to identify its fissures and expand them into new worlds. This is not reformism or apologism, it is recognizing the historical political-cultural reality in which we find ourselves and those with whom we would like to make revolution.
>
> *Starr (2010: 486)*

Yet let us not forget the costs that movements must pay when they seek, in Starr's words, to "take place in the market."

Returning to Slow Food: the article "The commoditization of products and taste" gives some good examples of these costs (Lotti 2010). In this article, the author argues quite convincingly "that commoditization makes the endeavors of Slow Food resemble the conventional agricultural system it is trying to oppose, as well as undermining the very agrobiodiversity the organization seeks to protect" (Lotti 2010: 71). One example of Slow Food given is Pélardon (a cheese). Producers of this slow cheese are expected to produce their particular Pélardon according to very specific standards, resulting in a high level of uniformity in their product. There are therefore enormous asymmetries in the incentive structure between slow Pélardon and nonslow Pélardon. Those select few in the former camp receive a premium price and considerable support from the Slow Food organization for the marketing of their product, while the rest struggle to make a living. Thus, as the author points out:

> Unless there are equal incentives elsewhere for the continued production of the whole variety of Pélardon cheeses, the actions of Slow Food actually risk having two negative effects. First, they risk eliminating biodiversity at the local scale. The eleven other Pélardon cheeses may cease to be produced, resulting in a decrease in biodiversity because where there were twelve different cheeses, there remains only one.
>
> *Lotti (2010: 79)*

Biodiversity is further reduced by the movement's emphasis on maintaining the genetic "purity" of the heritage variety. The article talks specifically of a pig farmer – Pedro – who maintains a breed included in the Ark of Taste:

> Pedro cannot cross his pigs with other breeds; this would ruin the purity of an autochthonous breed. It is this crossing between breeds and varieties, however, that maintains and creates the agrobiodiversity at the local level and upon which Slow Food depends to create the products it has identified. Pedro's pigs are preserved but the process from which future breeds might originate – that of crossing the Euskal Txerria with other breeds – has been stopped.
>
> *Lotti (2010: 79)*

Thus, while Slow Food may be seeking to preserve agrobiodiversity at the international level, it, somewhat paradoxically, is reducing it at the local level.

Elsewhere, criticism toward Slow Food has been more pointed. Some have labeled the movement as embracing a type of culinary romanticism, in that it fails to appreciate the democratization of food choice – maybe people actually want to eat fast food! – that has taken place over the last half of a century (Laudan 2001). Slow food has also been criticized for merely replacing global tyranny (by multinational food companies and retail firms) with a more regional form (by local elites) (Lauden 2004).

Earlier I mentioned how Slow Food implores us to slow down. Saying this, however, does nothing to change the socio-economic structures that make life for so many so fast. Until the economy is significantly altered many people will simply not have the time to do food (or much of anything) slowly. And *who*, if we suddenly started slowing down, would be burdened with this time consuming activity? Presumably, unless gender roles and expectations were to suddenly change too, women would be expected to engage in the majority of hidden domestic labor that makes Slow Food possible. Meneley (2004: 174), quoting from a newspaper article on Slow Food dishes, describes the Slow Food delicacy, capòn magro: "The dish consists of layers of mixed seafood, salsa verde, potato and smoked tuna. With all the boning, shelling, cleaning and chopping, it takes three people five hours to make a real capòn magro. It is worth every minute though [. . .]". Who – especially among the working class – has time for this?

Finally, there is concern that the movement does not aid small producers as much as its publications claim. In a study of Tuscan olive oil producers, Meneley (2004) notes that Slow Food's involvement in this industry largely resulted in the promotion of elite producers. What's the point of a food conservation movement that does not also conserve those ways of life upon which agri-*culture* is based, like small-scale (non-elite-owned) farms?

La Via Campesina and food sovereignty

Slow Food is arguably a Western, middle and upper class food movement. What alternatives, then, are available to the world's millions of peasant farmers? They have available to them peasant movements, one of the more famous of which is *La Via Campesina*.

The peasantry was targeted for extinction in the twentieth century. I don't mean to imply that peasants *themselves* were literally targeted for elimination. But their way of life was placed in the cross hairs of international development policy (McMichael 2003). *La Via Campesina* emerged as a direct response to the perceived injustices borne by the world's peasants.

La Via Campesina is viewed by many as the world's most important transnational social movement (Desmarais 2008; McMichael 2006; Edelman 2005; Martínez-Torres and Rosset 2010), even though most of the general public (at least in developed nations) have never heard of the organization. As explained on their website:

> *La Via Campesina* is the international movement which brings together millions of peasants, small and medium-size farmers, landless people, women farmers, indigenous people, migrants and agricultural workers from around the world. It defends small-scale sustainable agriculture as a way to promote social justice and dignity. It strongly opposes corporate driven agriculture and transnational companies that are destroying people and nature.
>
> *http://www.viacampesina.org/en/*

La Via Campesina is made up of roughly 150 local and national organizations in 70 countries from Africa, Asia, Europe, and the Americas and collectively represents about 200 million farmers. The movement is founded on the belief that for far too long rural and food policies have been developed in the absence of those most affected, noting that peasants (including rural women; Box 11.3) have a unique place and critical role in shaping agricultural policies (Martínez-Torres and Rosset 2010). The class profile of *La Via Campesina* is estimated as follows: landless peasants, tenant farmers, sharecroppers, and rural workers largely in Latin America and Asia; small and part-time farms in Europe, North America, Japan, and South Korea; peasant farms and pastoralists in Africa; small family farms in Mexico and Brazil; middle class (and some affluent) farmers in India; and poor urban (and urban-fringe) dwellers in countries like Brazil and South Africa (Borras 2008).

La Via Campesina seeks to recast the term "peasant" as an identity to be embraced, as are other organizations, like the Federation of Indonesian Peasant Unions and the Peasant Movement of the Philippines. This is opposite to the dominant English use of the term, where "peasant" has long been linked with feudalism thus giving it a pejorative meaning. Perhaps this explains why delegates from the UK, when seeking to name the movement, initially resisted the term *La Via Campesina* – which literally translates into Peasant Way – out of concern about the derogatory connotations attached to the term "peasant" (and especially in light of England's feudalistic past). The delegate instead preferred the term "farmer." However, as Saturnino Borras, a founding member of *La Via Campesina*, recalls, many outside the UK delegation preferred the term "peasant," noting that "farmer" also had connotations "that did not capture the nature and character of the farm sector we do represent" (as quoted in Desmarais 2008: 140). A compromise was reached: keep the name *La Via Campesina* but do not translate it into English (Desmarais 2008). Others also suggest that the title *La Via Campesina* pays homage to Latin America's founding role in the movement. Long a hotbed of peasant discontent, this region of the world has been plagued with grossly unequal distributions of land and income as well as a sharp decline in standards of living in the 1980s – the so-called "lost decade" – due to failed neo-liberal policies (Martínez-Torres and Rosset 2010).

The term "peasant" – as opposed to, say, "farmer" – is meant to imply a distinct way of life that is shared by many, in both developed and developing countries. Take the following quote from Nettie Wiebe, who at the time of the interview was regional coordinator of a Canadian branch of *La Via Campesina*:

BOX 11.3 *LA VIA CAMPESINA* AND THE PLIGHT OF RURAL WOMEN

A notable feature of *La Via Campesina* involves its striving for greater gender equality. At an organizational level, the movement has a man and a woman representative from each region. This has led to some other peasant organizations – most notably the Landless Workers' Movement in Brazil (in Portuguese *dos Trabalhadores Rurais Sem Terra*, or simply MST) – to make similar changes to their internal structure. As the movement has released a number of public statements about its views on the injustices women face around the world, I will let *La Via Campesina* speak on behalf of itself on the subject. The following is taken from the document titled "Declaration of Maputo" (Maputo, the capital and largest city of Mozambique, marks where this declaration was made):

> [A]ll the forms of violence that women face in our societies – among them physical, economic, social, cultural and macho violence, and violence based on differences of power – are also present in rural communities, and as a result, in our organizations. [. . .] We recognize the intimate relationships between capitalism, patriarchy, machismo and neo-liberalism, in detriment to the women peasants and farmers of the world. All of us together, women and men of *La Via Campesina*, make a responsible commitment to build new and better human relationships among us, as a necessary part of the construction of the new societies to which we aspire. [. . .] We recognize the central role of women in agriculture for food self-sufficiency, and the special relationship of women with the land, with life and with seeds. In addition, we women have been and are a guiding part of the construction of *La Via Campesina* from its beginning. If we do not eradicate violence towards women within our new gender relations, we will not be able to build a new society.
>
> La Via Campesina *(2008)*

> If you actually look at what "peasant" means, it means "people of the land." [. . .] We too are peasants and it's the land and our relationship to the land and food production that distinguishes us. We're not part of the industrial machine. We're much more closely linked to the places where we grow food and how we grow food, and what the weather is there. [. . .] It begins to make us understand that "people of the land" – peasantry everywhere, the millions of small subsistence peasants with whom we think we have so little in common – identifies them and it identifies us. They're being evicted from

> their land, and that decimates their identity and their community. And we're also being relocated in our society. [. . .] As long as you keep us in separate categories and we're the highly industrialized farmers who are sort of quasi-business entrepreneurs and they're the subsistence peasants, then we can't see how closely we and all our issues are linked.
>
> *quoted in Edelman (2003: 187)*

Tired of being spoken for, *La Via Campesina* emerge to give peasants *themselves* a voice. The movement therefore does not allow for the joining up of groups that are not actual, grassroots-based peasant organizations. This link to the peasant identity has been a source of tremendous strength; a common glue, as indicated in the above quote of Nettie Wiebe, that holds millions of people – from different countries, of different religions, speaking different languages – together (Martínez-Torres and Rosset 2010).

La Via Campesina is also responsible for making the issue of food sovereignty a topic of discussion around which activists within and outside the movement are rallying. The term was first injected into the international public debate during the World Food Summit in 1996. Used to encapsulate an alternative paradigm to food and food production, food sovereignty has become part of the popular lexicon among actors within nongovernmental organizations, academia, and the peasant community. As detailed in Table 11.1, food sovereignty speaks to a way of life that is in many ways diametrically opposed to the dominant view that presently dictates conventional food and agricultural policy. The centrality of food sovereignty is reflected, for example, in a unique ritual practiced by the organization. *La Via Campesina* engages in seed exchanges at many of their gatherings, where representatives will bring seeds from their homeland to share with others. This practice not only signifies the centrality of seed in the production of food but also (as also discussed in Chapter 7) in the very reproduction of culture itself.

Transition . . .

This chapter reviewed a number of alternatives or challenges to the current dominant food system, except local food. While mentioned briefly, I have yet to talk about the movement that gave rise to the word *locavore* – the word of the year for 2007 in the Oxford American Dictionary. The wait is over . . .

Discussion questions

1 Why do you buy (or not buy) organic food? Having learned about process-based standards, does the "organic" label mean what you thought it meant?
2 By embracing "virtuous globalization" are Slow Food proponents essentially advocating the "conventionalization" of Slow Food? Are "virtuous globalization" and "conventionalization" referring to identical processes?

TABLE 11.1 Dominant versus food sovereignty models

Issue	*Dominant model*	*Food sovereignty model*
Free trade	Everything ought to be governed by the market	Exempt food from trade agreements, as it is fundamentally different from, say, cars
Production priorities	Agroexports	National and local markets
Subsidies	Claims to favor market logics yet relies heavily on government subsidies for the largest farms	Subsides are okay if they level the playing field and do not unduly harm small-scale farms in developing economies
Food	A commodity to be traded (fundamentally no different from any other commodity)	A unique commodity that everyone has a right to
Agriculture	Increasingly an occupation for those with access to significant amounts of credit, capital, land, and labor	A right of rural peoples
Hunger	A technological/production problem	A problem of access (of food, water, and land to produce food)
Food security	Achieved through trade and by adopting green revolution principles	Improved by enabling the hungriest to produce food and by embracing Rainbow Evolution policies (see Chapter 12)
Seeds	Can be privatized (like land); a commodity with no cultural significance	Common heritage of humankind that everyone has a right to; an artifact that allows for the reproduction of culture
Overproduction	Good (leads to cheap food)	Bad (erodes food security around the world)
Peasantry	A holdover from feudalism (which no one wishes to return to); a pejorative term	"People of the land"; a proud identity for hundreds of millions around the world

Constructed with the help of Desmarais (2008) and Martínez-Torres and Rosset (2010).

3 One of the world's most important transnational social movements – *La Via Campesina* – has emerged around the issue of food sovereignty. Why would food sovereignty have such resonance among the world's peasants?
4 Does the Slow Food movement seem to be concerned about food sovereignty?
5 Does it surprise you that peasants are in some respects leading the charge to reform international food and agricultural policies?

Suggested reading: Introductory level

Desmarais, A. 2007. Where have all the peasants gone? Long time passing, In *La Via Campesina: Globalization and the Power of Peasants*. Ann Arbor, MI: Pluto Press, pp. 5–29.

Gussow, J. 1997. Can an organic Twinkie be certified? In *For All Generations: Making World Agriculture More Sustainable*, edited by P. Madden. Glendale, CA: O.M. Publishing, pp. 143–53.

Pollan, M. 2003. Cruising on the Ark of Taste. *Mother Jones* May/June, http://motherjones.com/politics/2003/05/michael-pollan-turkey, last accessed May 15, 2011.

Suggested reading: Advanced level

Aertsens, J., W. Verbeke, K. Mondelaers, and G. Van Huylenbroeck. 2009. Personal determinants of organic food consumption: A review. *British Food Journal* 111(10): 1140–67.

Hjelmar, U. 2011. Consumers' purchase of organic food products: A matter of convenience and reflexive practices. *Appetite* 56: 336–44.

Lotti, A. 2010. The commoditization of products and taste: Slow Food and the conservation of agrobiodiversity. *Agriculture and Human Values* 27: 71–83.

Martínez-Torres, M. and P. Rosset. 2010. La Vía Campesina: The Birth and Evolution of a Transnational Social Movement. *Journal of Peasant Studies* 37: 149–75.

Notes

1 http://www.organic-europe.net/country_reports/finland/default.asp, last accessed May 14, 2011.
2 http://www.slowfood.com/international/1/about-us, last accessed May 15, 2011.
3 http://www.slowfoodusa.org/index.php/programs/ark_product_detail/tuscarora_white_corn/, last accessed May 15, 2011.

References

Adams, D. and M. Salois. 2010. Local versus organic: A turn in consumer preferences and willingness-to-pay. *Renewable Agriculture and Food Systems* 25(4): 331–41.

Aertsens, J., W. Verbeke, K. Mondelaers, and G. Van Huylenbroeck. 2009. Personal determinants of organic food consumption: A review. *British Food Journal* 111(10): 1140–67.

Arbindra, P., W. Moon, and S. Balasubramanian. 2005. Agro-biotechnology and organic food purchase in the United Kingdom. *British Food Journal* 107(2): 84–97.

Ashmole, A. 1993. The organic values of agriculture. PhD thesis, University of Edinburgh, Edinburgh.

Baker, S., K. Thompson, J. Engelken, and K. Huntley. 2004. Mapping the values driving organic food choice: Germany vs. the UK. *European Journal of Marketing* 38(8): 995–1012.

Borras, S. 2008. La Via Campesina and its global campaign for agrarian reform. In *Transnational Agrarian Movements Confronting Globalization*, edited by S. Borras, M. Edelman, and C. Kay. Malden, MA: Blackwell, pp. 91–122.

Botonaki, A., K. Polymeros, E. Tsakiridou, and K. Mattas. 2006. The role of food quality certification on consumers' food choices. *British Food Journal* 108(2–3): 77–90.

Burton, M., Rigby, D., and T. Young. 1999. Analysis of the determinants of adoption of organic horticultural techniques in the UK. *Journal of Agricultural Economics* 50(1): 47–63.

Chen, M.F. 2009. Attitude toward organic foods among Taiwanese as related to health consciousness, environmental attitudes, and the mediating effects of a healthy lifestyle. *British Food Journal* 111(2): 165–78.

Chryssohoidis, G. and A. Krystallis. 2005. Organic consumers' personal values research: Testing and validating the list of values (LOV) scale and implementing a value-based segmentation task. *Food Quality and Preference* 16(7): 585–99.

Connolly, William. 2003. *Neuropolitics: Thinking, Culture, Speed.* Minneapolis, MN: University of Minnesota Press.

Darnhofer, I., W. Schneeberger, and B. Freyer. 2005. Converting or not converting to organic farming in Austria: Farmer types and their rationale. *Agriculture and Human Values* 22: 39–52.

Desmarais, A. 2008. The power of peasants: Reflections on the meanings of *La Vía Campensina. Journal of Rural Studies* 24: 138–49.

Dreezens E., C. Martijn, P. Tenbult, G. Kok, and N. Vries. 2005. Food and values: an examination of values underlying attitudes toward genetically modified- and organically grown food products. *Appetite* 44: 115–22.

Duram, L. 1999. Factors in organic farmers' decision making: Diversity, challenge, obstacles. *American Journal of Alternative Agriculture* 14(1): 2–9.

Edelman, M. 2003. Transnational peasant and farmer movements and networks. In *Global Civil Society Yearbook 2003*, edited by M. Kaldor, H. Anheier, and M. Glasius. London: London School of Economics, Centre for the Study of Global Governance, pp. 185–220.

Edelman, M. 2005. Bringing the moral economy back in . . . to the study of 21st-century transnational peasant movements. *American Anthropologist* 107(3): 331–45.

Fisher, P. 1989. Barriers to adoption of organic farming in Canterbury. MSc thesis, Center for Resource Management, Lincoln University, Lincoln, New Zealand.

Flaten, O., L. Gudbrand, M. Ebbesvik, M. Koesling, and P. Valle. 2006. Do the new organic producers differ from the "old guard"? Empirical results from Norwegian dairy farming. *Renewable Agriculture and Food Systems* 21(3): 174–82.

Fotopoulos, C., A. Krystallis, and M. Ness. 2003. Wine produced by organic grapes in Greece: Using means-end chains analysis to reveal organic buyers' purchasing motives in comparison to non-buyers. *Food Quality and Preference* 14: 549–66.

Gallons, J., U. Toensmeyer, J. Bacon, and C. German. 1997. An analysis of consumer characteristics concerning direct marketing of fresh produce in Delaware: A case study. *Journal of Food Distribution Research* 28(1): 98–106.

Geen, N. and Firth, C. 2006. The committed organic consumer. *Joint Organic Congress.* Denmark: Odense.

Gracia, A. and T. de Magistris, 2007. Organic food product purchase behaviour: A pilot study for urban consumers in the south of Italy. *Spanish Journal of Agricultural Research* 5(4): 439–51.

Guthman, J. 2004. The trouble with "Organic Lite" in California: A rejoinder to the "conventionalisation" debate. *Sociologia Ruralis* 44(3): 301–16.

Hjelmar, U. 2011. Consumers' purchase of organic food products: A matter of convenience and reflexive practices. *Appetite* 56: 336–44.

Howard, P. 2009. Organic industry structure. *Media-N: Journal of the New Media Caucus* 5(3): online, http://www.newmediacaucus.org/html/journal/issues.php?f=papers&time= 2009_winter&page=howard, last accessed May 14, 2011.

Hughner, R., P. McDonagh, A. Prothero, C. Schultz, and J. Stanton. 2007. Who are organic food consumers? A compilation and review of why people purchase organic food. *Journal of Consumer Behavior* 6: 94–110.

Kropotkin, P. 2011. The Conquest of Bread. online, Chapter 5, http://libcom.org/library/conquestofbread1906peterkropotkin5, last accessed May 18, 2011.

Laudan, R. 2001. A plea for culinary modernism: Why we should love new, fast, processed food. *Gastronomica* 1: 36–44.

Lauden, R. 2004. Slow Food, the French terroir strategy, and culinary modernism. *Food, Culture and Society* 7(2): 133–44.

La Via Campesina. 2008. Declaration of Maputo: V International Conference of La Via Campesina, October 19–22, http://viacampesinanorteamerica.org/en/viacampesina/conferencias/V%20conferencia%20Declaration%20of%20Maputo.pdf, last accessed May 17, 2011.

Lea, E. and T. Worsley. 2005. Australians' organic food beliefs, demographics and values. *British Food Journal* 107(11): 855–69.

Leitch, A. 2009. Slow Food and the politics of "virtuous globalization." In *The Globalization of Food*, edited by D. Inglis and D. Gimlin. New York: Berg, pp. 45–64.

Lockeretz, W. 1997. Diversity of personal and farm characteristics among organic growers in the Northeastern United States. *Biological Agriculture and Horticulture* 14(1): 13–24.

Lockie, S., K. Lyons, G. Lawrence, and J. Grice. 2004. Choosing organics: A path analysis of factors underlying the selection of organic food among australian consumers. *Appetite* 43(2): 135–46.

Lockie, S., K. Lyons, G. Lawrence, and D. Halpin. 2006. *Going Organic: Mobilizing Networks for Environmentally Responsible Food Production*. Wallingford, UK: CABI.

Lockie, S., K. Lyons, G. Lawrence, and K. Mummery. 2002. Eating "Green": Motivations behind organic food consumption in Australia. *Sociologia Ruralis* 42(1): 23–40.

Lotti, A. 2010. The commoditization of products and taste: Slow Food and the conservation of agrobiodiversity. *Agriculture and Human Values* 27: 71–83.

Loureiro, M., J. McCluskey, and R. Mittelhammer. 2001. Assessing consumer preferences for organic, eco-labeled, and regular apples. *Journal of Agricultural and Resource Economics* 26(2): 404–16.

Magnusson, M., A. Arvola, U. Koivisto Hursti, L. Aberg, and P. Sjödén. 2001. Attitudes towards organic foods among Swedish consumers. *British Food Journal* 103(3): 209–26.

Martínez-Torres, M. and P. Rosset. 2010. La Vía Campesina: The birth and evolution of a transnational social movement. *Journal of Peasant Studies* 37: 149–75.

McEachern, M. and P. McClean. 2002. Organic purchasing motivations and attitudes: Are they ethical? *International Journal of Consumer Studies* 26(2): 85–92.

McEachern, M. and J. Willock. 2004. Producers and consumers of organic meat: A focus on attitudes and motivations. *British Food Journal* 106(7): 534–52.

MacKenzie, Adrian. 2002. *Transductions*. London: Continuum.

McMichael, P. 2003. *Development and Social Change*. Thousand Oaks: Pine Forge.

McMichael, P. 2006. Peasant prospects in the neoliberal age. *New Political Economy* 11(3): 407–18.

Meneley, A. 2004. Extra virgin olive oil and Slow Food. *Anthropologica* 46(2): 165–76.

Mintz, S. 1986. *Sweetness and Power: The Place of Sugar in Modern History*. New York: Penguin.

O'Donovan, P. and M. McCarthy. 2002. Irish consumer preference for organic meat. *British Food Journal* 104: 353–70.

Padel, S. 2001. Conversion to organic farming: A typical example of the diffusion of an innovation? *Sociologia Ruralis* 41(1): 40–61.

Padel, S. and C. Foster. 2005. Exploring the gap between attitudes and behaviour: Understanding why consumers buy or do not buy organic food. *British Food Journal* 107(8): 606–25.

Padmavathy, K. and G. Poyyamoli. 2011. Alternative farming techniques for sustainable food production. In *Genetics, Biofuels and Local Farming Systems*, edited by Eric Lichtfouse. New York: Springer, pp. 367–424.

Pollan, M. 2003. Cruising on the Ark of Taste. *Mother Jones* May/June, http://motherjones.com/politics/2003/05/michael-pollan-turkey, last accessed May 15, 2011.

Roddy, G., C. Cowan, and G. Hutchinson. 1996. Consumer attitudes and behaviour to organic foods in Ireland. *Journal of International Consumer Marketing* 9(2): 1–19.

Schifferstein, H. and P. Oude Ophuis. 1998. Health-related determinants of organic food consumption in the Netherlands. *Food Quality and Preference* 9: 119–33.

Shepherd, R., M. Magnusson, and P. Sjoden. 2005. Determinants of consumer behavior related to organic foods. *Ambio* 34(4–5): 352–9.

Starr, A. 2010. Local Food: A social movement? *Cultural Studies, Critical Methodologies* 10(6): 479–90.

Storstad, O. and H. Bjorkhaug. 2003. Foundations of production and consumption of organic food in Norway: Common attitudes among farmers and consumers. *Agriculture and Human Values* 20: 151–63.

Thøgersen, J. 2010. Country differences in sustainable consumption: The case of organic food. *Journal of Macromarketing* 30(2): 171–85.

Toler, S., B. Briggeman, J. Lusk, and D. Adams. 2009. Fairness, farmers markets, and local production. *American Journal of Agricultural Economics* 91(5): 1272–8.

Tovey, H. 1997. Food, environmentalism and rural sociology: On the organic farming movement in Ireland. *Sociologica Ruralis* 37(1): 22–37.

Wandel, M. and A. Bugge. 1997. Environmental concerns in consumer evaluation of food quality. *Food Quality and Preferences* 8(1): 19–26.

Willer, Helga and Lukas Kilcher (eds). 2009. *The World of Organic Agriculture: Statistics and Emerging Trends 2009*. FIBL-IFOAM Report, IFOAM, Bonn; FiBL, Frick; ITC, Geneva.

Wolf, M., A. Spittler, and J. Ahern. 2005. A profile of farmers' market consumers and the perceived advantages of produce sold at farmers' markets. *Journal of Food Distribution Research* 36(1): 192–201.

Wolin, Sheldon. 1997. What time is It? *Theory and Event* 1: 1.

Zepeda, L. and J. Li. 2007. Characteristics of organic food shoppers. *Journal of Agricultural Applied Economics* 39(1): 17–28.

12

AVOIDING THE "TRAPS" IN AGRIFOOD STUDIES

KEY TOPICS

- While local food has incited much interest and excitement recently we must be careful not to fall into the local trap by believing that "the local" (and by extension local food) is clearly defined and inherently good.
- The global trap is also to be avoided.
- This does not mean either term – local or global – can no longer be used when talking about food and agriculture, though they must be used reflexively.

Key words

- Agriculture of the middle
- Alternative food networks
- Certification
- Civic agriculture
- Commodity agriculture
- Defensive localism
- Embeddedness
- Farm to School programs
- Food globalizations
- Foodshed
- Global trap
- Hollowing out of the state
- Local food
- Local trap
- Quality turn
- Rainbow evolution
- Reflexive localism
- Scale-neutral
- Scaling up
- Unreflexive globalism
- Unreflexive localism

This chapter reminds us that nothing should be taken at face value. The local, in conventional food-speak, has emerged to mean a more sustainable, just, equitable solution to the global. The chapter begins by reviewing literature that critically investigates this claim, leading some scholars to caution against falling into the "local trap" – speaking of the local as an objective state with inherent worth when in fact it is a social construction (Born and Purcell 2006). The fondness with which foodies, food activists, and food writers tend to speak of the local is often matched by their disdain for the global. Yet if there is a local trap is there not also a global trap? That is, if "the local" has multiple meanings and is open to contestation then shouldn't we expect the same to apply to the categorical box we refer to as "the global?" After discussing the problems associated with the uncritical adoption of "local" and "global" terminology, attention shifts to highlighting how we might avoid these traps.

The local trap and unreflexive localism

The last 20 years have given rise to some highly engaging, exceedingly thoughtful arguments in support of more local food systems, like Jack Kloppenburg *et al.*'s (1996) seminal essay "Coming to the foodshed." In this article, we find a compelling case to support the creation of food systems analogous to the watershed. The foodshed: "a place for us to ground ourselves in the biological and social realities of living on the land and from the land in a place that we can call home" (Kloppenburg *et al.* 1996: 33). The literature published between this article's publication and the first years of the new millennium represents the high water mark of pro-local food scholarship (see, e.g., Hassanein 1999; Hendrickson and Heffernan 2002; Murdoch *et al.* 2000; Norberg-Hodge *et al.* 2002). The message, which continues to be parroted by many food writers and food activists today, is that local food is not just *generally* better than its global alternative but *inherently* so. Yet that's an empirical statement. What, then, do the data say on the subject?

The data are clear: there is nothing inherently good (or bad) about local food. Issues of social justice, sustainability, and the like depend on a lot more than whether or not the system is local. And besides, just what does "local food" mean anyway?

Unwavering proponents of local food are said to have fallen into the "local trap," which they would do well to climb out of given that they are promoting what does not exist in any objectively real way (Born and Purcell 2006). As Born and Purcell (2006), the first to make the local trap argument, argue: "the local trap is the assumption that local is inherently good" (Born and Purcell 2006: 195). The data, however, say otherwise: "No matter what its scale, the outcomes produced by a food system are contextual: they depend on the actors and agendas that are empowered by the particular social relations in a given food system" (Born and Purcell 2006: 195–6). Furthermore, Local Food movements often fail "to address more intractable issues such as labor concerns, inequality, migration, [and] systemic patterns of social injustice, to name just a few" (Blue 2009: 69) (Box 12.1). Starr's (2010) account of a local food cooperative is a stark reminder that Local Food organizations can be as egregiously unjust as any global multinational firm: "I had

BOX 12.1 THE "TRAPPINGS" OF FARM TO SCHOOL PROGRAMS

Farm to School (FTS) programs seek to bring fresh, local produce to schools. At its face, it is a win-win: farmers, by selling directly to schools, receive a higher premium price for their labors; schools and school-aged children receive a seemingly safe, fresh, quality product. Not so fast, claim Allen and Guthman (2006). While not suggesting we campaign against FTS programs, these scholars point out a need to give these programs a more critical second look. Some points highlighted include:

- FTS programs have emerged as public resources are being slashed following a shift in governance toward neoliberalism (e.g. the market is providing for things previously provided by the state). Adapting to this economic climate, FTS organizers must find ways to make these programs profitable and competitive with other alternatives like allowing vending machines and restaurant chains into schools, which pay handsomely.
- FTS programs also have substantial start-up costs, compared with conventional school food that is made to "heat and serve" (and compared with vending machines which require no upfront capital investments). To deal with this problem, some schools have asked FTS programs to make available processed food, in an attempt to save on labor and equipment costs associated with preparing whole foods.
- FTS do not necessarily help small farms. For example, the largest organic vegetable producer in the world, Earthbound Farms, has partnered with All Saints Day School in Carmel Valley, California, in an FTS program.
- As local initiatives, FTS programs have greatest success in schools located in areas where capital (human, financial, natural, social, etc.) is greatest.

Allen and Guthman repeatedly highlight the inherent problematic nature of any movement that pushes responsibility and accountability "down" to the level of the local. Such a push fits well with neo-liberal philosophy, which seeks the devolution of the state in exchange for an emergence of local forms of governance. In the case of FTS programs, this trend is particularly disturbing as school food programs have historically been the responsibility of the state, which has helped guarantee their implementation for purposes of nationwide poverty alleviation (recognizing also the influence that agribusiness interests have had on these policies). Equally problematic, the devolution to the local "also abdicates responsibility on the part of the federal government and places it in the hands of those who happen to reside in a community, regardless of whether they have the will or wherewithal to act" (Allen and Guthman 2006: 408).

offered to help them identify local producers. They ignored me. They had also suppressed labor organizing and fired two workers, triggering concerns about racism. The board rebuffed members organizing on these issues" (Starr 2010: 486).

Calls by local food advocates to buy and grow local food, while well-intentioned, can also gloss over real constraints – like residing outside of well-serviced urban areas or lacking resources (e.g. money and time) – keeping people from adopting these prescriptions. Without calls for, and actions directed at, broader structural changes, the ability of people to practice localism will be concentrated in the hands of the more privileged, leaving an entire segment of the population out of the movement. Similarly, any shift toward local, in-season, home-produced foods necessitates an increase in work by a household's primary food provider, who continue to be women. A study of Canadian farmwomen notes that women were unable to fully engage in preferred local and self-provisioned food activities without the aid of children and extended family, as existing role expectations made them also responsible for house cleaning, child care, and food preparation activities (McIntyre and Rondeau 2011). The authors then proceed to critique Michael Pollan's unhelpful (and I might add nonsociological) "solution" to the additional time pressures that may be required as households switch over to more local, less processed, foods. According to Pollan, "If we decide cooking is important to our health and our family life, we'll find the time to do it" (as quoted in McIntyre and Rondeau 2011: 6).

So as not to diminish the path-breaking nature of prior research, literature dating back to the late 1990s has been warning us of the dangers of conflating local with ecologically sound and socially just (see, e.g., DuPuis and Goodman 2005; Hinrichs *et al.* 1998; Hinrichs 2000, 2003; Winter 2003). The novelty of the local trap argument, rather, lies in how it synthesizes a sizable literature on the social construction of scale – like Marston's 2000 article "The social construction of scale" – with the aforementioned agrifood literature where the ambiguous nature of "local" is highlighted. When taken as a whole, the local trap literature represents a powerful critique to those arguments that choose to embrace the concept of "local food" without a healthy dose of critical skepticism.

Yet just because "local" (like any scalar category) is political and therefore socially constructed does this therefore mean the term must be jettisoned from all future discussions about food and agriculture? Is the term such a free-floating signifier that it tells us nothing about reality? Might there be such a thing as, say, a local trap-trap? There may well be (Carolan 2011a; DuPuis *et al.* 2011).

To unpack this argument let's begin with a distinction that first appears in DuPuis and Goodman's 2005 important article "Should we go 'home' to eat?" between what they call "unreflexive localism" and "reflexive localism." Their point (or at least one of them): unfailingly pro-local food stances reflect not only poor scholarship but can be downright dangerous. According to these scholars, unreflexive localism comes with the following two negative consequences: "First, it can deny the politics of the local, with potentially problematic social justice consequences. Second, it can lead to proposed solutions, based on alternative standards of purity and perfection, that are vulnerable to corporate cooptation" (DuPuis and Goodman 2005: 360). Local, in the absence of critical reflection, risks being used as an ideological tool for reactionary politics, nativist sentiment, and corporate green washing (like Walmart's recent hijacking of the term in its push to be a major supplier of local food). I leave the concept of reflexive localism for later in the chapter, when I attempt to say something about how we can talk about the global and local without falling into any traps. The remainder of this section will discuss the trappings of unreflexive localism.

Let's start with the more obvious trappings, like who gets to define what *local* means? These definitions can vary significantly by corporation. For example, Walmart's definition of "local" means food produced in the same state. So: if you live near your state's border food produced a mile away, but just outside the state, would not be local while food produced hundreds of miles away within the state would (California, for instance, stretches 770 miles/1,239 km from north to south). Whole Foods Market, for a point of comparison, says food cannot be labeled as local unless it traveled to the store in 7 hours or less by car or truck (yet a truck traveling 7 hours on a freeway can go a great distance). And then: *who* gets to decide what local food is? For example, as discussed in Chapter 6, how this question is answered will shape the ethnic character of the available food thus determining such things as whether consumers can engage in, say, "eating black" (Slocum 2011: 4).

As a concept, "the local" is inherently political, in that it necessitates that creation of boundaries that only some people, places, and ways of life fall within. This is why local-talk is at its worse when it is unreflexive, acritical, and nondemocratic. When practiced as such, calls for the construction of regional or local food systems can be a backhanded way of lobbying for a "defensive politics of localization," leading to an elitist and reactionary movement that appeals through nativist sentiments (Hinrichs 2003: 37). This process has also been called "defensive localism," where the people, places, and ways of life defined as "not local" are cast in a negative light (if not outright demonized) (Winter 2003). Further examples of defensive localism have been documented by fair trade scholars, who note consumers that resist the fair trade label as it signifies helping "foreign" farmers (Long and Murray 2011). As Buller and Morris (2004: 1078) observe, "once territoriality becomes a component of value, it also becomes a commodity in itself, to protect and exploit." Under defensive localism only certain territory and the people are valued, to the explicit detriment of others.

It has also been argued that the sub-national and global scalar levels are gaining prominence in recent decades at the expense of the nation state. This has led some to view the resurgence of the local as a response toward the hollowing out of the state – the substitution of government for governance (Lawrence 2004). When viewed in this more critical light, local-talk is recast from representing a case of *resistance* to neo-liberal globalization to an *intrinsic* part of it (DuPuis and Goodman 2005). This is in recognition that, like neo-liberal reforms, local forms of governance ultimately endorse and foster "the self regulation of individuals and communities which, at the regional level, equates to the acceptance of programs, techniques and procedures that support market rule, productivism and global competition" (Lawrence 2004: 9).

The implementation of food localism to date has also remained primarily the responsibility of consumers (Allen and Hinrichs 2007; Blue 2009; DeLind 2002). Think about all the times you've come across the phrase "Buy Local." When was the last time you saw a sign that read (or heard a politician or activist proclaim), "Grow Local!"? At most, one-tenth of 1 percent of all domestic agriculture subsidies in the USA goes to supporting fruit and vegetable crops (Fields 2004: 822). Not long ago, I came across an article in *The Denver Post* celebrating that Feed Denver – an urban farming organization – was in the running for a US$4,000 grant (Robles 2010). I mention this as it is telling that a US$4,000 grant is cause for celebration for an urban agriculture organization while colleagues of mine at Colorado State University's College of Agriculture routinely obtain grants worth hundreds of thousands of dollars directed at more conventional food production endeavors.

In sum: until we understand why people do not (or cannot) buy or grow local – which are undeniably sociological topics – than we shouldn't expect people to do so in droves. And until those barriers are addressed, it is unfair of us to judge those who don't.

The global trap and unreflexive globalism

It is hard to locate the historical origins of globalization. There are a number of competing alleged starting points. Some locate its origins – of the full-blown spread of global capital – to as recently as the 1990s, after the fall of Communism and the rise of the information revolution (e.g. Friedman 2000). Marxists tend to prefer a later date; usually sometime during the nineteenth century, when global imperialism began to merge with Western industrial might (e.g. Brown 2009). While still others see a world system of sorts going back 5,000 years, following the start of trade across Eurasia (Frank and Gills 1994). Regardless of *when* it started, we all can agree that the globalizations of food are well underway. Yes, I wrote *globalizations* of food. The literature is quite clear on the trappings associated with talk of globalization in the singular – of globalization as if it were a monolithic, unequivocal process. There is also a tendency to get so caught on what stays the *same* with this process that we sometimes miss its complexity, nuance, and indeed even *creative novelty* (Box 12.2).

Moreover, not only are people around the world consuming many similar commodities, as either foods or as agrifood inputs (e.g. corn). Consumers themselves are increasingly taking the same form. A recent report by the Organization for Economic Co-operation and Development (OECD) published in *The Lancet* explains that 7 in 10 Mexican adults are overweight or obese, while approximately half of all adult Brazilians, Russians, and South Africans are also in this category (Cecchini *et al.* 2010: 1). Mexico is just behind the "leader," the USA, where 74 percent of adults are classified as "overweight" or "obese." In the case of Mexico,

BOX 12.2 THE GLOCALIZATION OF SPAM

While a lowbrow staple in the USA, SPAM (short for SPiced-hAM) has become a delicacy in the Philippines among the growing middle class. As Ty Matejowsky (2007: 38) explains, "SPAM personifies America in all of its perceived grandeur [. . .] [due to] its associations with the US military, its efficient packaging, and use as *pasalubong* (homecoming gifts), SPAM reflects an idealized version of American that many everyday Filipinos hold to be true." While studying Filipino cuisine, Matejowsky describes being frequently invited to dinner to Filipinos' households to be served fried SPAM. The hosts hoped to make their guest feel at home – the assumption being all Americans eat SPAM. Matejowsky goes on to describe the confusion experienced by the hosts upon learning that many in the USA go their entire lives never having eaten this product. Nevertheless, the vision of SPAM as an all-American food was successfully capitalized on by the founders of SPAMJAM: an all-SPAM Philippine fast-food restaurant. SPAMJAM claims to offer Filipinos an "authentic" American eating experience, even though it involves a food that many in the USA never eat. SPAMJAM represents a clear example of how elongated food networks rarely (if ever) result in the exportation of unequivocal "sameness." Though clearly connected to the USA, SPAMJAM reflects locally rooted understandings and applications of this spiced ham.

global trade has been linked to its recent waistline "prosperity." Since enacting the North American Free Trade Agreement (NFTA) in 1994, imports of processed food and drinks from the USA have increased rapidly. Mexico now consumes more Coca-Cola products per capita than any other country on the planet. Each Mexican today averages 635 8-ounce bottles annually – three times the amount consumed in 1988 (Grillo 2009).

When talking about global similarities, however, we must be careful not to overemphasize the degree to which globalizations homogenize. "Similarity" is not equal to "sameness" when talking about food globalizations. To say otherwise is to succumb to yet another trap.

Take the case of McDonald's. On the one hand, there are undeniable similarities across this global restaurant chain – a point famously detailed in Ritzer's (2010) *McDonaldization of Society* (Box 12.3). Nevertheless, while having learned there are substantial cost savings through standardization, the food giant also knows the importance of being able to adapt to local socio-cultural environments for its long-term success. There are a number of instances where McDonald's altered its menu due to local laws and customs. In certain stores in Israel, for instance, Big Macs are served without cheese, which allows for the separation of meat and cheese required of kosher restaurants. In India customers can order Vegetable McNugggets and a mutton-based Big Mac (called a Maharaja Mac), recognizing that

BOX 12.3 THE HOMOGENIZATION THESIS

While important not to overstate these processes, the homogenizing aspects of globalization are hard to deny. Marianne Elisabeth Lien (2009) describes how farmed salmon are essentially place-less entities. As opposed to wild salmon, whose biological characteristics tie them to particular environments (as they evolve within a particular place), farmed salmon are simultaneously both *everywhere* and *nowhere* as they are bred for the standardized salmon farm (Inglis and Gimlin 2009).

The industrialization process itself has helped to smooth out diversity. Difference is deviance when it comes to field corn, for example, where plants need to mature at even rates (so all can be simultaneously harvested), remain standing by fall (so kernels are accessible to the combine), and have a standardized starch, fiber, and protein ratio (for later agri-industrial processes). Similarly, grocery stores tend to be populated with the same selection of fruits and vegetables – namely, those that withstand the brutalities of mechanical harvesting and transportation. And as those transportation networks continue to expand, the "fresh" produce able to survive the trip will shrink still further (Freidberg 2009).

Hindus and Muslins do not eat beef and pork, respectively, while other religious groups (like Jains) avoid meat entirely. Some other examples of Mcglocalization include beer in Germany, Teriyaki burgers in Japan, cold pasta in Italy, McSpaghetti in Philippines, and McLaks (grilled salmon sandwich) in Norway (Vignali 2001).

Richard Wilk (2009) argues that the categories "local" and "global" are actually part of the process of globalization itself. Historically, "when people are exposed to foreign goods, especially novel ones which challenge culturally important consumption practices, some will always react by inventing tradition" (Wilk 2009: 190–1). In other words, perceived external threats lead to certain practices being selected (usually by elites) and established as a cultural, local standard. In his study of Mayan ethnohistory, Wilk (1991) describes how tradition assumes a level of permanence during periods of cultural crisis when ways of life are viewed as under attack. The aforementioned defensive localism resembles something very similar to this phenomena described by Wilk. Similarly, "national" cuisine has been shown to emerge (often through mass produced and widely dispersed cookbooks) in response to globalization and in the accompaniment of nation-building. This is evident, for example, throughout Europe during the nineteenth century, as national cuisine and national identity grew in lockstep fashion (Appadurai 1988).

There is also a tendency to talk about globalization in the context of motion and speed. Terms like *fluid* capital, information *flows*, and time–space *compression* – which are sprinkled throughout the food globalization literature – imply a world where immobility is quickly being replaced by mobility. Yet this mobility *presupposes* structures of immobility. Lien (2009: 69), in her study of the global salmon, states this point plainly: "Too often in globalization studies, the empirical attention drifts toward that which moves, at the expense of structures of immobility that, in fact, make movement possible." Lien looks specifically at food standards and science that allows for the creation of "a 'generalized' knowledge of salmon 'everywhere': a notion of salmon as a universal biogenetic entity the qualities of which may be known through a set of scientifically informed and standardized techniques" (Lien 2009: 69). As Lien highlights, structures of immobility can take many forms. They can be structures in the conventional sense, like landing strips and satellites (which makes global air travel and GPS location technology possible). Yet structures of immobility also refers to "things" like international food standards, shared beliefs about how food ought to look, and standardized forms of knowledge.

To the best of my knowledge, the term "unreflexive globalism" has never been used before. It is not my intention to create yet another neo-logism by using it here. I introduce the term only to provide some symmetry to the earlier mentioned concept "unreflective localism." For like "the local," essentializing and monolithic treatments of "the global" massively understate the complexity of the world.

Reflexive glocalism

Does this mean that we as social scientists ought to reject the terms "local" and "global?" Not necessarily. They remain meaningful analytic concepts; we just need to be mindful of their limits (and the "traps" that arise when we overstep those limits). The surest road to food security comes from having many options on the table, not fewer (Box 12.4; Figures 12.1 and 12.2). I would therefore hate

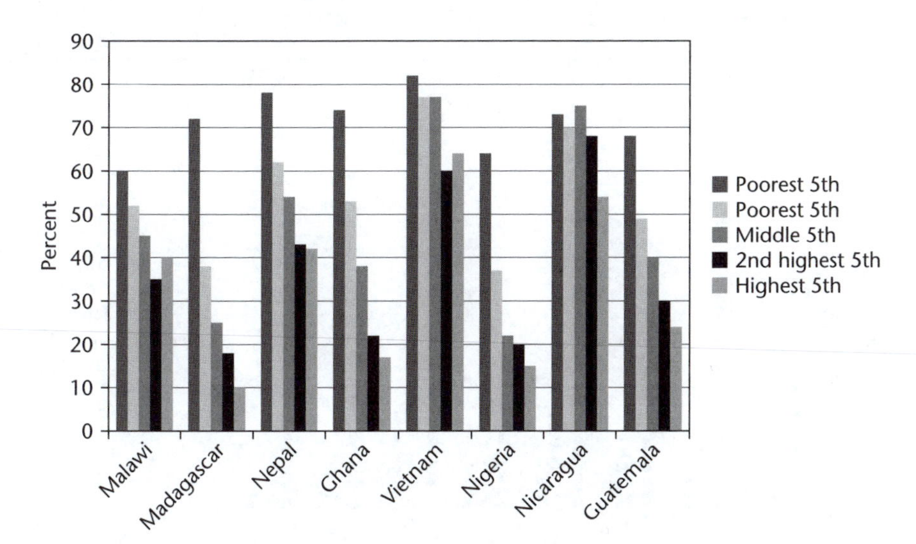

FIGURE 12.1 Percentage of households participating in urban farming for select countries, by quintile. Based upon data from Zezza and Tasciotti (2010: 269)

BOX 12.4 A STUDY OF URBAN AGRICULTURE ACROSS 15 COUNTRIES

Approximately 70 percent of the world's population is expected to live in urban areas in 2050 (in 2010 that figure is around 50 percent) (Rabobank 2010: 6). In the swiftly urbanizing developing nations, the poor are moving to the city at a faster rate than the population as a whole (Ravallion *et al.* 2007: 2). If current trends continue, the world's slum population – those living in irregular settlements, without access to suitable food, water, shelter, and sanitation – will increase 50 percent between now and 2020 (Mougeot 2005: 1).

Most urban poor in the developing world spend the majority of their incomes on food. An estimated 60 percent of all household income generated by the poor in Asia goes to buying food, compared with Canada and the USA where the figure is closer to 10 percent (Redwood 2010: 5). Food insecurity among urban poor is further exacerbated by diets composed heavily of tradable commodities (versus traditional foods) and a lack of space to grow their own food (Redwood 2010: 5). A recently published study looking at 15 less affluent nations located in Africa, Asia, Eastern Europe, and Latin America assesses the role of urban agriculture in promoting food and economic security and diet diversity among the urban poor (Zezza and Tasciotti 2010). Participation in urban agriculture among poor households in the countries studied was as high as 81 percent (Figure 12.1). Urban agriculture was also shown to be of major economic significance, making up over 50 percent of the income for poor households in certain countries (Figure 12.2). Two-thirds of the countries analyzed were shown to have a correlation between active participation in agricultural activities and greater household dietary diversity, even after controlling for economic welfare and other household characteristics. There was also a tentative link between urban agriculture participation and the consumption of fruits and vegetables (Zezza and Tasciotti 2010).

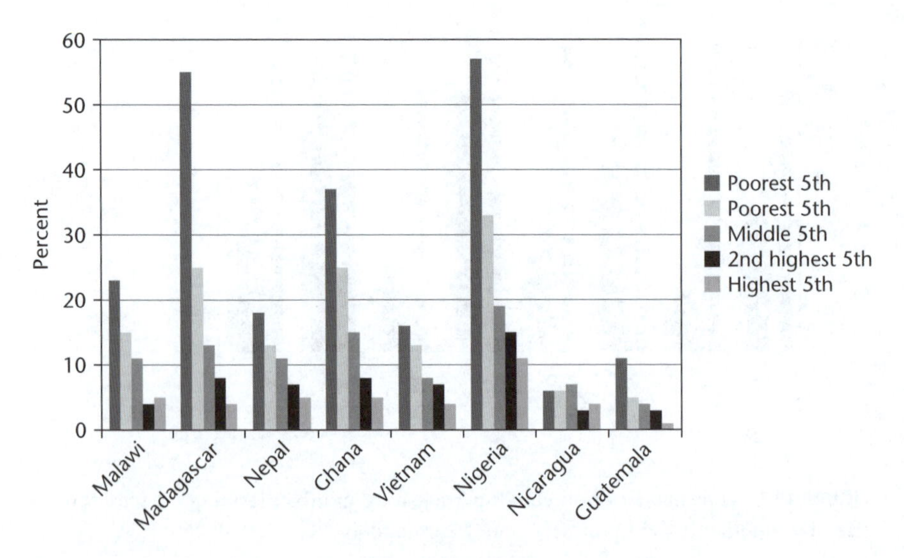

FIGURE 12.2 Share of household income from urban farming for select countries, by quintile. Based upon data from Zezza and Tasciotti (2010: 269)

to stop talking about, say, local food just because it is a problematic term – after all, what term isn't problematic? As others have argued, the key lies in reflexivity (see, e.g., DuPuis and Goodman 2005; DuPuis *et al.* 2011). Any effective food social movement, "must be reflexive and critical, rather than embraced as a panacea for global industrial agriculture" (McIntyre and Rondeau 2011). Let's look briefly at some examples of what it means to be more reflexive with our food.

Thinking about food reflexively means remembering consumers do not exist in a vacuum

For example, although the Local Food movement attempts to move beyond an ideology of nutritionism – the tendency, addressed in Chapter 4, of reducing food to (and organizing eating patterns around) nutritional components – calls to buy local, while often well-intentioned, risk reducing discussions about food to matters of geography (e.g. where was it grown?). Doing this, however, disassociates food from, among other things, the broader context in which household gatekeepers make decisions about what foods to grow, purchase, prepare, and consume. When food writers like Michael Pollan (2008: 197) instruct readers to "cook and, if you can, plant a garden" they ignore (and unintentionally delegitimize) the very real sociological barriers that keep many individuals and households, especially the underprivileged living in food deserts (Chapter 4), from purchasing (or growing) and preparing whole foods (McIntyre and Rondeau 2011).

Thinking about food reflexively means remembering that characteristics pertaining to the "quality" of food are sociologically produced

The move toward "the local" – and away from "the global" – has parallels with an equally recent turn toward "quality" among food and agriculture scholars (Goodman 2003: 1; see also, e.g., DuPuis and Goodman 2005; Mansfield 2003). This turn refers to growing desire among people to eat "better" – fresher, safer, more local, less environmentally harmful – foods as is reflected in the recent proliferation of "alternative agro-food networks" (D. Goodman 2003: 1). What makes quality-talk by scholars different from that articulated by, say, food writers resides in the fact that sociological understandings of quality are, well, sociological (Box 12.5). Rather than speaking of some inherent state, to talk about "quality" food sociologically is to look at what gives some foods this designate but not others. Mansfield (2003: 11–12) puts it this way: "Quality is neither a subjective judgment (what different people like), nor an objective measure (the characteristic of a commodity), but instead is produced within relations of commodity production and consumption." An excellent example of quality-as-a-relational-effect is offered by Freidberg (2009) through her term "industrial freshness" – a "mass-produced, nationally distributed, and constantly refrigerated" freshness (Freidman 2009: 2). Industrial freshness is a freshness that travels well; freshness *because* of transportation and processing rather than despite it.

BOX 12.5 FOOD SAFETY AND SCALE

Avoiding the local trap does not require that we avoid saying anything positive about smaller scale, more local forms of food production, distribution, and consumption on issues of food safety. We just need to be sure that any such discussions do not essentialize the local, as inferring (or outright stating) that local food is inherently safe. It most certainly is not. But the scope and scale of potential risks do differ between conventional and alternative food networks.

Writing on the 2006 outbreak of *E. coli* O157:H7, traced to bagged spinach from California, DeLind and Howard (2008) argue convincingly that scale *does* matter on issues of food system food safety. In their words, "local, more regionally based production and marketing systems have not perpetuated – and cannot perpetuate – crises of the nature or magnitude of the spinach scare" (DeLind and Howard 2008: 303), which negatively impacted the health of consumers in 26 US states (and resulted in industry losses of about US$100 million). When systems are more regionally based, any pathogen to enter into the food system is by definition limited to the regional level. In nationally and globally based food systems, conversely, there is no such thing as a "limited food outbreak." There are only national and international food scares.

In response to the outbreak, the US government fell into a "trap" as well. They did this by ascribing an inherent superiority to large-scale, conventional forms of food production and distribution. Moreover, the "solutions" proposed sought to prop up the very system that makes widespread outbreaks possible while creating further barriers to potentially safer food system alternatives. DeLind and Howard describe how the government's response to enhance food safety introduced the following two ironies:

> One such irony is the central role afforded scientific solutions and postincident, diagnostic technologies (e.g., vaccines, irradiation, DNA signatures), while leaving uninvestigated the naturally low incidence of *E. coli* O157:H7 in the production systems of developing nations and in pasture-based animal production [. . .] This, in turn, generates the further irony that the preferred solution – [. . .] e.g., state-of-the-art equipment, increased paperwork and fees, centralized inspection sites and uniform protocol, increased transportation costs [. . .] – privileges the very system responsible for the public health crisis in the first place.
>
> *DeLind and Howard (2008: 302)*

Reflexivity does not favor any one scale or political practice

Falling for the local trap means ascribing to the mistaken belief that localism is the natural solution to the ills of the globalization (in the singular) of food (DuPuis *et al.* 2011). Reflexivity means selecting political strategies based upon the case in question, rather than making *a priori* universal assertions about which solutions "work." Doing this reveals the complexity of the problems at hand while privileging no specific scale, mode of production, or food system. For example, when striving for social justice should we select bananas grown by smaller farmers in cooperatives in the Caribbean or those grown on plantations in Central American by unionized and politically empowered workers (DuPuis *et al.* 2011)? Don't worry if you are having problems answering this question. You should. That's the point.

Reflexive approaches emphasize means rather than predefined ends

Instead of proclaiming what an ideal food system(s) looks like, a reflexive approach chooses instead to focus on the *processes* used that will ultimately arrive at this end. A key term here is "inclusivity," as one goal should be to include stakeholders beyond the commodity chain. As DuPuis *et al.* (2011) note, "Those we sell and buy from are not the only country people worthy of our thoughts." Local Food movement activists must also be cautious about making alliances with those not guided by principles of inclusivity (e.g. those who practice defensive localism), while similarly being mindful to include those who lack the resources to participated in localism (DuPuis *et al.* 2011).

Reflexive approaches reject one-size-fits-all solutions

This goes for solutions aimed at any scale (Box 12.6). The green revolution's single magic bullet style has been handily critiqued on the grounds that it *increased* global food insecurity (Carolan 2011b). In 2002, then-UN secretary general Kofi Annan asked an appointed panel of experts from such countries as Brazil, China, Mexico, and South Africa how a green revolution could be achieved in Africa. A year later the group came back with their answer. Foremost, they questioned the one-size-fits-all approach to food security that underlies the green revolution ideology: "The diverse African situation implies that no single magic 'technological bullet' is available for radically improving African agriculture" (InterAcademy Council 2003: xviii). The panel's strategic recommendations explain that "African agriculture is more likely to experience numerous 'rainbow evolutions' that differ in nature and extent among the many systems, rather than one Green Revolution as in Asia" (InterAcademy Council 2003: xviii). The term reframes the discussion away from looking for that magic bullet while also emphasizing how agricultural development cannot be divorced from the social, economic, and agroecological conditions of place.

BOX 12.6 A FAILED CASE OF DIRECT MARKETING IN NAIROBI, KENYA

While extensive in terms of available literature on the subject, the majority of research looking at shorter food supply chains remains narrow in its geographic scope, as much of it focuses on Europe, North America, and the wealthier parts of Asia-Pacific. Are conditions similar enough to generalize from developed to developing countries? It could be argued that they are; after all, most certified fair trade networks source most of their materials from less affluent nations. Yet, and this brings us back to a point made earlier about the structures of immobility of globalization, these networks depend upon a wide range of technologies and infrastructures that make fair trade commodities mobile. Yet these structures of immobility are not everywhere. For instance, no distance is "short" for farmers and consumers lacking easy access to motorized vehicles and/or well-maintained roads.

Freidberg and Goldstein (2011) make just this point in their study of a failed direct marketing initiative in Kenya. They note how a community's soil and water can be dangerously polluted (as in the case of Nairobi) and its public sanitation and infrastructure in disarray – points that ought to make people think twice about the virtues and viability of local food networks in some parts of the world. In the context of their case study, they found that getting the produce to the city of Nairobi was even costlier and more difficult than anticipated as the vans used to transport produce often broke down. When this happened, more expensive buses and minibuses had to be used. Meeting customers' quality standards was also a challenge, partly because of the rough roads and lack of refrigeration. But also this was partly because farmers lacked knowledge about the kind of quality their customers expected. As many of the small farmers had never even been to Nairobi, most had never seen the displays of uniform, unblemished produce in the supermarkets where many of their affluent customers shopped.

Social justice in particular is hard to come by through magic bullet prescriptions, as what's "right" and "just" in *one* situation for *one* population is rarely generalizable to *all* situations and *all* people. In California, for example, workers' compensation law regulates the long-standing cultural practice of unpaid labor sharing among extended family members. As a result, all workers on a farm, paid or unpaid, must be covered by workers' compensation insurance. One study recently looked at the effects of this law on small-scale Hmong (an Asian ethnic ground from the mountainous regions of China) farmers in California (Minkoff-Zern *et al.* 2011). The cost of this insurance for the average California Hmong farmer is US$445 a year – well beyond what most can afford. Originally designed to protect workers from exploitation by growers (and thus enhance social justice), this one-size-fits-all piece of otherwise progressive legislation has had a crippling effect on this ethnic group.

For farmers who do not carry the insurance and who are caught illegally employing extended family members (which is a common cultural practice among the Hmong), fines can range between US$14,500 and 25,000. Legislative attempts have been made to amend workers' compensation insurance requirements to exempt farms with less than US$100,000 a year in taxable income. Labor unions opposed this on the grounds that the US$100,000 ceiling would have exempted far more than the Hmong and Southeast Asian grower community.

The messy reality of alternative food networks

Alternative food networks have been defined as: "New and rapidly mainstreaming spaces in the food economy defined by – among other things – the explosion of organic, Fair Trade, and local, quality, and premium specialty foods," noting also that "the politics and practices of alternative food networks have more recently come under critical scrutiny [...] as a narrow and weakly politicized expression of middle- and upper-class angst" (Goodman and Goodman 2008: 208). The creation of these networks involves a complexity that is missed in overly simplistic calls to "buy local." Any successful alternative food network must still be embedded within the very material conditions shaped by the economic, political, and social dominance of conventional food networks (Hinrichs 2000). Alternative food networks must therefore still play by the same rules as the globalizations of food, recognizing further that some of those very rules could well have been written by actors whose interests lie in supporting conventional food networks. What we find, then, is that alternative food networks often create what have been called "hybrid spaces" (Smithers and Joseph 2010). An example of this is when producers, seeking to make a living through local sales, repeatedly report having to also sell to conventional markets to dispose of surplus product (Oberholtzer *et al.* 2010; Ilbery *et al.* 2010).

Yet this embeddedness is a Catch 22. It is inevitable and necessary for long-term success – even the most ecologically, socially conscious producer needs to make at least some profit. But it could also undermine the future viability of any alternative food movement – after all, they are called *alternative* food movements. Much of the value of these shortened food networks lie in their representation as an "alternative" to the status quo (DuPuis and Gillon 2009; Seyfang 2008). So while embedded within many of the same networks as conventional food chains – the same market economy, the same web of rules and regulations, etc. – alternative food networks must simultaneously avoid the appearance of morphing into that which they are said to be an alternative to (Talyor 2004). It's a fine line to walk.

Alternative food networks also cannot be dependent upon niche pricing in the long term, as any "niche" will eventually be saturated with the entry of additional producers. Something other than price premiums must therefore keep producers involved in shortened food networks. For DeLind (2006), for example, that "something" is nurtured through emotional embodied relationships.

Then there's the thorny question of how to scale up alternative food networks, especially those premised upon shortened commodity chains. As Friedmann (2007: 391) notes in her study of alternative food networks in Toronto, Canada: "Local community food organizations believe they have reached the limits of scale in both supply and delivery." This has caused them to find ways to scale up and expand their networks beyond that of the "local." How one goes about doing this, however, is far from straightforward. For example, as "value-added" for local markets often means "labor-added," how can small and mid-sized farms be recruited into alternative food networks when most of those farmers also work off the farm? Where will they find the time to provide this additional labor? And *who* will be providing this extra labor – will it be, for example, women?

Scholars who have argued for an "agriculture of the middle" note that mid-sized family farms have become trapped between conventional and alternative food networks, unable to compete successfully in either: "mid-sized farms are the most vulnerable in today's polarized markets, since they are too small to compete in the highly consolidated commodity markets, and too large and commoditized to sell in the direct markets" (Kirschenmann *et al.* 2008: 3). These mid-sized farms may be well positioned to allow for the scaling up of alternative food networks. Yet this will likely only happen if those markets can be supplied without increasing these operations' labor requirements, as most mid-sized family farms are supported by off-farm income. For example, a major operational challenge for Farm to School (FTS) programs (as discussed in Box 12.1) is getting the produce from the farm to the school. The logistics of having multiple farmers delivering to schools makes the practice unfeasible for many schools. Mid-sized farms might be just the right scale for these programs, as one could credibly supply an entire school, thereby overcoming the logistical problems associated with sourcing from multiple smaller growers (Allen and Guthman 2006).

Finally, what role does certification play in the scaling up of alternative food networks? Certification schemes seek to broadly differentiate a given product from its conventionally produced counterpart based on certain "qualities" ascribed to it, usually through a third party. In responding to the growing demand within affluent nations for foods that are produced in environmentally sustainable and socially just ways, certification schemes (among other things) seek to re-embed consumers to the distant food products that they consume (M. Goodman 2004).

However, certification schemes are not without their problems (Higgins *et al.* 2008). First, smallholders (especially those in less affluent nations) often have restricted access to certification, due largely to the costs associated with this process. Certification also often requires additional inputs to ensure that certain standards are met; inputs that smallholders often cannot afford (remember, from Chapter 11, the immense expenses associated with "local" growers supplying Walmart). Furthermore, consumers might not understand the message of a label, leading to insufficient market demand while creating additional work for those seeking to sell their certified produce. There is also the risk that powerful private actors might hijack the certification process for their own ends, which is what some contend happens to the "organic" label (Guthman 2004).

There is also the point that, while proxies for so-called local, face-to-face relationships, certification schemes are not identical substitutes for these relationships (Box 12.7). Thus, as local, alternative food networks are scaled-up, what

BOX 12.7 CIVIC AGRICULTURE

The late distinguished sociologist Thomas Lyson (2004: 63) wrote that civic agriculture:

> [E]mbodies a commitment to developing and strengthening an economically, environmentally, and socially sustainable system of agriculture and food production that relies on local resources and serves local markets and consumers. The imperative to earn a profit is filtered through a set of cooperative and mutually supporting social relations. Community problem solving rather than individual competition is the foundation of civic agriculture.
>
> *Lyson (2004: 63)*

Juxtaposed to civic agriculture is commodity agriculture, composed of farms, "large-scale, absentee-owned, [and] factory.-like," that rely on large amounts of migrant labor or labor-saving capital and which sell their produce for global markets. These farms, according to Lyson, while not entirely devoid of civic spirit, nevertheless can "not be deemed very civic" (Lyson 2004: 62). Lyson, a tireless proponent of rural communities and human-scale agriculture, saw in civic agriculture a way to promote both ends.

Moreover, civic agriculture is not scale-neutral. It seems to work best with farms of a certain limited scale. Civic agriculture is nevertheless seen by its proponents as a way to promote alternative food networks without, if done right, the traps described earlier. In recent years, the concept of civic agriculture has been extended to a variety of phenomena, from, for example, Farm to School programs (Bagdonis *et al.* 2009) to dietetic practices (Wilkins 2009). It is worth noting that some have constructively criticized the concept of civic agriculture, suggesting that it places too much emphasis on building economic capacity within rural communities to sustain alternative markets (see e.g., Delind 2002). While this is important, critics contend we must equally value and seek to nurture alternatives to market-oriented relationships between producers and consumers if we are serious about building a truly *civic* agriculture and food system.

happens to those qualities and attributes that arise out of the social relationships maintained through shorter-chained food networks? Labels cannot substitute entirely for those. Can they?

Transition . . .

If this were a meal, the following, and final, chapter represents a light desert. The heavy dishes have already been dispensed with, as the "meat" (or, for the vegans out there, the "tofu") of the subdiscipline has been served over the span of 12 chapters. As any good desert should, the next chapter still brings something new to the table. Yet more than anything else, it gives us a chance to reflect on all that we've dined on over the previous pages.

Discussion questions

1 Does the Slow Food movement fall into any "traps?" Is there such a thing as unreflexive slowness?
2 What does "local food"/"global food" mean to you? Has this chapter altered these definitions at all?
3 What forms of knowledge might certification schemes miss as local, alternative food networks are scaled up?
4 As all food systems are embedded in national and international market logics and regulatory structures, is it even possible to have an entirely local food system?

Suggested reading: Introductory level

Guthman, J. 2007. Commentary on teaching food: Why I am fed up with Michael Pollan *et al. Agriculture and Human Values* 24(2): 261–4.

Kleiman, J. 2009. Local food and the problem of public authority. *Technology and Culture* 50(2): 399–417.

Suggested reading: Advanced level

Born, Branden and Mark Purcell. 2006. Avoiding the local trap: Scale and food systems in planning research. *Journal of Planning Education and Research* 26(2): 195–207.

DuPuis, E., J. Harrison, and D. Goodman. 2011. Just food? In *Cultivating Food Justice: Race, Class and Sustainability*, edited by A. Alkon and J. Agyeman. Cambridge, MA: MIT Press, pp. 283–308.

Taylor, P. 2004. In the market but not of it: Fair Trade coffee and Forest Stewardship Council Certification as market-based social change. *World Development* 33(1): 129–47.

Winter, Michael. 2003. Embeddedness, the new food economy and defensive localism. *Journal of Rural Studies* 19(1): 23–32.

References

Allen, P. and J. Guthman. 2006. From "old school" to "Farm to School": Neoliberalization from the ground up. *Agriculture and Human Values* 23(4): 401–15.

Allen, P. and C. Hinrichs. 2007. Buying into "Buy Local": Engagements of United States local food initiatives. In *Alternative Food Geographies*, edited by D. Maye, L. Holloway, and M. Kneafsey. Oxford: Elsevier, pp. 255–72.

Appadurai, A. 1988. How to make a national cuisine: Cookbooks in contemporary India. *Comparative Studies of Society and History* 30(1): 3–24.

Bagdonis, J., C. Hinrichs, and K. Schafft. 2009. The emergence and framing of Farm-to-School initiatives: Civic engagement, health and local agriculture. *Agriculture and Human Values* 26: 107–19.

Blue, G. 2009. On the politics and possibilities of locavores: Situating food sovereignty in the turn from government to governance. *Politics and Culture* 9: 68–79.

Born, Branden and Mark Purcell. 2006. Avoiding the local trap: Scale and food systems in planning research. *Journal of Planning Education and Research* 26(2): 195–207.

Brown, T. 2009. The time of globalization: Rethinking primitive accumulation. *Rethinking Marxism* 21(4): 571–84.

Buller, H. and C. Morris. 2004. Growing goods: The market, the state, and sustainable food production. *Environment and Planning A* 36: 1065–84.

Carolan, M. 2011a. *Embodied Food Politics*. Burlington, VT: Ashgate.

Carolan, M. 2011b. *The Real Cost of Cheap Food*. London: Earthscan.

Cecchini, M., F. Sassi, J.A. Lauer, Y. Lee, V. Guajardo-Barron, and D. Chisholm. 2010. Tackling of unhealthy diets, physical inactivity, and obesity: Health effects and cost-effectiveness. *The Lancet* 6736(10): 1–10, www.oecd.org/dataoecd/31/23/46407986.pdf, last accessed 29 December 2010.

DeLind, L. 2002. Place, work, and civic agriculture: Common fields for cultivation. *Agriculture and Human Values* 19: 217–24.

DeLind, L. 2006. Of bodies, place, and culture: Re-situating local food. *Journal of Agricultural and Environmental Ethics* 19: 121–46.

DeLind, L. and P. Howard. 2008. Safe at any scale? Food scares, food regulation, and scaled alternatives. *Agriculture and Human Values* 25: 301–17.

DuPuis, E.M. and S. Gillon. 2009. Alternative modes of governance: Organic as civic engagement. *Agriculture and Human Values* 26: 43–56.

DuPuis, E. Melanie and David Goodman. 2005. Should we go "home" to eat?: Toward a reflexive politics of localism. *Journal of Rural Studies* 21(3): 359–71.

DuPuis, E., J. Harrison, and D. Goodman. 2011. Just food? In *Cultivating Food Justice: Race, Class and Sustainability*, edited by A. Alkon and J. Agyeman. Cambridge, MA: MIT Press, pp. 283–308.

Fields, S. 2004. The fat of the land: Do agricultural subsidies foster poor health? *Environmental Health Perspectives* 112(14): 820–3.

Frank, A. and B. Gills. 1994. *The World System: Five Hundred Years or Five Thousand?* London: Routledge.

Freidberg, S. 2009. *Fresh: A Perishable History*. Cambridge, MA: Harvard University Press.

Freidberg, S. and L. Goldstein. 2011. Alternative food in the global south: Reflections on a direct marketing initiative in Kenya. *Journal of Rural Studies* 27: 23–34.

Friedman, T. 2000. *The Lexus and the Olive Tree*. New York: HarperCollins.

Friedmann, H. 2007. Scaling up: bringing public institutions and food service corporations into the project for a local, sustainable food system in Ontario. *Agriculture and Human Values* 24: 389–98.

Goodman, David. 2003. Editorial: The quality "turn" and alternative food practices: Reflections and agenda. *Journal of Rural Studies* 44: 1, 3–16.

Goodman, Michael. 2004. Reading fair trade: Political ecological imaginary and the moral economy of fair trade foods. *Political Geography* 23: 891–915.

Goodman, D. and Goodman, M. 2008. Alternative food networks. In *International Encyclopedia of Human Geography*, edited by R. Kitchin and N. Thrift. Oxford: Elsevier, pp. 208–20.

Grillo, I. 2009. Mexico's growing obesity problem. *Tucson Sentential* September 5, www.tucsonsentinel.com/nationworld/report/090509_mexico_obese, last accessed 29 May 2010.

Guthman, J. 2004. *Agrarian Dreams: The Paradox of Organic Farming in California*. Los Angeles, CA: UCLA Press.

Hassanein, N. 1999. *Changing the Way America Farms: Knowledge and Community in the Sustainable Agriculture Movement*. Lincoln, NE: University of Nebraska Press.

Hendrickson, M. and W. Heffernan. 2002. Opening spaces through relocalization: Locating potential resistance in the weaknesses of the global food system. *Sociologia Ruralis* 42(4): 347–69.

Higgins, V., J. Dibden, and C. Cocklin. 2008. Building alternative agri-food networks: Certification, embeddedness, and agri-environmental governance. *Journal of Rural Studies* 24: 15–27.

Hinrichs, Clare. 2000. Embeddedness and local food systems: Notes on two types of direct agricultural markets. *Journal of Rural Studies* 16(3): 295–303.

Hinrichs, Clare. 2003. The practice and politics of food system localization. *Journal of Rural Studies* 19: 33–45.

Hinrichs, Clare, Jack Kloppenburg, George Stevenson, Sharon Lezberg, John Hendrickson, and Kathryn DeMaster. 1998. *Moving Beyond Global and Local*. United States Department of Agriculture, Regional Research Project NE-185 working statement, October 2, <http://www.ces.ncsu.edu/depts/sociology/ne185/global.html> last accessed May 21, 2011.

Ilbery, B., P. Courtney, J. Kirwan, and D. Maye. 2010. Marketing concentration and geographical dispersion: A survey of organic farms in England and Wales. *British Food Journal* 112(9): 962–75.

Inglis, D. and D. Gimlin. 2009. Food globalizations. In The *Globalization of Food*, edited by D. Inglis and D. Gimlin. New York: Berg, pp. 3–42.

InterAcademy Council. 2003. Realising the promise and potential of African agriculture: Science and technology strategies for improving agricultural productivity and food security in Africa. Amsterdam, The Netherlands, http://www.cgiar.org/pdf/agm04/agm04_iacpanel_execsumm.pdf, last accessed June 7, 2011.

Kirschenmann, F., S. Stevenson, F. Buttel, T. Lyson, and M. Duffy. 2008. Why worry about the agriculture of the middle? In *Food and the Mid-Level Farm: Renewing an Agriculture of the Middle*, edited by T.A. Lyson, G.W. Stevenson, and R. Welsh. Cambridge, MA: MIT Press, pp. 3–22.

Kloppenburg, J., J. Hendrickson, and G. Stevenson. 1996. Coming to the foodshed. *Agriculture and Human Values* 13(3): 33–42.

Lawrence, G. 2004. Promoting sustainable development: The question of governance. Plenary Address, XI World Conference of Rural Sociology, Trondheim, Norway, July 26, http://www.irsa-world.org/prior/XI/program/Lawrence.pdf, last accessed May 23, 2011.

Lien, M. 2009. Standards, science and scale: The case of Tasmanian Atlantic salmon. In *The Globalization of Food*, edited by D. Inglis and D. Gimlin. New York: Berg, pp. 3–42.

Long, Michael and Doug Murray. 2011. *Ethical Consumption, Values, Convergence/Divergence and Community Development*. Center for Fair and Alternative Trade, Sociology Department, Colorado State University, Fort Collins, CO, USA.

Lyson, T. 2004. *Civic Agriculture: Reconnecting Farm, Food and Community*. Lebanon, NH: Tufts University Press.

Mansfield, Becky. 2003. Fish, factory trawlers, and imitation crab: The nature of quality in the seafood industry. *Journal of Rural Studies* 19(1): 9–21.

Marston, S. 2000. The social construction of scale. *Progress in Human Geography* 24(2): 219–42.

Matejowsky, T. 2007. SPAM and fast food globalization in the Philippines. *Food, Culture and Society* 10(1): 23–41.

McIntyre, L. and K. Rondeau. 2011. Individual consumer food localism: A review anchored in Canadian farmwomen's reflections. *Journal of Rural Studies* 27(2): 116–24.

Minkoff-Zern, L., N. Peluso, J. Sowerwine, and C. Getz. 2011. Race and regulation: Asian immigrants in California agriculture. In *Cultivating Food Justice: Race, Class and Sustainability*, edited by A. Alkon and J. Agyeman. Cambridge, MA: MIT Press, pp. 65–86.

Mougeot, Luc. 2005. Introduction: Urban agriculture and millennium development goals. In *Agropolis: The Social, Political and Environmental Dimensions of Urban Agriculture*, edited by Luc Mougeot. London: Earthscan, pp. 1–29.

Murdoch, J., T. Marsden, and J. Banks. 2000. Quality, nature and embeddedness: Some theoretical considerations in the context of the food sector. *Economic Geography* 72(6): 107–25.

Norberg-Hodge, H., T. Merrifield, and S. Gorelick. 2002. *Bringing the Food Economy Home: Local Alternatives to Global Agribusiness*. Bloomfield, CT: Kumarian Press.

Oberholtzer, L., K. Clancy, and J.D. Esseks. 2010. The future of farming on the urban edge: Insights from fifteen US counties about farmland protection and farm viability. *Journal of Agriculture, Food Systems, and Community Development* 1(2): 59–75.

Pollan, Michael. 2008. *In Defense of Food: An Eater's Manifesto*. New York: Penguin.

Rabobank. 2010. Sustainability and security of the global food supply chain. Rabobank Group, The Netherlands, http://www.rabobank.nl/images/rabobanksustainability_29286998.pdf?ra_resize=yes&ra_width=800&ra_height=600&ra_toolbar=yes&ra_locationbar=yes, last accessed June 7, 2011.

Ravallion, Martin, Shaohua Chen, and Prem Sangraula. 2007. New evidence on the urbanization of global poverty. Policy Research Working Paper 4199. World Bank, Washington, DC, http://siteresources.worldbank.org/INTWDR2008/Resources/2795087-1191427986785/RavallionMEtAl_UrbanizationOfGlobalPoverty.pdf, last accessed June 8, 2011.

Redwood, M. 2010. Commentary: Food price and volatility and the urban poor. In *Urban Agriculture: Diverse Activities and Benefits for City Society*, edited by C. Pearson. Earthscan, London, pp. 5–6.

Ritzer, G. 2010. *McDonaldization of Society*. Thousand Oak, CA: Pine Forge Press.

Robles, Y. 2010. Urban-farming group in running for $4,000 grant. *The Denver Post* September 30, wwwdenverpostcom/news/ci_16211365?source=email, last accessed June 2, 2010.

Seyfang, G. 2008. Avoiding Asda? Exploring consumer motivations in local organic food networks. *Local Environment* 13(3): 187–201.

Slocum, R. 2011. Race in the study of food. *Progress in Human Geography* 35(3): 303–27.

Smithers, J. and A.E. Joseph. 2010. The trouble with authenticity: Separating ideology from practice at the farmers' market. *Agriculture and Human Values* 27(2): 239–47.

Starr, A. 2010. Local food: A social movement. *Cultural Studies, Critical Methodologies* 10(6): 479–90.

Talyor, P. 2004. In the market but not of it: Fair Trade coffee and forest stewardship council certification as market-based social change. *World Development* 33(1): 129–47.

Vignali, C. 2001. McDonald's: Think global, act local – the marketing mix. *British Food Journal* 103(2): 97–111.

Wilk, R. 1991. *Household Ecology: Economic Change and Domestic Life among the Kekchi Maya of Belize*. Tucson, AZ: University of Arizona Press.

Wilk, , R. 2009. Difference on the menu. In *The Globalization of Food*, edited by D. Inglis and D. Gimlin. New York: Berg, pp. 185–96.

Wilkins, J. 2009. Civic dietetics: Opportunities for integrating civic agriculture concepts into dietetic practice. *Agriculture and Human Values* 26: 57–66.

Winter, Michael. 2003. Embeddedness, the new food economy and defensive localism. *Journal of Rural Studies* 19(1): 23–32.

Zezza, Alberto and Luca Tasciotti. 2010. Urban agriculture, poverty and food security: Empirical evidence from a sample of developing countries. *Food Policy* 35: 265–73.

13

LOOKING BACK . . . AND FORWARD

KEY TOPICS

- As "free trade" is often talked about in the context of food security, it is useful to turn our sociological imaginations back onto this widely used, but often poorly defined, concept.
- In light of how far the sub-discipline has come, gaps in the literature remain, such as around issues of gender and social justice.
- The re-peasantization of the countryside will reawaken an interest in peasant agriculture in years to come.
- The book concludes with what I hope is an instructive map of our journey.

Key words

- Fair trade
- Food democracy
- Food labels
- Freedom from/freedom to
- Free trade
- Neo-liberalism
- Public sociology of food and agriculture
- Re-peasantization
- Resiliency

An exciting future awaits the sociology of food and agriculture (and what is known as "agrifood studies" more generally). The popularity of the sociological study of food, from seed to stomach, has never been greater. I will conclude this book first with some general comments; the hope being they will be taken with you as you utilize your sociological imaginations to think critically about food and agriculture in the future. In this vein, I begin by opening up the concept of "free markets" for a brief but necessary critical reflection. Other subjects discussed include the gendered politics of food and social justice, both of which deserve more thorough treatment by scholars of food and agriculture. I will also say a little more about peasant agriculture. As the eradication of the peasantry is no longer a formal goal of the international development community, future food and agriculture scholars, politicians, and activists will need to better understand peasants and their once disparaged mode of food production. To conclude, both the chapter as well as the book, I make an honest attempt at mapping the terrain I spent the previous chapters discussing – an exercise started by Buttel (2001a,b) over a decade ago. While imperfect (the first principle of map making is that maps are, by definition, imperfect representations of what they are said to represent), I find such an exercise pedagogically useful as it helps make visible links that might have otherwise been missed. I also take the liberty to offer some tentative suggestions forward, based upon current blind spots in the literature.

Free trade

Any current and future discussions about food and agriculture must involve at least some talk about markets. Even alternative food networks, as stated in the previous chapter, do not operate outside conventional markets. So what *are* free markets? According to Google dictionary, free markets are "an economic system in which prices are determined by unrestricted competition between privately owned businesses." *Unrestricted competition*: that term alone ought to raise red flags. As discussed repeatedly in the previous chapters, some economic actors have clear advantages over others in the food system – due to, among other things, subsidies (see Chapter 2) and monopoly and monopsony conditions (see Chapter 3). Moreover, as long as we continue to value things like the environment, worker rights, and consumer welfare we are going to continue to regulate business and demand certain standards when it comes to how food is produced and processed. This is not to suggest that regulation has been antithetical to agrifood corporate interests. As in the case of the organic label, regulations and standards have certainly not emptied the coffers of agribusiness.

But I ask you: do we even want unbridled "free" competition? Clearly, proponents of neo-liberalism do (or at least they claim to want it). As David Harvey (2005: 2) explains in *Brief History of Neoliberalism*, "Neoliberalism is in the first instance a theory of political economic practices that proposes that human well-being can best be advanced by liberating individual entrepreneurial freedoms and skills within an institutional framework characterized by strong private property rights, free markets, and free trade." As opposed to individual liberties (the Holy Grail in classic liberalism), neo-liberalism sees salvation only through free enterprise – and a really no-holds-barred, dog-eat-dog approach to free enterprise at that. This explains, for example, the rapid push in recent decades to hollow out the state by eliminating regulations and privatizing anything and everything (even, as mentioned in Chapter 10, water).

To answer the question of whether totally unbridled competition is what we really want when we talk about "free markets" I would like to take a step back and ask the following question. What does "freedom" mean (as understandings of a free market fundamentally hinge, I assume, on our collective understandings of this concept)? I have found twentieth century English philosopher Isaiah Berlin's dual understanding of freedom extremely useful for making sense of this over-used, under-understood, concept (see also Carolan 2011).

Conventional understandings of free markets are dangerously narrow, as evidenced by the simple fact that free markets are rarely ever fair (Box 13.1).

BOX 13.1 FAIR (AS OPPOSED TO FREE) TRADE

Fair trade represents a new producer–consumer relationship through a supply chain that attempts to distribute economic benefits more fairly between all stakeholders (Raynolds *et al.* 2007). This is in contrast to the conventional supply chain model, which primarily seeks the maximization of return somewhere in the middle of the food system hourglass (often at the expense of producers, as discussed in Chapter 3). Fair trade makes trade more fair through a number of practices. Some of these practices include the following (taken from Nicholls and Opal 2005: 6–7).

- Agreed upon minimum prices that are often higher than those set by the market. This is in recognition that agricultural commodity market prices, thanks to things like subsidies and monopoly and monopsony conditions, are rarely fair. Farmers are therefore given a living wage for their work.
- An additional social premium is paid on top of the fair trade price. This allows producers and farm laborers to collectively implement larger development projects (like the building of new schools). How the money is spent is usually decided democratically, through cooperatives.
- Purchasing directly from producers. This reduces the number of profit-taking "middleman" in the commodity chain and ensures that more of each dollar spent on the goods return to producers.
- Making credit available to producers. As importers from affluent countries typically have greater access to credit than developing country producers, importers must pre-finance a significant portion of the year's harvest.
- Farmers and workers are democratically organized. This helps minimize labor abuses (e.g. child and slave labor) and ensures socially responsible production practices.

That is because most of this talk is heavily infused with, to draw from Berlin, "freedom *from*" rhetoric – freedom *from* domestic protectionist policies, freedom *from* other countries' production-oriented subsidies, freedom *from* government regulation, freedom *from* the state, and so forth. Freedom *from* talk is especially pervasive in neo-liberalism discourse. Isaiah Berlin (1969 [1958]: 127) describes this as "negative freedom," which refers to the "absence of interference." It seems as though when most people are asked about "freedom," negative freedom is the freedom most have in mind – a freedom where we are allowed to pursue actions unimpeded (see also Bell and Lowe 2000: 287–9).

Yet in addition to its negative component freedom also has a positive side. "Positive freedom" – or freedom *to* – refers to the ability "to lead one prescribed form of life [. . .] that derives from the wish on the part of the individual to be his [or her] own master" (Berlin 1969 [1958]: 127). Without the active pursuit of positive freedom most people would not – indeed *could not* – feel free. To explain what I mean by this I evoke the words of the old English aphorism, "freedom for the pike is death for the minnow." Without the pursuit of positive freedom for all – pikes and minnows alike – the pikes of the world would be clearly advantaged.

How do we actively pursue positive freedom for all? This is where the concept of *constraint* comes into the picture. There must be some constraints if all minnows are to prosper. Positive freedom, in other words, allows us to talk about constraint in the context of freedom without any sense of paradox.

Let's now apply this thinking to free markets. In order for me to be and feel free (and trade freely), I need some assurances that the pikes of the world will not freely have their way with me. This is why freedom – and, yes, free markets – *require* some level of government intervention.

What are some of the consequences of this understanding of free markets? First, it gives force to arguments that favor opening up the middle of the hourglass figure detailed in Chapter 3, so as to allow for greater competition between firms. Market monopoly, oligopoly, and monopsony conditions are an inevitable outcome of food policies overly influenced by one-sided understandings of freedom – namely, freedom *from* (Bell and Lowe 2000).

A Berlinian understanding of free markets also gives justification to spending initiatives to build food security *capabilities* in less developed countries. To play upon an ancient Chinese proverb: rather than giving other countries fish (or more specifically grain, as the USA does with food aid), so they can eat for a day, affluent nations can instead assist less affluent nations build the capabilities to fish, which would allow them to eat for a lifetime. This is not, however, a call to *dictate* to developing nations what these capabilities ought to look like. They need to decide this path on their own.

The rise of neo-liberal forms of governance over the last decades has brought with it the transference of responsibility for food security from government to the market – a process some have called the "marketization of food security" (Zerbe 2009: 172). A Berlinian understanding of free markets could act as a philosophical counterweight to neo-liberalism. For it provides rational justification for the reversal of its policies (Box 13.2; Table 13.1).

Gendered politics of food and agriculture

The subject of gender has come up a number of times in the preceding chapters, so it might surprise some to learn that "gender analysis remains on the margins of the sociology of agriculture" (Allen and Sachs 2007: 4). This omission no doubt

BOX 13.2 LABELING STRATEGIES

How do food labels fit into all of this talk about free markets and social justice? The literature does not supply a clear-cut answer. Some, like Barham (2002: 350; see also Barham 1997) view labeling as "one historical manifestation of social resistance to the violation of broadly shared values by systemic aspects of 'free market capitalism'." Others, however, take a notably more critical stance toward labels, arguing, for instance, that the food label strategy "shifts the responsibility for social reforms from the state and manufactures to individual consumers, bringing with it important social justice implications" (Roff 2007: 511). Perhaps the best answer, then, to the question "How do food labels fit into a broader discussion about free markets and social justice" is "It depends."

The alleged protections labels are said to offer are terribly uneven and often directed to those least in need of protection. This point echoes Guthman (2007: 462), who writes "among the various claims made about labels, some are meant to protect or preserve existing arrangements while others are meant to improve them." In making this argument, Guthman attempts to categorize some of the differences among the various food labels. This categorization scheme is reproduced in Table 13.1. As detailed, the intent, focus of verification, barriers to entry, and distributional effect of labels vary considerably depending upon label type.

Thus, for example, labels – like Protected Designation of Origin (PDO) – that aim to preserve existing conditions have (not surprisingly) steep barriers to entry, as label-gatekeepers are usually those already trained in the craft being protected. These designations offer strong protections to existing producers, even in the face of efforts by agribusiness to produce cheaper substitutes. While these labels are not redistributive, they quite successfully protect certain nonconventional forms of food production. Other labels, like the organic label, are based on barriers that are quite easy to overcome (as evidenced by infiltration of capital into organics). Then there is the Fair Trade movement/label, which is re-distributive in its intent; though its low barriers to entry have led to the involvement of multinational corporations (e.g. Starbucks) which has undermined the label's credibility.

TABLE 13.1 Taxonomy of food labels

Type	*Example*	*Intent*	*Focus of verification*	*Barriers to entry*	*Distributional effect*
Process based	Organic	Encourage ecologically sustainable practices	Inputs used, land used for production	Meeting standards, verification, 3-year transition	After transition subject to competition, price declines
	PDO, craft cheese	Safeguard taste/traditional methods	Labor process	Entry into apprentice/craft	Value retained by producer
Place based	AOC–terroir	Safeguard taste/specific qualities	Territorial qualities of land	Access to designated land	Monopoly ground rent
	Local	Reduce food miles, support regional development	Within designated boundary	Marketing agreement	Inter-regional competition
Distributive	Fair trade	Redistribution of commodity prices	Revenue allocation along supply chain	Verification only	Redistributional between consumers and producers

Based on Guthman (2007: 463).

AOC, *Appellation d'origine contrôlée* (controlled designation of origin); PDO, Protected Denomination of Origin.

reflects, at least in part, the discipline's long-standing interest in the production "end" of the food system; a focus that dates back to its rural sociological roots. As women have traditionally made decisions about the "inside" (the home), leaving decisions about the "outside" (the farm) to men (Rickson *et al.* 2006: 126), social scientists have historically restricted their gaze to those whose role included "farmer" – namely, men.

Interest in gender among agrifood scholars will likely continue to grow. For one thing, especially with the rise of more sustainable forms of agriculture, more women than ever are self-described "farmers" in affluent countries. Similarly, food and agriculture scholars and practitioners are becoming wise to the fact that the majority of farm labor conducted in less affluent nations is being (and has long been) done by women. A changing structure of agriculture in affluent countries (which was discussed extensively earlier in the book) means also that an increasing percentage of farm land is being rented. In the USA, for instance, close to 50 percent of private agricultural land is being utilized by someone other than the landowner. And a growing number of these landowners are women (Box 13.3) (Duffy and Smith 2008; Petrzelka and Marquart-Pyatt 2011).

Moreover, as Allen and Sachs (2007: 4) note, "as the sociology of agriculture moves more towards a focus on consumption, it would seem that gender relations would emerge as an obvious key problematic." As discussed previously, food purchasers and especially food preparers continue to be, typically, women; a role that comes with a host of competing expectations and interests, while also giving those involved at the consumption "end" of the food system the potential for considerable agency (Wilk 2010). On the whole, the literature continues to view subjects that populate our food system as largely ungendered. This, though, appears to be changing (see, e.g., McIntyre and Rondeau 2011).

Those in affluent countries are also increasingly eating outside the home. Approximately half of all the money US households spend on food, for example, is spent on food away from home (Carolan 2011). Recognizing that "women" is not a unified category – discussions of gender cannot be divorced from its intersections with race, ethnicity, and class – we must ask how this shift to food "away from home" will impact women differently, as women hold most of the jobs in the food service industry; though, interestingly, the "top" positions of this industry continue to be overwhelmingly male-dominated – for example, 87 percent of the head cooks and chefs in the USA are men (US Bureau of Labor Statistics 2008). Those jobs dominated by women are not high paying. Most pay only minimum wage, which means they will be filled by women of a lower socio-economic status, immigrants, and by women of color. The majority of these jobs are also part-time, with few (if any) benefits (Allen and Sachs 2007).

There will undoubtedly be growth in certain sectors of the food system in the near future. Yet are these the kinds of jobs we want to see (as most will likely not even pay a living wage)? If not, who are benefiting from these low wages? And, at the other extreme, who will be most negatively impacted by them?

BOX 13.3 CHANGING LAND TENURE AND GENDER: A BRIEF REVIEW

An early study looking at gendered elements of the tenant–landlord relationship found female landlords less likely than their male counterparts to be involved in decisions related to fertilizer, herbicide, and pesticide applications on their land, though age seemed to help mitigate this effect as younger landlords (both male and female) reported greater involvement (Rogers and Vandeman 1993). A few years later, a team of researchers found being male, having a higher dependence upon rent for income, and coming from a farming background were strong predictors among landlords of involvement in decisions about farming practices on their land (Constance *et al.* 1996). More recently, notable differences were found between male and female landlords in a study examining the adoption of conservation practices on rented land in Iowa (Carolan 2005). For example, among female landlords interviewed who had a desire that sustainable agricultural practices be adopted on their land, self-censorship was reported out of concern they would "scare away good tenants" (Carolan 2005: 396). Similarly, validating earlier research (e.g. Gilbert and Beckley 1993), the study reports that "all of the female landlords described inequitable power relations between themselves and their male tenants" (Carolan 2005: 402). More recent still, looking at gender relations among landlords and tenants in the US states of Wisconsin, Michigan, and New York, it was found little has changed in the last 20 years for female landowners (Petrzelka and Marquart-Pyatt 2011). In the authors' own words:

> While the number of female landlords in the US has changed, this study's findings suggest that the power issues which exist for women in agriculture have perhaps not. Female landlords appear to have the least power due to their relationship with their tenant, their relationships with their siblings, or a combination of these situations. [. . .] Women may hold title to the land, but whether the land is truly theirs in terms of control over what occurs on it is called into question with our study's findings.
>
> *Petrzelka and Marquart-Pyatt (2011: 558)*

Social justice and public sociology of food and agriculture

"It is clear," as Patricia Allen (2008: 158) wrote a few years back, "that our food system does not meet the fundamental criteria of social justice such as freedom from want, freedom from oppression, and access to equal opportunity." It's a curious development. The growing local food market, the introduction of the "food miles" concept to millions of consumers around the world, and the conventionalization of organic (and/or the organicification of industrial food), each indicates a rising awareness among consumers, food activists, producers, and policy makers of issues relating to the ecological sustainability of our food system. The rise of vegetarianism and veganism indicates a growing sensitivity to issues of animal welfare (and among many vegans a belief that animals, like human animals, have inalienable rights). There is also increasing talk about building regional food systems in order to foster community development. Yet, as we strive toward these admirable goals, the incorporation of equally admirable social justice outcomes often get lost in the process. Occasionally, social justice is *sacrificed* – either intentionally or unintentionally – as another goal is doggedly pursued, like when social justice standards were dropped during the formation of the US National Organic Program.

Even when a tacit goal, social justice outcomes can be undermined by other socio-economic factors. We have seen this with community-supported agriculture (CSA). As others have described through their research (see, e.g., Alkon 2008; Hinrichs 2000; Slocum 2008), even when CSAs and farmers' markets are originally organized to help low-income minority groups they nevertheless often end up serving groups with plenty of access already to whole, locally and/or regionally produced foods – namely, affluent, educated, European-American consumers.

It is also possible to spend too much time talking about the creation of alternative food networks (e.g. CSAs, regional food systems), ignoring the many injustices that continue to exist within the dominant food system. One of the arguments for alternative food networks is that, once available, they will allow consumers to choose between different food system models. Yet it's not entirely clear that consumer-oriented action can even resolve problems of social justice. Leaving social change up to consumers, while leaving the status quo intact, risks creating a release value for any discontent toward the dominant food system, as political action is reduced to individual consumer choice (versus action directed at structurally oriented social change). This sets up a situation where the affluent are the ones most likely to participate in these alternative spaces where they can feel safe. In doing this, however, they risk becoming dissuaded from leveraging their social, cultural, political, and economic capital to meaningfully change the dominant food system (Szasz 2007).

We must also remember, as discussed in earlier chapters, that social justice is not inherently a problem tied to scale. Labor injustices are not only committed in confined animal feeding operations (CAFOs) but can arise wherever the employer–employee relation exists, including, for example, small-scale farms, "mom and pop" grocery stores, and local food co-ops. The sociology of food and agriculture literature has a history of conflating farm scale and social justice, where "small" is often equated with "just." As future agrifood scholars, it is up to you to make sure the conflation of scale and justice/sustainability does not continue.

And it is not just farm workers. Injustices occur throughout the food system, as touched on previously in Chapter 5 when discussing the health risks associated with working in industrial feed lots and with pesticides. Focusing only on farmers also risks perpetuating a subtle agrarian idealism that underlies much of the academic and popular writing on the subject. This idealism holds the family farm as the core of democracy and care while ignoring all the other laborers that feed us.[1]

This leads me to ask: do sociologists have an obligation to make the world better or are we merely charged with its study? The broader discipline has been grappling with the practice of what has come to be known as "public sociology" for well over a century. Indeed, a recent President-Elect of the American Sociological Association, Michael Burawoy, spent the better part of his term actively campaigning to make sociology (and sociologists) more socially engaged; more, in a word, relevant. As he wrote in his 2004 American Sociological Association Presidential Address:

> Economics, as we know it today, depends on the existence of markets with an interest in their expansion, political science depends on the state with an interest in political stability, while sociology depends on civil society with an interest in the expansion of the social. [. . .] In time of market tyranny and state despotism sociology – and in particular its public face – defends the interests of humanity.
>
> *Burawoy (2005: 287)*

In other words, Burawoy is arguing that sociology has an inherent obligation to protect the interests of actors within society. I recognize that this is an immensely complex task, as often actors have competing interests, which is certainly the case when talking about something as ubiquitous as food. This point was articulated by Neva Hassanein (2003) in her discussion of food democracy, recognizing that something as foundational to alternative food movements as "sustainability" means something different to everyone. Consequently,

> it is difficult to apply this – or any – definition as a practical guide for action. What does it really mean in practice to equitably balance concerns for environmental soundness, economic viability, and social justice among all sectors of society? How should each dimension be evaluated in relation to the others? How should society weigh, for example, the protection of water quality from agricultural runoff against the possibility that additional regulation of farming practices might make it even more difficult for small agricultural producers to operate in an economically viable manner? [. . .] How do we make judgments about the needs, wants, and rights of current generations in light of considerations for future generations? Perhaps most importantly, who gets to decide where the "equitable balance" lies?
>
> *Hassanein (2003)*

At their core, many of the issues being debated in the context of food and agriculture are around competing values. And as such, there is no independent authority – not science, not the law, not politicians – that can tell us which (and whose) values trump all the others. This very point, however, is sociologically significant. For if we can better understand which questions are empirical and which are about competing values then perhaps we can better focus conversations about food and make sure all stakeholders have a place at the table.

But is that enough? Not so, according to a number of prominent scholars who passionately argue for a praxis dimension in their work (e.g. Allen 2004; Friedland 2008; Lyson 2004; Tanaka and Mooney 2010). According to this argument, sociologists of food and agriculture need to do more than just help stakeholders understand the world better. We must actively work to change it, with an aim toward justice. Sociologist Douglas Constance, past-President of Agriculture, Food, and Human Values Society, is quite firm on this point:

> As a community of critical agrifood scholars, we must embrace a praxis and volunteer in service of this movement. We must commit to a praxis of public social science that moves beyond the comfortable, conservative confines of the academy. Whatever the focus of our work, the long-term goal remains the same; emancipatory change to end injustice.
>
> *Constance (2008: 154)*

Peasant agriculture

A slow, but noticeable, shift in attitudes is occurring among academics and developmental practitioners when it comes to understanding the peasantry. For much of the twentieth century, as detailed in Chapter 5, "peasant" was all too often equated with "backward," "poverty," and "inefficient." There is a growing call to abandon this view. If anything, these labels say more about those casting aspersions than they do about peasants and peasant agriculture. The assumptions that have turned peasant into a pejorative term in the West are based more in ideology than empirical fact. McMichael puts this point succinctly:

> Colonizers and developers routinely conflate "frugality" with "modernized poverty", by projecting the latter onto the non-European world. Naturalizing the African woman as a baseline for a "development ladder" is as misleading as implying that agro-industrialization is inevitable, and desirable, at a time when low-carbon livelihoods are often more appropriate and possibly more just. The poverty trope informs, and legitimates, the development trajectory with which the [World] Bank works.
>
> *McMichael (2009: 236)*

Conventional economic theory legitimates the disappearance of peasants as "growth" is reduced to purely quantitative measures with predefined assumptions that severely undervalue peasant agriculture. One such assumption is its fixation upon monocultures, which, among other things, hastens the substitution of labor for capital and necessitates the use of chemical inputs – a logic that only makes economic sense, I might add, when oil, technology, and the inputs in question and are incorrectly priced (Carolan 2011). When monoculture agriculture is naturalized – when assumed to be the only option – the agroecology practiced by peasants finds itself in a poor competitive position to conventional industrial agriculture as the latter's inputs (chemical poisons that immediately kill pests) are stronger in the short term than the former's. This may be why organic farming in affluent countries is less likely to out-yield conventional agriculture, while in the Global South peasant agroecological systems consistently average higher total levels of productivity than conventional monocultures (Altieri and Koohafkan 2008; Rosset *et al.* 2011). For decades, an unbridled faith in conventional agriculture has led to the de-peasantization of developing economies. In more recent years, with agroecology principles once again gaining favor (described in Chapter 9) and the rise of such peasantry movements as *La Via Campesina* (described in Chapter 11), we are witnessing a reversal of this trend, toward what van der Ploeg (2008, 2010) calls re-peasantization.

Moreover (and linking this section with the one it proceeds), there is also evidence that this process of re-peasantization – a transition to diversified agroecological farming – reduces gender inequities inside peasant families; a fact that goes against the commonly held view that peasant families are highly patriarchal. As described by one peasant woman, in conventional agriculture "the man is the king. [. . .] [The] crop belongs to the man. He drives the tractor, plants, applies chemicals, harvests and sells the crop. And all the money goes to him" (quoted in Rosset *et al.* 2011: 183). In more diversified, peasant-style modes of production, the roles and income earning opportunities are equally as diverse. While men may still manage the row crops, the animals and medicinal plants are the responsibility of the woman and she in turn receives the income from these activities (Rosset *et al.* 2011).

In a reversal of positions, peasant agriculture may in fact turn out to be what saves us from future food systems' collapse, as climate change in particular causes us to redefine understandings of "resiliency." Scholars and practitioners of agroecology have long known that peasant agriculture is highly adaptable to climate change and therefore more resilient to extreme climate events than conventional agriculture (Altieri and Koohafkan 2008; Mercer and Perales 2010). Research comparing the farms of peasants utilizing agroecological principles with those farming more conventionally in Central America in the wake of Hurricane Mitch in 1998 found that the agroecological farms suffered considerably less erosion and fewer landslides as a result of the hurricane (Holt-Gimenez 2002). In another study, the effects of Hurricane Ike in 2008 on Central American provinces revealed far higher levels of crop loss in areas of industrial monoculture than agroecological peasant farms with multi-storied agroforestry farming systems (Rosset *et al.* 2011). In the peasant farms, lower story annual and perennial crops were compensating for any losses with exuberant growth, thus "taking," in the words of the researchers, "advantage of the added sunlight when upper stories were tumbled or lost leaves and branches" (Rosset *et al.* 2011: 182). The research team "also saw tremendous new leaf growth on branches that had been stripped. And perhaps most impressive of all, a substantial portion of the trees that had been blown down had been saved by peasant families who stood them back up and covered their roots the first morning after the storm" (Rosset *et al.* 2011: 182).

A map of our journey

It has been over 10 years since the late Fred Buttel "mapped" the sociology of food and agriculture subdiscipline; a decade that brought with it an explosion of interest among scholars, activists, and the general public toward agrifood issues. Building upon this previous work, I update this "map" to reflect recent changes in the literature. Figure 13.1 represents the outcome of this labor, while also providing an illustrative overview of the terrain covered over the previous chapters.

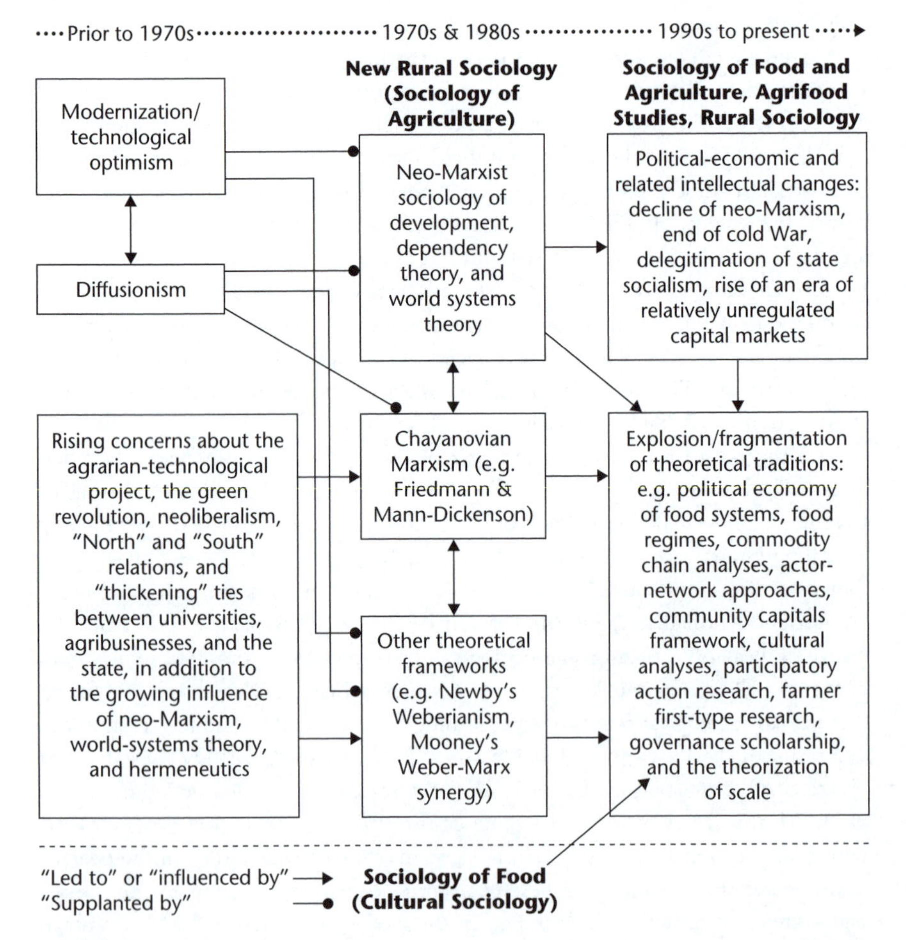

FIGURE 13.1 A geneology of transitions in the sociology of food and agriculture. Inspired by Buttel (2001a,b)

Over the last decade, we have seen a continuation of the long-standing political economy of food systems tradition, which, while still popular, has seen its theoretical supremacy weakened as competing frameworks – like actor-network approaches, cultural analyses, participatory action research, and theorizations of scale – gain followers. When Buttel mapped the sub-discipline back in 2001, analytic attention was still primarily centered on production, as that was the "side" viewed as being of most consequence to the shape and structure of the food system and by extension to consumer "choice" (though consumption was beginning to be viewed as a legitimate site of political action even at this time; e.g. DuPuis 2000). Since then, we have witnessed a reorientation in the literature, where considerably more attention is now paid to actors further along in the "hourglass." Most notably, retail (as discussed in Chapter 3) has emerged as perhaps the most powerful sector of the entire food system.

As for consumption: the sociology of food and agriculture is now addressing this subject head-on. This omission up until the new millennium was a product of a long-held division of labor within sociology. Sociologists have long been interested in consumption; they just didn't, until recently, happen to be rural sociologists or sociologists of agriculture. As explained by the subtitle of a leading text from the early 1990s titled *The Sociology of Food*, sociologists of food have historically been interested in "eating, diet and culture" (Mennell *et al.* 1992); an orientation clearly different from the agrarian political economic approach that guided the majority of early sociologists of agriculture.

A similar point can be made about gender. Allen and Sachs (2007: 4), as quoted earlier, note that "gender analysis remains on the margins of the sociology of *agriculture*" (my emphasis). There is undoubtedly a larger literature in the sociology of *food* that has focused on gender, which has been discussed throughout the book (and in Chapter 6 in particular). Consequently, as implied by Figure 13.1, gender is becoming a more dominant subject in the literature by nature of this coming together of food and agriculture sociological scholarship.

This brings me to my final point, which I will express as a question: Has the sociology of food and agriculture outgrown its subdisciplinary title? I admit that my own uncomfortability with the title reveals itself in my writing through the occasional injection of such terms as "agrifood studies" and "agrifood scholarship" when referring to this book's literature base. I have justified using the term "sociological" throughout the book on the basis that while the scholarship presented may not be entirely by sociologists they nevertheless employ their sociological imaginations to investigate the complexities of our food system. The growing multidisciplinarity of the literature might also be an artifact of its rural sociological roots (though, as Hinrichs [2008] notes, rural sociology's radical turn in the 1970s arguably sidetracked this multidisciplinary spirit for a time). William Freudenburg, in a paper based upon his Presidential Address to the Rural Sociological Society, notes the following:

> Rather than seeking ivory-tower isolation, members of the Rural Sociological Society have always been distinguished by a willingness to work with

> specialists from a broad range of disciplines, and to work on some of the world's most challenging problems. What is less commonly recognized is that the willingness to reach beyond disciplinary boundaries can contribute not just to the solution of real-world problems, but also to the advancement of the discipline itself.
>
> *Freudenburg (2006: 3)*

He goes on to explain how this point is increasingly being illustrated in studies of environment–society relationships. Yet I think an equally illustrative case can be found in studies relating to food and agriculture. The vitality of the sub-discipline lies precisely in its willingness to be driven by the problems rather than disciplinary dogma, recognizing that a (sub)disciplinary narrative only becomes dogmatic by failing to learn, change, and adapt (which necessitates the occasional crossing of disciplinary lines).

This cross-disciplinarity is one of the things that first attracted me to the field. And it is, among other things, what keeps me nourished as I continue to feed upon its rich literature. *Bon appétit!*

Discussion questions

1 What are your thoughts about public sociology? Do you think sociologists of food and agriculture should just study the world or study *and* actively work to change it?
2 What points most surprised you in this book? Any points that you strongly disagree with?
3 In light of all that you have just read, what do you believe to be the most pressing agrifood issues confronting us today?

Suggested reading: Introductory level

Constance, D. 2008. The emancipatory question: The next step in the sociology of agrifood sytems. *Agriculture and Human Values* 25: 151–5.

Hassanein, N. 2003. Practicing food democracy: A pragmatic politics of transformation. *Journal of Rural Studies* 19: 77–86.

Niles, D. and R. Roff. 2008. Shifting agrifood systems: The contemporary geography of food and agriculture – an introduction. *Geojournal* 73(1): 1–10.

Suggested reading: Advanced level

Allen, P. and C. Sachs. 2007. Women and food chains: The gendered politics of food. *International Journal of Sociology of Food and Agriculture* 15(1): 1–23.

Bell, Michael and Philip Lowe. 2000. Regulated freedoms: The market and the state, agriculture and the environment. *Journal of Rural Studies* 16: 285–94.

Burawoy, Michael. 2005. 2004 American Sociological Association Presidential Address: For Public Sociology. *British Journal of Sociology* 56(2): 259–94.

Hinrichs, C. 2008. Interdisciplinarity and boundary work: Challenges and opportunities for agrifood studies. *Agriculture and Human Values* 25: 209–13.

Rosset, P., B. Sosa, A. Jaime, and D. Lozano. 2011. The Campesino-to-Campesino agroecology movement of ANAP in Cuba: Social process methodology in the construction of sustainable peasant agriculture and food sovereignty. *Journal of Peasant Studies* 38(1): 161–91.

Note

1 Thanks to the anonymous reviewer who inspired this reference to agrarian idealism.

References

Alkon, A. 2008. Paradise or pavement: The social constructions of the environment in two urban farmers' markets and their implications for environmental justice and sustainability. *Local Environment* 13(3): 271–89.

Allen, P. 2004. *Together at the Table: Sustainability and Sustenance in the American Agrifood System*. University Park, PA: Penn State University Press.

Allen, P. 2008. Mining for justice in the food system: Perceptions, practices, and possibilities. *Agriculture and Human Values* 25: 157–61.

Allen, P. and C. Sachs. 2007. Women and food chains: The gendered politics of food. *International Journal of Sociology of Food and Agriculture* 15(1): 1–23.

Altieri, M. and P. Koohafkan. 2008. *Enduring Farms: Climate Change, Smallholders and Traditional Farming Communities*. Penang: Third World Network.

Barham, E. 1997. Social movements for sustainable agriculture in France: A Polanyian perspective. *Society and Natural Resources* 10(3): 239–49.

Barham, E. 2002. Towards a theory of values-based labeling. *Agriculture and Human Values* 19(4): 349–60.

Bell, Michael and Philip Lowe. 2000. Regulated freedoms: The market and the state, agriculture and the environment. *Journal of Rural Studies* 16: 285–94.

Berlin, Isaiah. 1969 (1958). *Four Essays on Liberty*. London: Oxford University Press.

Burawoy, Michael. 2005. 2004 American Sociological Association Presidential Address: For Public Sociology. *British Journal of Sociology* 56(2): 259–94.

Buttel, Fredrick. 2001a. Some reflections on late-twentieth century agrarian political economy. *Sociologia Ruralis* 41(2): 165–81.

Buttel, F. 2001b. Some reflections on late-twentieth century agrarian political economy. *Cadernos de Ciência and Tecnologia, Brasília* 18(2): 11–36.

Carolan, M. 2005. Barriers to the adoption of sustainable agriculture on rented land: An examination of contesting social fields. *Rural Sociology* 70: 387–413.

Carolan, M. 2011. *The Real Cost of Cheap Food*. London: Earthscan.

Constance, D. 2008. The emancipatory question: The next step in the sociology of agrifood systems. *Agriculture and Human Values* 25: 151–5.

Constance, D., J. Rikoon, and J. Ma. 1996. Landlord involvement in environmental decision making on rented Missouri cropland: Pesticide use and water quality issues. *Rural Sociology* 61: 577–605.

DuPuis, M. 2000. Not in my body: rBGH and the rise of organic milk. *Agriculture and Human Values* 17: 285–95.

Duffy, M. and D. Smith. 2008. Farm ownership and tenure in Iowa, 2007. Iowa State University Extension, Ames, IA, http://www.econ.iastate.edu/sites/default/files/publications/papers/paper_13013.pdf, last accessed June 10, 2011.

Freudenburg, W. 2006. Environmental degradation, disproportionality, and the double diversion: Reaching out, reaching ahead, and reaching beyond. *Rural Sociology* 71(1): 3–32.

Friedland, W. 2008. "Chasms" in agrifood systems: Rethinking how we can contribute. *Agriculture and Human Values* 25: 197–201.

Gilbert, J. and T. Beckley. 1993. Ownership and control of farmland: Landlord–tenant relations in Wisconsin. *Rural Sociology* 58: 569–79.

Guthman, J. 2007. The Polanyian Way? Voluntary food labels as neoliberal governance. *Antipode* 39(3): 456–78.

Harvey, D. 2005. *Brief History of Neoliberalism*. New York: Oxford University Press.

Hassanein, N. 2003. Practicing food democracy: A pragmatic politics of transformation. *Journal of Rural Studies* 19: 77–86.

Hinrichs, C. 2000. Embeddedness and local food systems: Notes on two types of direct agriculture market. *Journal of Rural Studies* 16: 295–303.

Hinrichs, C. 2008. Interdisciplinarity and boundary work: Challenges and opportunities for agrifood studies. *Agriculture and Human Values* 25: 209–13.

Holt-Gimenez, E. 2002. Measuring farmers' agroecological resistance after Hurricane Mitch in Nicaragua: A case study in participatory, sustainable land management impact monitoring. *Agriculture, Ecosystems and Environment* 93(1–3): 87–105.

Lyson, T. 2004. *Civic Agriculture: Reconnecting Farm, Food, and Community*. Medford, MA: Tufts University Press.

McIntyre, L. and K. Rondeau. 2011. Individual consumer food localism: A review anchored in Canadian farmwomen's reflections. *Journal of Rural Studies* 27(2): 116–24.

McMichael, P. 2009. Banking on agriculture: A review of the World Development Report 2008. *Journal of Agrarian Change* 9(2): 235–46.

Mennell, S., A. Murcott, and A. van Otterloo. 1992. *The Sociology of Food: Eating, Diet and Culture*. London: Sage Publications.

Mercer, K. and H. Perales. 2010. Evolutionary response of landraces to climate change in centers of crop diversity. *Evolutionary Applications* 3: 480–93.

Nicholls, A. and C. Opal. 2005. *Fair Trade: Market Drive Ethical Consumption*. Thousand Oaks, CA: Sage.

Petrzelka, P. and S. Marquart-Pyatt. 2011. Land tenure in the US: Power, gender, and consequences for conservation decision making. *Agriculture and Human Values* 28: 549–60.

Raynolds, L., D. Murray, and J. Wilkinson. 2007. *Fair Trade: The Challenges of Transforming Globalization*. New York: Routledge.

Rickson, S., R. Rickson, and D. Burch. 2006. Women and sustainable agriculture. In *Rural Gender Relations: Issues, Case Studies*, edited by B. Beck and S. Shortall. Wallingford, UK: CABI Publishing, pp. 119–35.

Roff, R. 2007. Shopping for change? Neoliberalizing activism and the limits of eating non-GMO. *Agriculture and Human Values* 24: 511–22.

Rogers, D. and A. Vandeman. 1993. Women as farm landlords: Does gender affect environmental decision making on leased land? *Rural Sociology* 58: 560–8.

Rosset, P., B. Sosa, A. Jaime, and D. Lozano. 2011. The Campesino-to-Campesino agroecology movement of ANAP in Cuba: Social process methodology in the construction of sustainable peasant agriculture and food sovereignty. *Journal of Peasant Studies* 38(1): 161–91.

Slocum, R. 2008. Thinking race through feminist corporeal theory: Divisions and intimacies at the Minneapolis farmers' market. *Social and Cultural Geography* 9(8): 849–69.

Szasz, A. 2007. *Shopping Our Way to Safety: How We Changed from Protecting the Environment to Protecting Ourselves* Minneapolis, MN: University of Minnesota Press.

Tanaka, K. and P. Mooney. 2010. Public scholarship and community engagement in building community food security: The case of the University of Kentucky. *Rural Sociology* 75(4): 560–83.

US Bureau of Labor Statistics. 2008. Highlights of Women's Earnings in 2004. Washington, DC: US Department of Labor, http://www.bls.gov/cps/cpswom2008.pdf, last accessed June 11, 2011

van der Ploeg, J. 2008. *The New Peasantries: Struggles for Autonomy and Sustainability in an Era of Empire and Globalization*. London: Earthscan.

van der Ploeg, J. 2010. The peasantries of the twenty-first century: The commoditization debate revisited. *Journal of Peasant Studies* 37(1): 1–30.

Wilk, R. 2010. Power at the table: Food fights and happy meals. *Cultural Studies, Critical Methodologies* 10(6): 428–36.

Zerbe, Noah. 2009. Setting the global dinner table: Exploring the limits of the marketization of food security. In *The Global Food Crisis: Governance Challenges and Opportunities*, edited by Jennifer Clapp and Marc Cohen. Waterloo, Canada: Wilfrid University Press, pp. 161–77.

INDEX